Y0-BYU-814

Photobiological Techniques

NATO ASI Series

Advanced Science Institutes Series

A series presenting the results of activities sponsored by the NATO Science Committee, which aims at the dissemination of advanced scientific and technological knowledge, with a view to strengthening links between scientific communities.

The series is published by an international board of publishers in conjunction with the NATO Scientific Affairs Division

A	**Life Sciences**	Plenum Publishing Corporation
B	**Physics**	New York and London
C	**Mathematical and Physical Sciences**	Kluwer Academic Publishers Dordrecht, Boston, and London
D	**Behavioral and Social Sciences**	
E	**Applied Sciences**	
F	**Computer and Systems Sciences**	Springer-Verlag
G	**Ecological Sciences**	Berlin, Heidelberg, New York, London,
H	**Cell Biology**	Paris, and Tokyo

Recent Volumes in this Series

Volume 210—Woody Plant Biotechnology
edited by M. R. Ahuja

Volume 211—Biophysics of Photoreceptors and Photomovements in Microorganisms
edited by F. Lenci, F. Ghetti, G. Colombetti, D.-P. Häder, and Pill-Soon Song

Volume 212—Plant Molecular Biology 2
edited by R. G. Herrmann and B. Larkins

Volume 213—The Midbrain Periaqueductal Gray Matter: Functional, Anatomical, and Neurochemical Organization
edited by Antoine Depaulis and Richard Bandler

Volume 214—DNA Polymorphisms as Disease Markers
edited by D. J. Galton and G. Assmann

Volume 215—Vaccines: Recent Trends and Progress
edited by Gregory Gregoriadis, Anthony C. Allison, and George Poste

Volume 216—Photobiological Techniques
edited by Dennis Paul Valenzeno, Roy H. Pottier, Paul Mathis, and Ron H. Douglas

Series A: Life Sciences

Photobiological Techniques

Edited by

Dennis Paul Valenzeno
The University of Kansas Medical Center
Kansas City, Kansas

Roy H. Pottier
The Royal Military College of Canada
Kingston, Ontario, Canada

Paul Mathis
CEN Saclay
Gif-sur-Yvette, France

and

Ron H. Douglas
The City University
London, United Kingdom

Plenum Press
New York and London
Published in cooperation with NATO Scientific Affairs Division

Proceedings of a NATO Advanced Study Institute
on Photobiological Techniques,
held July 1-14, 1990,
in Kingston, Ontario, Canada

Library of Congress Cataloging-in-Publication Data

NATO Advanced Study Institute on Photobiological Techniques (1990 :
Kingston, Ont.)
Photobiological techniques / edited by Dennis Paul Valenzeno ...
[et al.].
p. cm. -- (NATO ASI series. Series A, Life sciences ; vol.
216)
"Proceedings of a NATO Advanced Study Institute on Photobiological
Techniques, held July 1-14, 1990, in Kingston, Ontario, Canada"-
-T.p. verso.
"Published in cooperation with NATO Scientific Affairs Division."
Includes bibliographical references and index.
ISBN 0-306-44057-1 (hardbound). -- ISBN 0-306-43778-3 (paperbound)
1. Photobiology--Technique--Congresses. I. Valenzeno, Dennis
Paul. II. North Atlantic Treaty Organization. Scientific Affairs
Division. III. Title. IV. Series: NATO ASI series. Series A, Life
sciences ; v. 216.
QH515.N17 1990
574.19'153--dc20 91-34489
CIP

ISBN 0-306-44057-1

A Division of Plenum Publishing Corporation
233 Spring Street, New York, N.Y. 10013

Printed in the United States of America

PREFACE

The first edition of the *Science of Photobiology* edited by Kendric C. Smith (Plenum Press, 1977) was a comprehensive textbook of photobiology, devoting a chapter to each of the subdisciplines of the field. At the end of many of these chapters there were brief descriptions of simple experiments that students could perform to demonstrate the principles discussed. In the succeeding years some photobiologists felt that a more complete publication of experiments in photobiology would be a useful teaching tool. Thus, in the 1980s the American Society for Photobiology (ASP) attempted to produce a laboratory manual in photobiology. Cognizant of these efforts, Kendric Smith elected to publish the second edition of *The Science of Photobiology* (1989) without experiments, anticipating the completion of the ASP laboratory manual. Unfortunately, the initial ASP efforts met with limited success, and several years were to pass before a photobiology laboratory manual became a reality.

One of the major stumbling blocks to production of an accurate and reliable laboratory manual was the requirement that the experiments be tested, not just by the author who is familiar with the techniques, but by students who may be quite new to photobiology. How could this be accomplished with limited resources? Many ideas were considered and discarded, before a workable solution was found. The catalyst that enabled the careful screening of all experiments in this book was a NATO Advanced Study Institute (ASI) devoted entirely to this purpose. In some cases lecturers personally transported smaller pieces of specialized scientific equipment from various countries and set them up so that NATO ASI participants could perform each experiment at the school.

A total of 36 experiments were thus set up by the faculty and successfully tested by the participants at a NATO ASI entitled *Photobiological Techniques*, held at the Royal Military College of Canada, 1-14 July 1990. Working in small groups of 4 to 8 people, participants performed all 36 experiments in turn. Their criticisms were invaluable in producing the final revised version that is found in this book. Further, three participants submitted experiments that were successfully tested by the faculty, and these experiments have also been included here (Chaps. 3, 6, and 9).

• • • • •

The book is organized into 21 chapters. Each chapter addresses one of the major subdisciplines of photobiology: photophysics, photochemistry, photosensitization, photosynthesis, UV effects, environmental photobiology, vision, photomorphogenesis, photomovement, chronobiology, bioluminescence, and photomedicine. Some subdisciplines are covered in more than one chapter. With the exception of extraretinal photobiology, all subdisciplines covered in Smith's *The Science of Photobiology* are covered here.

Each chapter contains a general overview of the subdiscipline, this overview being based on the lecture material of the contributing author. The overview is not intended to be a comprehensive coverage of the field, but rather serves to emphasize the salient points required to understand the experiments that follow.

Following the overview in each chapter are one or more experiments, including a comprehensive list of materials required to perform the experiment. Each experiment was successfully carried out during a two-hour period at the NATO ASI in Kingston. A discussion follows each experiment. The discussion may contain a critical review of the various experimental parameters that can influence the results of the experiments, suggestions for expansion of the experiment for longer laboratory periods, and/or typical experimental results. At the end of the experiments is a list of supplementary reading, judiciously chosen to guide the student towards a fuller understanding of basic principles. Further, review questions with detailed answers are provided.

• • • • •

The level of the experiments in this book is quite varied. The original intent was to develop a series of "mini" or "simplified" experiments at the senior undergraduate/graduate student level. However, the enthusiasm of the lecturers led to more detailed experiments in many cases.

As an aid to the instructor attempting to select experiments, we offer the following listing of the experiments in which we have attempted to indicate the complexity of equipment requirements for the 39 experiments tested at the ASI. We expect that most colleges and universities should have the necessary equipment to perform the experiments in the first category without trouble. Similarly, we anticipate that many colleges and universities should be able to assemble the equipment to do many of the experiments in the second category, perhaps with some advance work or modification. The experiments in the last category require special equipment (listed in parenthesis) that may not be available at all institutions. We have assumed that a standard scanning absorbance spectrophotometer is available in most laboratories. Some photomedicine experiments are listed as having significant equipment requirements because of the complexities in the use of animals and/or human volunteers.

Experiments with minimal equipment requirements: Photophysics - 2; Photochemistry - 9, 10; Photosensitization - 11, 12, 13; Environmental Photobiology -21; Vision - 22; Photomorphogenesis - 24 and 25; Chronobiology - 29; Bioluminescence - 30, 32, 33; Photomedicine - 39

Experiments with moderate equipment requirements: Photophysics - 1, 4, 5; Photosensitization - 14; UV Effects - 17, 18, 19; Chronobiology - 28

Experiments with significant equipment requirements: Photophysics - 3 (fluorescence lifetime measurement system); Photochemistry - 6, 7, 8 (HPLC); Photosynthesis - 15, 16 (flash absorption and fluorescence system); Vision - 23 (electrophysiology equipment); Photomovement - 26 and 27 (photodetectors and power supplies); Bioluminescence - 31 (luminometer); Photomedicine - 34, 35, 36 (human volunteers), 37, 38 (mice)

Aside from the 39 detailed experiments tested during the NATO ASI, the original experiments published in the first edition of *The Science of Photobiology* have also been reproduced here. We are thus indebted to Dr. Kendric Smith and to Plenum Press for permission to reprint the experiments which constitute Chapter 21 of this book. These experiments have already proven to be very useful as a pedagogical tool in the teaching of photobiology and are a welcome addition to this text. In general, they require very little equipment and would be included in the first category in the ranking of equipment requirements given above.

• • • • •

The editors wish to express their thanks to NATO for providing the funding which made this book and the ASI possible. Participants came from 19 countries, most from western Europe, Canada and the United States. However, Australia, Austria, Israel, Japan, Lithuania, New Guinea, Yugoslavia and the U.S.S.R. were also represented.

Both the American and European Societies for Photobiology made significant financial contributions towards the ASI, *Photobiological Techniques*. Members of both Societies share a common interest in the promotion of photobiology, and it is our hope that students and research scientists throughout the world will benefit from the support of photobiology provided by these Societies.

The authors and editors are also indebted to the Royal Military College of Canada for its hospitality and support of this project through the use of its facilities. In addition we wish to acknowledge the support of the following companies who contributed to the success of our efforts:

The Optikon Corporation

Millipore (Canada) Ltd.

SLM Instruments

PTI, Photon Technology International Inc.

Munksgaard Publishing

Cadna Medical Diagnostic Inc.

Solar Light Company

• • • • •

Finally, although each chapter has received thorough peer review, the editors retain sole responsibility for any errors or omissions. In order to improve the text for possible future editions, readers are asked to submit any criticisms, corrections or suggested experimental modifications to Dr. Dennis Paul Valenzeno, Department of Physiology, The University of Kansas Medical Center, 39th and Rainbow Boulevard, Kansas City, KS 66103, U.S.A. Fax: (913) 588-7430.

Dennis Paul Valenzeno

Roy Pottier

Paul Mathis

Ron Douglas

CONTENTS

Photophysics

Photochemistry

Photosensitization

Photosynthesis

UV Effects

Environmental Photobiology

Vision

Photomorphogenesis

Photomovement

Chronobiology

Bioluminescence

Photomedicine

Other Experiments

The Nature of the Light Field in Biological Media

Brian C. WILSON
McMaster University and
Ontario Laser and Lightwave Research Center
Canada

INTRODUCTION

Fundamentals

Photobiology is the science of the interactions of light with living systems. "Light" is that part of the electromagnetic spectrum encompassing the ultraviolet, visible and infrared. Here we will discuss the fundamental properties of light and the propagation of light in biological media, specifically mammalian tissues. This latter is particularly important in many photobiological and photomedical applications, since all photobiological effects result, in the first instance, from the absorption of optical energy by tissue components, so that the spatial, and in some cases also the temporal, distribution of light in tissue is a major factor in the final biological result.

When propagating in space or in a dielectric medium such as tissue, light may be considered either as an electromagnetic wave or as discrete photons (Ishimaru, 1978). In the former, at any position and time, the wave is specified by the amplitude and phase of its electric and magnetic vectors, as illustrated in Fig. 1a. The wavelength, λ, speed, c, and frequency, υ, of the wave are related by

$$c = \lambda \cdot \upsilon \qquad [1]$$

In free space, $c = c_o = 3 \times 10^8\ m \cdot s^{-1}$, whereas in tissue $c' = c_o/n$, where n = average refractive index of the tissue (typically in the visible range, $n \sim 1.4$).

In the *particle* description, the photon energy is discrete and depends only on the wavelength: $E = 1240/\lambda$ in units of eV/nm. Thus, a photon of, say, green light at 500 nm has a quantum energy of about 2.5 eV, so that in a single absorption interaction, this is the maximum energy available to excite a molecule to a higher energy level. Table I summarizes the wavelength, wave number, and photon energy ranges for the UV, visible and infrared. Note that 1 Joule = 6.24×10^{18} eV, so that 1 watt ($1\ W = 1\ J \cdot s^{-1}$) of green light represents a photon flux of about 2.5×10^{18} photons per second.

BRIAN C. WILSON, Hamilton Regional Cancer Center, McMaster University, Hamilton and Ontario Laser and Lightwave Research Center, Toronto, Ontario, Canada

Photobiological Techniques, Edited by D.P. Valenzeno *et al.*
Plenum Press, New York, 1991

Table 1

Wavelength, photon energy and wavenumber ranges for the ultraviolet, visible and infrared regions of the electromagnetic spectrum. The defined ranges are approximate, and vary slightly in different applications and standards. The wavenumber is the inverse of the wavelength and is often used in spectroscopy. The photon energy is given by the relationship $E = h \cdot c/\lambda$, where λ is the wavelength and c is the speed of light *in vacuo*; hc = 1240 when E is in eV and λ is in nm.

	ULTRAVIOLET			VISIBLE	INFRARED
	UVC	UVB	UVA		
Wavelength Range	100-280 nm	280-315 nm	315-400 nm	400-700 nm	0.7-20 μm
Energy Range per Photon	12.4-4.4 eV	4.4-3.9 eV	3.9-3.1 eV	3.1-1.8 eV	1.8-0.6 eV
Wavenumber	100 000-36 000 cm^{-1}	36 000-32 000 cm^{-1}	32 000-25 000 cm^{-1}	25 000-14 000 cm^{-1}	14 000-500 cm^{-1}

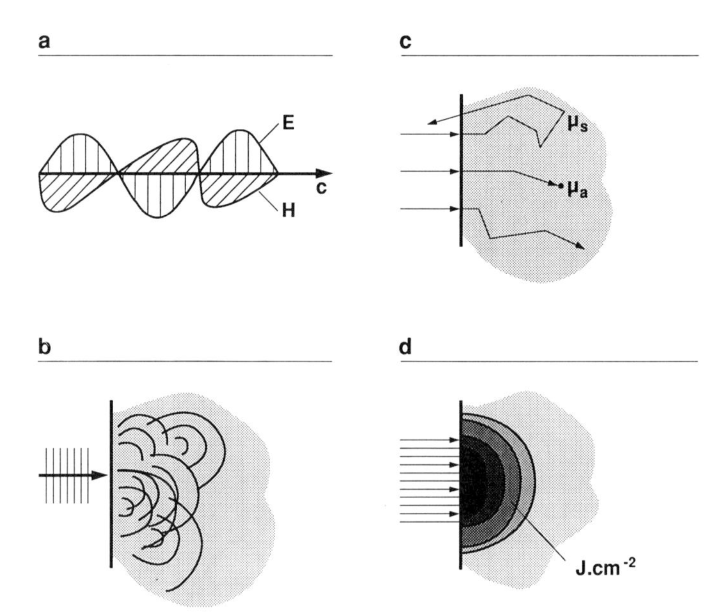

Figure 1. Electromagnetic wave (a,b) and photon transport (c,d) models of light propagation in tissue. a) Electric (E) and magnetic (H) wave vectors; b) Plane wave incident on tissue, with subsequent phase and amplitude shifting due to attenuation; c) Scattering and absorption of discrete photons; d) Resulting fluence distribution in tissue.

The wave picture allows the description of effects such as polarization, diffraction, reflection and refraction of light. As illustrated in Fig. 1b, a large collimated light beam incident on tissue is represented as a plane wave. Upon encountering the tissue, which comprises a medium with random or pseudo-random microscopic variations of the dielectric constants, this simple wave is 'decomposed' into a complex pattern of multiple wavelets, propagating in all directions with different amplitudes and phases.

Fortunately, for most purposes in photobiology, we are primarily interested in the local energy or power density (fluence [Joules·m^{-2}] and fluence-rate [W·m^{-2}]) within or outside the tissue, rather than in the component amplitudes and phases of the light field. For example, the rate of energy absorption by the tissue at any point is proportional simply to the local fluence rate, and does not depend separately on the amplitude and phases of each of the wave components at the point. Similarly, in transmission or reflection spectroscopy, the relevant quantities are the fluence rates of light which have interacted with, and then propagated out of the tissue. As illustrated in Figs. 1c and d, this allows the light propagation to be described *empirically*, as equivalent to the propagation of descrete photons, whose number density and flow rate define the local fluence rate. The interaction of the light field with the tissue is then specified by the absorption and scattering coefficients (μ_a, μ_s) of the tissue, which are measures of the probability of the photons being locally absorbed or locally scattered, i.e., undergoing a change in direction of propagation. The angular dependence of this scattering is described by the *phase function* which is a measure of the relative probability of scattering through a particular angle. Clearly, this so-called photon transport model does not incorporate wave-dependent effects, but it has, nevertheless, been successful and useful as a model to describe the light fluence from and within tissues (Star et al., 1987; Wilson et al., 1990), and will, therefore, be used in the experiments described below.

It should be noted that there is only a poor understanding of the relationship between the parameters of the transport model (absorption and scattering coefficients and scattering phase functions) and the dielectric microstructure of tissues. Rigorous electromagnetic wave descriptions have been derived for the interaction of light waves with dielectric particles of known size, shape and refractive index, so that, if the integrated and directional outgoing fluence rates are calculated from these, then the equivalent absorption and scattering coefficients and scattering phase functions can be obtained. For example, for particles much smaller than the wavelength (Rayleigh scattering), the scattering coefficient varies roughly inversely as the fourth power of the wavelength, and the phase function is approximately constant, i.e., the scattering is isotropic, with equal scattered fluence in all direction. At the other extreme (Mie scattering) for large particles, the scattering probability goes as $1/\lambda$ and the phase function is forward peaked, i.e., most photons are scattered through a small angle, or, equivalently, the wave intensity is high at small angles to the incoming wave. It is now known that, at least in the visible and near-infrared wavelength range, the scattering coefficients of tissue are "Mie-like", i.e., decrease slowly with wavelength and are forward peaked. However, tissue is clearly not a suspension of independent simple dielectric spheres floating in a uniform dielectric medium, so that it is not clear how to relate this observation to the actual microstructure of tissues, where wave-related processes such as reflection and refraction at inter- and intra-cellular structures result in the observed scattering behavior.

The total absorption coefficient in a tissue is simply the sum of the absorption coefficients of the constituent chromophores, which in turn are given by the product of the tissue concentration and the extinction coefficient. Different chromophores contribute in different wavelength regions: in the UV, the absorption by biomolecules such as nuclei acids dominates; in the visible and near-IR, the absorption is due mainly to hemoglobin and pigments such as melanin and bilirubin, but other chromophores (e.g. cytochrome aa) are also important for specific applications, even though their contribution to the total absorption coefficient of the tissue may be small; in the mid and far-IR, water accounts for most of the optical absorption.

Tissue Optical Properties

The linear (i.e., one-photon) absorption and scattering properties of several mammalian tissues have been measured using a variety of *in vitro* and, more recently, *in vivo* techniques: i) directly in optically thin tissue samples (Flock et al., 1987; Marchesini et al., 1989); ii) from the diffuse transmittance and reflectance through optically thick samples; iii) by *doping* homogenized tissue with known concentrations of an exogenous absorber (Wilson et al., 1986); iv) through *mapping* of the fluence distributions within intact, bulk tissue (Marijnissen and Star, 1987); and v) by measuring the spatial and/or temporal distribution of diffusely reflected or transmitted light. Methods (iv) and (v) may be done *in vivo*, while all the techniques except (i) require the analysis of the measured data by some radiation transport model(s) of light propagation in order to derive the fundamental optical coefficients.

In optically turbid tissue (excluding the transparent structures of the eye), the nature of the light field, i.e., the spatial/temporal distribution of light within and from the tissue, depends primarily on the relative contributions of absorption and scattering. This relative contribution is very wavelength dependent. First, the total absorption coefficient of tissues

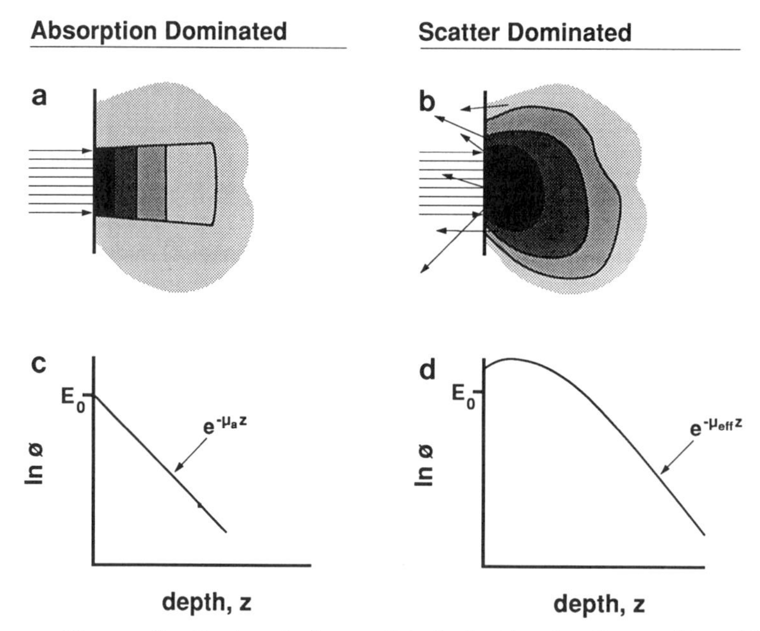

Figure 2. Fluence distributions in tissue. E_0 is the incident irradiance. (a) and (b) show the fluence contours for absorption >> scattering and scattering >> absorption, respectively. (c) and (d) show the corresponding fluence-depth curves: for the absorption-dominated case (c), the fluence has a simple exponential decrease with depth, according to the absorption coefficient. For high scattering (d), the curve is more complex, but eventually becomes exponential, where the exponent (effective attenuation) depends on both absorption and scattering.

is spectrally complex, since many of the individual chromophores have complex extinction spectra, some of which vary with the physiologic or metabolic status of the tissue: for example the hemoglobin spectrum is altered markedly between the reduced and oxygenated states. Second, the scattering coefficient, and scattering anisotropy, g (the average cosine of the scattering angle), also vary with wavelength, although the spectra are much simpler, without the detailed structure seen in the absorption spectra: in the wavelength ranges where measurements have been done, i.e., in the visible and near-IR, the scattering coefficient decreases roughly inversely with increasing wavelength, while the anisotropy is nearly constant (Parsa et al., 1989).

Finally, the average refractive index of tissues, which determines both the speed of light propagation and the specular reflection properties at tissue boundaries, has also been found to be fairly constant with wavelength, although measurements have only been done over a limited range in the visible.

It should be emphasized that, in the visible and near-IR, different tissues have very different optical properties. At any given wavelength, at least an order of magnitude difference in absorption coefficients can be found, while 2- or 3-fold differences in the scattering coefficients have been measured. The anisotropy is also tissue dependent.

Fluence Distributions in Tissue

In reality the spatial distribution of light fluence in an irradiated tissue volume clearly depends not only on the absorption and scattering coefficients of the tissue at any point, but also on heterogeneity of these properties, irradiation geometry, boundary conditions and the shape and size of the tissue volume. In order to illustrate some underlying principles, consider the simplified case of a large volume of optically homogeneous tissue irradiated by a collimated beam incident normally on the flat tissue surface, as shown in Fig. 2a. The light enters the tissue, where the photons may be absorbed or elastically scattered (i.e., scattered without loss of energy), as governed, respectively, by the absorption coefficient, μ_a, and the scattering coefficient and anisotropy, μ_s and g.

At wavelengths where absorption dominates over scattering, the fluence distribution follows Beer's Law, and at any depth z in the tissue (Fig. 2a, c) is given by

$$\phi(z) = \phi_o e^{-\mu_a z} \qquad [2]$$

where ϕ_o is the fluence at the surface, which in this case is equal to $E_o(1-r)$, where E_o is the incident beam irradiance and r is the specular reflection coefficient ($r = \{[n-1] / [n+1]\}^2 \approx 3\%$, for tissue irradiated in air). Since there is little (in the limit, zero) multiple scattering of the light, the incident beam profile is maintained in the tissue with no sideways spreading of the fluence, and the diffuse reflectance, i.e., the amount of light which propagates back through the tissue surface after scattering in the tissue, is low. This is the situation in soft tissues in the UV and mid/far-IR regions of the spectrum.

At the other extreme, when scattering dominates over absorption (Fig. 2b, d), the beam spreads (diffuses) rapidly in the tissue and the diffuse reflectance is high. In soft tissues this situation occurs between about 600 and 1300 nm, the so-called *optical window*, where the absorption of tissue pigments has fallen from the high values at shorter visible wavelengths and that of water has not started to rise as it does at longer wavelengths (Anderson and Parrish, 1982).

The depth distribution of fluence in this case is no longer a simple exponential, but has the following characteristics:

a) There is a sub-surface peak, which is most pronounced when the tissue is irradiated through a refractive index-matched fluid: with index mismatch, as for irradiation in air, there is substantial total internal reflection at the surface which increases the fluence close to the surface.

b) The fluence can be several times higher than the incident beam irradiance, due to the *backscattered* flux of photons. At first sight this might appear to violate energy conservation, but it should be noted that we are here describing the steady-state condition of continuous irradiation, and that the fluence is simply a measure of the number of photons passing through a tissue element per unit time. The same photon may thus contribute several times to the fluence at any point as it undergoes multiple scattering. The total optical energy in a given time interval, representing both propagating photons inside and outside the tissue and energy absorbed in the tissue, is conserved.

c) The fluence-depth distribution eventually reaches an exponential region, given by

$$\phi(z) \propto e^{-\mu_{eff} z}. \qquad [3]$$

In this case the so-called effective attenuation coefficient, μ_{eff}, depends on both the absorption and scattering properties of the tissue. The inverse is referred to as the (effective) penetration depth. Note that this is the depth increment over which the fluence falls by the factor 1/e (37%). It is *not* the depth below the surface at which the fluence is 37% of the incident irradiance, as can be seen from Fig. 2d.

The penetration depth (and the subsurface behavior) depends also on the size of the incident beam. In the limiting case of a very large beam, diffusion theory can be applied to obtain an approximate analytic expression for μ_{eff},

$$\mu_{eff} = [3\mu_a (\mu_a + \mu_s [1-g])]^{1/2} \approx [3\mu_a \mu_s']^{1/2} \qquad [4]$$

where $\mu_s' = [1-g]\, \mu_s$ (μ_s' is called the transport scattering coefficient).

Note that, since $\mu_s' >> \mu_a$, μ_{eff} is always greater than μ_a, i.e., the final rate of decrease in a tissue is always increased by scattering compared to the case where absorption is the dominant process. (This is true even if the scatter is not large compared with the absorption, but Eq. 4 cannot be applied for this case).

In the transition case, where neither absorption nor scattering dominate the other, there is no simple expression to describe the fluence-depth distribution nor the diffuse reflectance (see below), and numerical or computational models have to be used to study the dependency of the light field on the tissue properties. This situation holds at shorter visible wavelengths (≤600 nm) and also in a region around 1300-2000 nm for soft tissue. Note, however, that there is not a clean separation between these various absorption/scattering zones, and that the boundaries between them vary between tissues.

Diffuse Reflectance/Transmittance

As the scattering in tissue increases, the probability of photons being backscattered through the irradiated surface after multiple scattering also increases. This is measured by the diffuse reflectance. The total diffuse reflectance, R_t, is the ratio of the photon flux integrated over the irradiated surface to the incident flux, after accounting for the loss due to specular reflection. In the limit of pure absorption, $R_t = 0$. For the scattering-dominated case it can be shown from diffusion theory that R_t depends only on the transport albedo, $a' = \mu_s' / \{\mu_a + \mu_s'\}$, of the tissue and, in the case of a refractive index mismatch at the surface, also on the tissue refractive index. Table 2 shows the dependence of R_t on the transport albedo in the range 0.9 - 0.9999, i.e., for scattering-to-absorption ratios varying from approximately 10:1 to 10 000:1. For lower scattering, R_t falls monotonically to zero, but the value is not derived simply from diffusion theory (Wilson and Patterson, 1990).

The total diffuse transmittance, T, through a slab of tissue clearly also depends on the scattering and absorption, as well as on the thickness. T increases as the absorption and scattering decrease. Table 2 gives some representative values for different tissue thicknesses.

As well as the total diffuse reflectance, the local diffuse reflectance contains information on the tissue absorption and scattering when the surface is irradiated by a small diameter collimated beam, for example, the direct beam from a laser. It can be shown (Patterson et al., 1989) that, in the case of high albedo (i.e., scattering-dominated tissue), the dependence of local reflectance $R(\rho)$ on the source-detector distance on the surface is

$$R(\rho) \propto \frac{1}{\rho^2} \cdot e^{-\mu_{eff}\rho}. \qquad [5]$$

Table 2

Diffuse reflectance and transmittance for a scattering/absorbing slab. Thickness is optical (e.g., $5 \equiv e^{-5} \equiv 6\%$ transmission of unscattered primary beam: $1 \equiv e^{-1} \equiv 37\%$): values for $R_t(\infty)$ at $a' \geq 0.83$ from diffusion theory (Wilson and Patterson, 1990): other values from numerical solutions (van de Hulst, 1980).

Surface	Slab Thickness	%	μ_s'/μ_a								
			0.5	1	2	5	10	10^2	10^3	10^4	∞
	∞	R_t	7	11	20	32	45	77	93	98	100
		T	----------	----------	----------	----------	0	----------	----------	----------	----------
Index Matched	5	R_t	7	11	20	31	43	68			74
		T	1	1.5	2.5	5	9	22			26
	1	R_t	6	10	16	23	27	33			34
		T	41	45	50	56	60	65			66
Mismatched (n = 1.4)	∞	R_t				17	30	63	85	95	100
	a'		0.33	0.5	0.67	0.83	0.91	0.99	0.999	0.9999	1.0

Thus, after correcting for the *inverse square law* term $1/\rho^2$, the radial dependence of local reflectance is the same as the depth dependence of the fluence when the tissue is irradiated by a broad beam. This can be used as a non-invasive method to determine the effective penetration depth.

Time-Resolved Photon Propagation

In the above, we have assumed steady-state conditions such as one would obtain for continuous irradiation or for pulsed irradiation where the pulse length is large compared to the propagation time of light in tissue. However, with picosecond lasers and ultrafast light detectors such as multichannel plate photomultiplier tubes and streak cameras, it is now both feasible and of practical interest in tissue spectroscopy to examine the time evolution of the light field in tissue. The current cost and complexity of the equipment precludes inclusion of time-resolved light propagation experiments in this course, but it is worthwhile to consider a qualitative description of the main features of time-dependent light propagation .

In the absorption-dominated case, little insight or information is obtained. (We exclude here the use of high power, short pulse lasers where 2-photon absorption may occur and for which the time spectrum of fluence in the tissue is important.) The light simply propagates as a pulse of decreasing energy density at speed $c' = c_o/n$.

However, with significant light scattering the situation is quite different. At any time, t, following a short pulse, the total distance (optical path length) travelled by a photon is $l = c' \cdot t$. For unscattered, primary photons this corresponds to the depth below the tissue surface. However, for scattered photons, at time t there is a spread in the spatial distribution of the photons, and the pattern of photon distribution *evolves* with time. Ultimately, after the light has been completely diffused, the fluence-time dependence is governed by the absorption. On the tissue surface, the time dependence of the local and the total diffuse reflectance also depends initially on the scatter and absorption, but at later times is absorption dominated.

Note that in the visible/near-IR optical window, photons can be detected in soft tissue at times of several nanoseconds after the input pulse. Since 1 ns corresponds to an optical path length of 21 cm (for n = 1.4) and the typical scattering coefficient in this region is ~ 50 mm^{-1}, such photons have, on average, undergone some 10 000 scattering interactions. However, despite the long path length, photons can be detected at the tissue surface even at a distance from the input source position, due to the random-walk nature of their propagation.

EXPERIMENT 1: LIGHT FLUENCE AND REFLECTANCE IN A TISSUE-SIMULATING MEDIUM

OBJECTIVE

This exercise provides an opportunity to compare observations of light fluence and reflectance in a tissue-simulating medium with the predictions of theory. The exercises are meant to serve as a hands-on introduction for someone new to the photophysical research field.

This laboratory is made up of four parts:

Part A: Light fluence distribution in a scattering and absorbing medium resulting from a narrow beam source.

Part B: Light fluence distribution in a scattering and absorbing medium resulting from a broad beam source.

Part C: Diffuse reflectance from a scattering and absorbing medium.

Part D: Subsurface fluence peak in a scattering medium.

The scattering and absorbing medium will be a solution of Intralipid®, a fat emulsion with more or less pure scattering properties, and ink which has almost purely absorbing characteristics. Such different solutions, of appropriate concentrations, are often used to model tissues in experimental studies of the interaction of light with matter. They are useful because they allow the experimenter to differentiate and quantify the effects of scattering and absorbing elements. Part B may be omitted to reduce the length of this laboratory experiment.

LIST OF MATERIALS

Small laser. A 10 mW HeNe is ideal
Small lens for broadening the laser beam
Optical fiber*, isotropic tipped, large core (100-400 μm)
Photometer, nW sensitivity, which will couple to the optical fiber (the photodiode based Photodyne 88XL is a good example)
Two or three axis stereotaxic stand
Liquid tight tank with:
- at least one thin transparent face
- at least 10 cm width, height and breadth (volume ~ 1liter)
Glassware for measuring liquid quantities
Pipetting system for measuring to 0.5 ml accuracy

PREPARATIONS

Intralipid Suspension: Prepare a 10% or 20% Intralipid-water suspension (fat emulsion) by mixing 20 ml or 40 ml, respectively, of Intralipid with water, to a final volume of 200 ml. If this is not available, it may be replaced with cream or non-dairy creamer diluted in water at the highest concentration possible.

---

*If an isotropic tipped fiber is not available, a mock-up may be created by dipping the end of a cleaved fiber, with 1 mm of the cladding removed, in typing correction fluid.

India Ink Solution: Prepare a 1% and a 0.1% solution of India ink by dissolving 0.1 ml of India ink in 9.9 ml and 99.9 ml of water, respectively.

EXPERIMENTAL PROCEDURE

Part A: Light fluence distribution in a scattering and absorbing medium resulting from a narrow beam source.

1. Fill the tank to a depth of at least 10 cm with a measured volume of water.
2. Align the laser so that the beam enters the tank through, and perpendicular to, a transparent tank face. The beam should enter and cross the tank at least 4 or 5 cm below the surface of the water.
3. Suspend the isotropic end of the optical fiber from the stereotaxic stand. Arrange the stand so that the isotropic tip falls in the center of the laser beam within the water tank. The stand should be set up so that one of the position adjustments moves the fiber tip exactly along the beam, and another moves it perpendicularly across the beam. The equipment should appear as in Fig. 3.

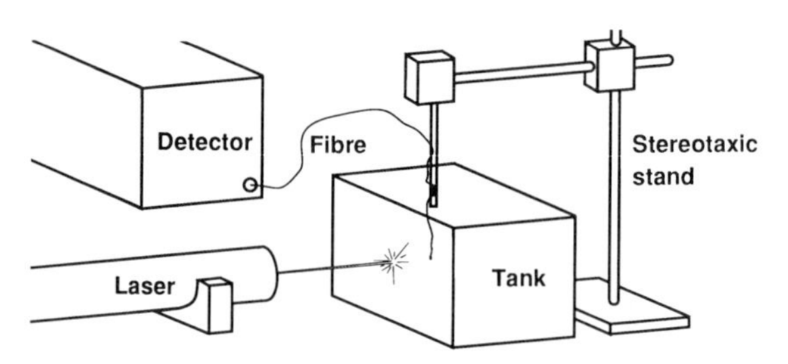

Figure 3. General layout of the experiment.

4. Couple the other end of the fiber into the light detector.
5. It is most likely that the stereotaxic stand will not trace the fiber tip exactly along the beam. A set of reference/normalizing measurements must be taken to determine how well this has been accomplished.
    a) Starting with the fiber tip centered in the beam and very near the entry point of the beam into the tank, measure and record the beam fluence and the fiber position. Call this position (0,0,0).
    b) Translate the fiber tip along the beam. At 1 cm intervals record the fiber position and measured beam fluence (i.e., at points (0,0,1.0), (0,0,2.0), etc.). Continue this to a depth of between 5 and 10 cm.
6. With the fiber tip at a depth of 5 cm, (0,0,5.0), add measured amounts of the 0.1% ink solution until the beam fluence is cut in half. It should require about 0.2% or 1 part in 500 of the 1% ink solution, or 1 part in 50 of the 0.1% ink solution.
7. a) Repeat the beam fluence vs. depth measurements made in step 5.
    b) Normalize these measurements using the no-ink reference measurements taken in step 5. That is, divide each measurement by the reference measurement *at the same position.*
    c) Plot the normalized results on a graph of $\ln(\phi)$ vs z (depth).

8. Now measure a transverse beam fluence profile at a depth of 1 cm:
   a) Move the fiber tip to (1.0,0,1.0); 1 cm deep and 1 cm off the beam axis.
   b) Move the fiber across the beam making periodic measurements which reveal a good picture of the beam fluence profile.
   c) Plot these results on a graph of ϕ vs. radial distance (ρ).
9. Add scattering medium to the tank:
   a) If Intralipid is available, make a 0.2% fat solution in the tank by adding 1 part in 100 by volume of the 20% lipid solution. If cream is being used as a substitute, add measured amounts until the signal as measured at one half centimetre depth, (0,0,0.5), is attenuated to about 1/25 of its original value. (Note: before doing this make sure that the fibre is actually centered in the beam by maximizing the measured signal).
   b) Repeat the depth vs. fluence measurements made in step 5. Plot these results *without normalizing*, on the fluence vs. depth graph of step 7.
   c) Make transverse beam profile measurements like those made in step 8. However, to properly observe the beam profile, measurements will have to be made as far as 2 or 3 cm on either side of the beam axis. Plot these results on the beam profile graph started in step 8.
10. a) Increase the amount of scattering medium in the tank to five times that used in step 9 (i.e., add 4 more parts).
    b) Repeat the depth and transverse profile fluence measurements. Add these plots to the graphs.

Part B: Light fluence distribution in a scattering and absorbing medium resulting from a broad beam source.

1. Empty the tank from part A, refilling it with clean water to the same level. Wash the fiber tip. Keep the equipment arranged as in part A.
2. Introduce the beam broadening lens into the laser beam. The beam should obtain a diameter of at least 2 cm where it enters the tank. The beam should be reasonably collimated, diverging at a rate of no more than about f/8.
3. Place the fiber tip in the center of the beam inside the tank, as close to the tank wall as possible, as in part A. Call this position (0,0,0).
4. The broad beam now being used may be diverging, and the translational equipment again may not track the beam exactly. To account for this, make measurements for a reference vs. depth profile as in part A, measuring at 1 cm intervals. See part A, step 5.
5. Add an identical quantity of ink to the tank as that added in part A, step 6.
6. a) Remeasure the beam depth profile as in step 4, in 1 cm increments.
   b) Normalize these measurements using the reference values taken in step 4.
   c) Plot these normalized values in a ln(ϕ) vs z (depth) graph.
7. a) Add a quantity of scatterer equivalent to that added on the first occasion in part A, step 9.
   b) Make measurements of the light fluence at several depths along the beam axis.
   c) Plot the fluence vs. depth profile on the graph of step 6.

8. a) Increase the amount of scattering medium in the tank to five times that used in step 7 (add 4 more parts as in part A, step 10).

   b) Repeat the fluence-depth profile measurements. Plot these data on the graph of step 6.

Part C: Diffuse Reflectance From a Scattering and Absorbing Medium.

1. Empty the tank and fill it with the same volume of clean water.
2. Remove the lens which was added in part B.
3. Rearrange the stereotaxic stand such that the fiber tip now just touches the outside of the transparent tank wall through which the beam enters. The transverse adjustment should slide the fibre tip along the tank wall, and its path should intersect the beam. The point where this occurs will be (0,0,0).
4. Add to the tank one half the amount of ink as was added in parts A and B (See part A, step 6).
5. Calculate the size of one unit of scattering material: Divide by 4 the quantity of scatterer originally added in parts A and B (see part A, step 9). The resulting quantity shall be known as one unit.
6. Move the fiber tip to (0.5→1.0,0,0); against the outer tank wall between 0.5 and 1.0 cm from the beam entry point. Ensure that the fiber does not intersect with the beam at any point.
7. a) Add 1 unit of scattering material to the tank. Measure and record the diffuse reflectance signal.

   b) Continue to add quantities of scatterer in quantities of 1 or 2 units recording the reflectance signal at each addition, to a total of about 20 units. At this point, the measured reflectance should level off or diminish. If this phenomenon is not observed, continue to add scatterer until it is seen.

   c) Plot the observations on a graph of Reflectance (R) vs. scatterer concentration.
8. a) Move the fiber to a position about 1 mm from the beam entrance, (0.1,0,0). Measure and record the reflectance at this point.

   b) Move the fiber radially away from the beam entrance in 1 mm increments until the reflectance signal drops to about 1% of its maximum value. Measure the reflectance at each radial distance, ρ.

   c) Plot the data on a graph of $\ln(\rho^2 R)$ vs. ρ.
9. a) Adjust the concentration of ink and scatterer in the tank to match one of those used in parts A and B (do not choose the case in which no scatterer was in the tank).

   b) Repeat the measurements of reflectance vs. radial distance made in step 8.

   c) Add a plot of this data to the graph of step 8.

   d) Calculate the slope of the tail of this graph. Compare it to the fall of the fluence (ϕ) vs. depth plot resulting from a broad beam source for the same scatterer and ink concentration.

   e) Calculate μ_{eff} for the medium used in step 9 d. Compare this with the measured slopes.

Part D: Subsurface Fluence Peak in a Scattering Medium

1. Set up the tank, laser and fiber as in Part A with the tank filled with water and the fiber at (0,0,0).
2. After checking that the fiber tip is collinear with the laser beam, move the fiber laterally until it is just out of the laser beam and the signal is less than 1% of the maximum in the beam.
3. Add 1 ml of scatterer and move the fiber in depth, looking for the depth of the maximum reading (z_m) and the value of the maximum reading (ϕ_m).
4. Repeat step 3 until z_m is constant or zero.
5. Determine how z_m and ϕ_m depend on the scattering of the medium.

DISCUSSION

Apart from the specific information obtained on the behavior of light in scattering and absorbing media, this experiment has provided experience in making optical measurements. Problems which may have been encountered include: a) varying signal due to background lighting in the laboratory, b) fluctuations or drift in the laser or lamp output, c) errors in positioning of the fiber, or sensitivity to movement on the laboratory bench. Experience will also show short-cuts in some of the experiments: for example, plotting the fluence values as each measurement is made avoids *over-sampling* in regions where there is little change in the signal.

In Part A it may be noticed that the accuracy of alignment of the fiber movement along the laser beam optical axis can have substantial effect on the fluence-depth plots, but that this becomes less important as the scattering is increased. The reason can be seen by visually observing the light distribution in the medium, viewing this from the top surface. The alignment is also less critical for a broad beam, as in Part B. The plots obtained in A7 should be straight lines, the (negative) slope of which equals the absorption coefficient, μ_a. The transverse plots in A8 should show the laser beam sharply delineated. These findings should be checked by visual observation of the laser beam as it propagates in the medium. The final slopes of the graphs in A9 (b) will give the values for μ_{eff}. These are "narrow beam" values which should be smaller than "broad-beam" values found in Part B7.

In Part C, it is important again to *observe* and *comment on* what is happening when scatterer and absorber are added to the tank. The *glow* of backscattered light produced by the scattering should be obvious. Compare what is seen at the front face (diffuse reflectance) with the distribution of light seen at the top surface. If the front transparent face of the tank is not thin enough, a pattern of rings may be seen as scatterer is added. This is due to multiple reflections within the face, and may lead to erroneous readings in the plots constructed in C8, C9.

In Part D, the readings are made slightly off-axis, with the fiber tip out of the laser beam. This enables the subtle subsurface effects to be measured. Again, visual observation of the objective readings should be made.

REVIEW QUESTIONS

1. For the hypothetical tissue absorption spectrum shown in the figure, calculate and plot the diffuse reflectance spectrum which would result from tissues in which the (transport) scattering coefficient, μ_s', is

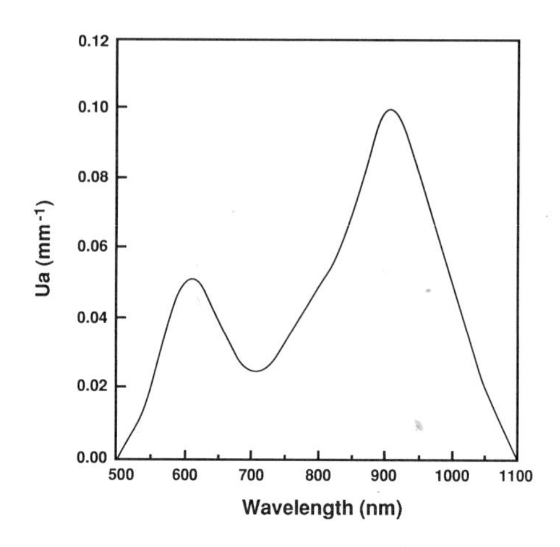

   a) constant at 20 mm^{-1}

   b) linearly decreasing from 20 mm^{-1} at 500 nm to 2 mm^{-1} at 1200 nm.

   Assume index-matching. Discuss the influence of scattering on the relative size of the dips in the reflectance spectra corresponding to the two chromophore peaks.

2. Draw the shapes of the fluence rate-depth profiles in tissue irradiated at normal incidence with a collimated beam of irradiance 1 $W{\cdot}cm^{-2}$ for:

   a) a large area beam with $\mu_a = 1\ mm^{-1}$, $\mu_s' = 0$
   b) a large area beam with $\mu_a = 0.5\ mm^{-1}$, $\mu_s' = 0$
   c) a large area beam with $\mu_a = 0.5\ mm^{-1}$, $\mu_s' = 0.5\ mm^{-1}$
   d) a large area beam with $\mu_a = 0.5\ mm^{-1}$, $\mu_s' = 5\ mm^{-1}$
   e) a small area beam with $\mu_a = 0.5\ mm^{-1}$, $\mu_s' = 0$
   f) a small area beam with $\mu_a = 0.5\ mm^{-1}$, $\mu_s' = 5\ mm^{-1}$

   Explain the main features of these profiles. The graphs should all be drawn on the same scale, and, where possible, should be quantitative in both fluence rate and depth values.

3. Draw qualitatively the fluence-depth profiles for the following, non-homogeneous, tissue model irradiated by a large area beam. Explain the features of the graph.

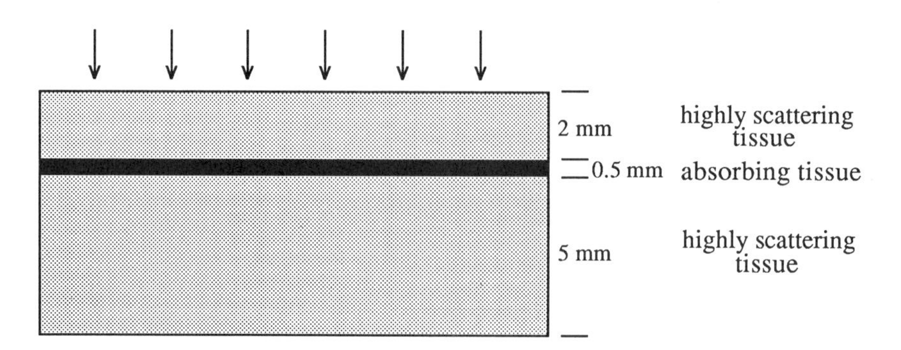

4. Draw qualitatively the iso-fluence distributions for the heterogeneous tissue shown, irradiated by a large area beam. Explain the features of the distribution.

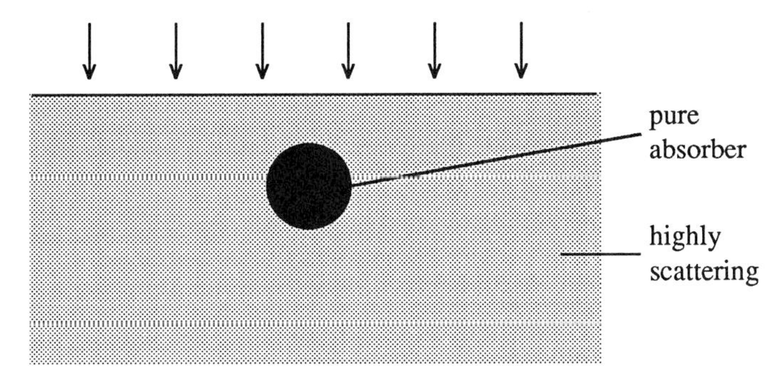

ANSWERS TO REVIEW QUESTIONS

1. The results in the table below are calculated as follows:

Column 4: Graphically, from the straight-line plot of μ_s' versus λ with $\mu_s' = 20$ at $\lambda = 500$ and $\mu_s' = 2$ at $\lambda = 1200$, or from the equivalent formula $\mu_s' = 20 - [(\lambda - 500)/700] \cdot 18$.

Columns 5,6: From columns 3 and 2, and columns 4 and 2, respectively.

Columns 7,8: Determine the R values by drawing a graph of R versus μ_s'/μ_a' using the data given in Table 2 (index-matched, infinite slab thickness).

As can be seen from the resulting reflectance spectra, the relative sizes of the reflectance dips at 600 and 900 nm for cases (a) and (b) are significantly different. The main effect of the wavelength-dependent scatter is to decrease the diffuse reflectance at longer wavelengths, as the scattering albedo falls, so that measuring the relative reflectance to estimate the relative magnitude of the absorption at 600 and 900 nm gives false results. If we define the absorbance ratio, $A = \log_{10}(R_{peak}/R_{background})$ as shown on the figure at thetop of the next page, then $A_{900}/A_{600} = 1.48$ for case (a) and 2.28 for case (b), whereas the true absorption ratio $\mu_a(900)/\mu_a(600)$ is 2.0.

1	2	3	4	5	6	7	8
Wavelength (nm)	μ_a (mm^{-1})	μ_s' (mm^{-1}) (a)	(b)	μ_s'/μ_a (a)	(b)	R_t(%) (a)	(b)
500	0.0		20	∞	∞	100	100
550	0.020		18.7	1000	935	93	92.5
600	0.050		17.4	400	348	86.5	86
650	0.040		16.1	500	402	88	86.5
700	0.025		14.9	800	596	91	89
750	0.032		13.6	625	425	89.5	86.5
800	0.050	20	12.3	400	246	86.5	83.5
850	0.070		11.0	286	159	84.5	80.5
900	0.100		9.7	200	97	82	76
950	0.080		8.4	250	105	83.5	78
1000	0.050		7.1	400	142	86.5	80
1050	0.018		5.9	1111	328	93	85
1100	0.0		4.6	∞	∞	100	100

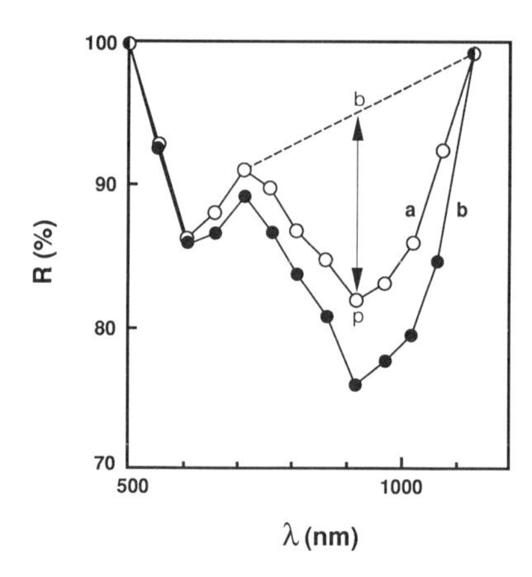

2. For no scattering (cases a,b,e), the fluence curve is obtained simply from $\phi = \phi_o e^{-\mu_a z}$, where $\phi_o = 1$ W·cm^{-2} and z = depth in tissue. In this case the beam size makes no difference, since there is no light scattered into or out of the beam. For example, for $\mu_a = 0.1$ mm^{-1} and z = 20 mm, $\phi = e^{-2} = 0.135$; for $\mu_a = 1$ mm^{-1} and z = 5 mm, $\phi = e^{-5} = 0.0067$. For the cases with both scattering and absorption and large beam size (cases c and d), the final exponential part of the curve is given by Eq.3 in the text. The slope of the line on a log-linear plot is $-e^{-\mu_{eff}}$, where the effective attenuation coefficient is calculated as $\mu_{eff} = \sqrt{3\mu_a \mu_s'}$ (Eq. 4): $\mu_{eff} = 0.17$ mm^{-1} for case c and 0.55 mm^{-1} for case d. For case c the scattering is fairly small so that the curve is close to exponential everywhere and the peak is approximately equal to ϕ_o: at z = 20 mm, $\phi = e^{-3.4} = 0.033$. For case d, there is a sub-surface peak, as in Fig. 2d, and the fluence is greater than 1 W·cm^{-2} at the surface due to backscattered light. Qualitatively, we have shown the peak at about 2 mm depth and twice the incident fluence. Thus, at z = 12 mm (10 + 2), $\phi = 2\phi_o e^{-10\times 0.55} = 0.0082$. For a small beam, case f, the fluence is less than in case d everywhere, and the slope of the curve is greater, due to decreased scatter into the center of the beam. This is plotted qualitatively in the diagram.

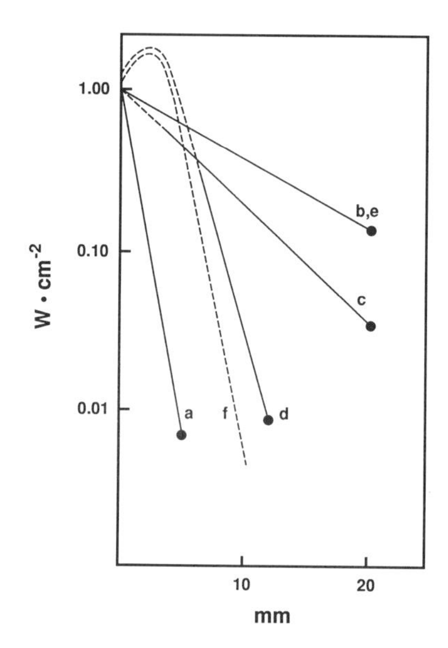

3. The figure below shows two different cases: A) when the absorption of the middle layer is greater than the effective attenuation in the upper and lower layers and B) vice versa. In case A, the fluence falls more rapidly in the absorbing layer. The exponential fluence distribution governed by the effective attenuation becomes re-established in the lower layer after the transition through the purely absorbing layer.

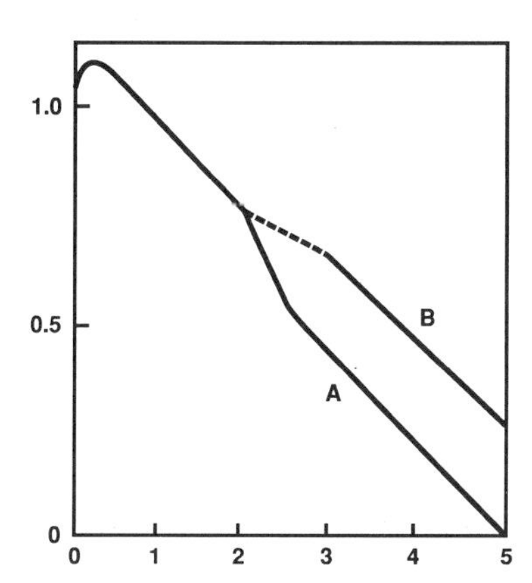

4. The effect of the absorbing structure is to reduce the fluence "down stream" of the structure. The contour lines marking regions of equal fluence curve around the structure are as shown in the figure below. The degree of "filling in" of the fluence past the structure depends strongly on the degree of scattering of the medium. At one extreme, for very low scatter, the shadow of the structure is maintained, whereas at very high scatter the effect of the optical inhomogeneity is felt only locally.

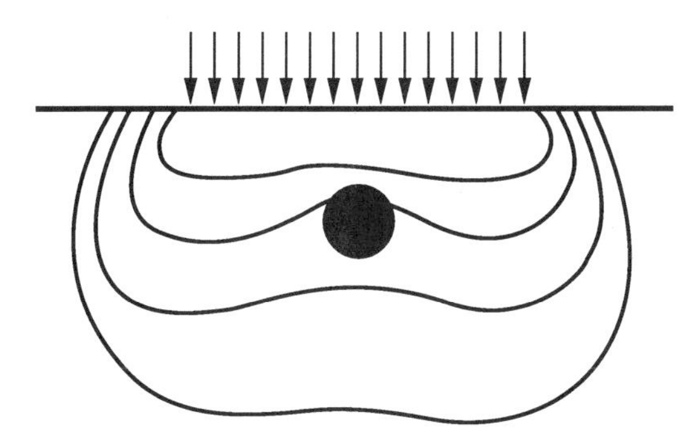

SUPPLEMENTARY READING

Grossweiner, L.I. (1989) Photophysics. In *The Science of Photobiology*, 2nd edition (Edited by K.C. Smith), pp. 1-45. Plenum Press, New York.

REFERENCES

Anderson, R.R. and J.A. Parrish (1982) Optical properties of human skin. In: *The Science of Photomedicine* (Edited by J.D. Regan and J.A. Parrish), pp 147-194, Plenum Press, New York.

Flock, S.T., B.C. Wilson and M.S. Patterson (1987) Total attenuation coefficients and scattering phase functions of tissues and phantom materials at 633 nm. *Med. Phys.* **14**, 835-842.

van de Hulst, H.C. (1980) *Multiple Light Scattering Tables, Formulas and Applications*, Academic Press, New York.

Ishimaru, A. (1978) *Wave Propagation and Scattering in Random Media*, Academic Press, New York.

Marchesini, R., A. Bertoni, S. Andreola, E. Melloni and A.E. Sichirollo (1989) Extinction and absorption coefficients and scattering phase functions of human tissues *in vitro*. *Appl. Opt.* **28**, 2318-2324.

Marijnissen, J.P.A. and W.M. Star (1987) Quantitative light dosimetry *in vitro*. *Laser Surg. Med.* **7**, 235-242.

Parsa, P., S.L. Jacques and N.S. Nishioka (1989) Optical properties of rat liver between 350 and 2200 nm. *Appl. Opt.* **28**, 2325-2330.

Patterson, M.S., E. Schwartz and B.C. Wilson (1989) Quantitative reflectance spectrophotometry for the noninvasive measurement of photosensitizer concentration in tissue during photodynamic therapy. *Proc. S.P.I.E.*, **1065**, 115-122.

Star, W.M., J.P.A. Marijnissen and M.J.C. van Gemert (1987) Light dosimetry: Status and prospects. *J. Photochem. Photobiol. B* **1**, 149-168.

Wilson, B.C., M.S. Patterson, S.T. Flock and D.R. Wyman (1990) Tissue optical properties in relation to light propagation models and *in vivo* dosimetry. In: *Photon Migration in Tissues* (Edited by B. Chance), pp 25-42, Plenum Press, New York.

Wilson, B.C. and M.S. Patterson (1990) The determination of light fluence distributions in photodynamic therapy. In: *Photodynamic Therapy of Neoplastic Disease* (Edited by D. Kessel), **Vol. 1**, pp 129-168, CRC Press, Boca Raton.

Wilson, B.C., M.S. Patterson and D.M. Burns (1986) Effect of photosensitizer concentration in tissue on the penetration depth of photoactivating light. *Lasers Med. Sci.* **1**, 235-244.

Measurement of Fluoresecence Decays

Alessandra ANDREONI
C.E.O.S.–C.N.R. and University of Naples
Italy

INTRODUCTION

This chapter discusses both the excited state techniques that can be implemented in a non-specialized laser laboratory and the identification of instruments that would allow experiments on large classes of dyes. Systems based on both time- and frequency-domain analysis of the fluorescence decay will be examined. Time- and frequency-domain analysis based on fast detection and averaging of the detector output or on single-photon timing will be reviewed. Criteria for choosing among commercially available equipment, as well as for implementing home-made systems, will be examined.

Basic Photophysics

The first step in the activation process of a photochemical reagent or a photosensitizer is the photoexcitation of the reagent to its lowest excited singlet state, S_1. From there, it can be transformed by a chemical reaction, undergo intersystem crossing to the triplet state, T_1, or decay back to the ground state, S_o, via nonradiative decay pathways or fluorescence emission. It is generally assumed that for most photosensitizers, the lifetime of S_1 is too short for reaction with a substrate. However, since the lifetime of T_1 is relatively longer, photoreactions, as well as photosensitized reactions, can occur from the triplet state. This is normally true, however, only in the case of dilute substrate solutions where long lived states of the photosensitizer can react due to diffusion limits of the reaction rate. At very high concentrations of a particularly reactive substrate, the sensitizer singlet may be intercepted and undergo reactions within its short lifetime. Thus, in principle, in biological systems, where binding of sensitizer to substrate often occurs, the direct reactions (Type I) of excited sensitizer and substrate are particularly favored compared to Type II reactions which require oxygen.

Fluorescence Decay

Recent fluorescence decay studies of chromophores (either as photosensitizers or fluorescent labels to stain specific biomolecules) have shown this method to be a valuable

ALESSANDRA ANDREONI, C.E.O.S.-C.N.R. and Department of Biology and Cellular and Molecular Pathology, University of Naples, Napoli, Italy

tool for the investigation of the interactions underlying a variety of photophysical and photochemical phenomena. Furthermore, the recent technological improvements of systems with sub-ns time resolution have made this approach rather common. In fact, for both fluorescent labels and photosensitizers, interactions with the substrate, that can easily be investigated by means of time-resolved fluorescence spectroscopy, occur on such a time scale.

In general, to determine the most suitable approach to measure fluorescence decays, the following points have to be considered:

(i) the range of excitation wavelengths of interest

(ii) the maximum exposure in terms of light dose and/or intensity that can be delivered to the sample without producing changes in the fluorescence signal, or the maximum time that is judged to be convenient to obtain reasonably good, i.e. statistically significant, results

(iii) the time-resolution desired.

Unfortunately, the requirements that arise from these points cannot be satisfied independently of each other. For instance, adopting a lamp as the excitation source gives maximum wavelength flexibility (i), but places serious limitations on time resolution (iii) (Pesek *et al.*, 1988). In addition, even when a time resolution of 0.2 ns is satisfactory (feasible with reasonably unsophisticated lamp systems), the requirements of point (ii) may not be fulfilled. As far as point (ii) is concerned, the measurement of fluorescence decay by single-photon timing techniques offers the highest reliability. This technique may involve impractically long durations of the experiment unless excitation sources with very high repetition rates, (e.g. mode-locked lasers) are available in the required spectral range. On the other hand, commercial systems utilizing mode-locked lasers and single-photon timing techniques easily achieve time resolutions as short as ~0.1 ns. Alternatively, and advantageously in terms of duration of the experiment, when a low fluorescence level originates from a low absorbance of the sample and point (ii) does not set a limit on the excitation intensity that can be safely delivered to the sample, either thyrathron-switched gas-filled ns lamps or pulsed laser sources may be adopted in conjunction with fast detectors and averagers such as either fast-transient digitizers and boxcar integrators. Since these systems allow fluorescence decays to be measured with time resolutions as short as 0.1 ns, they can compete well with those utilizing single-photon timing. However, only single-photon timing systems can detect pulses as short as 20 ps. Such resolution requires optimum conditions and a mode-locked dye laser as the excitation source (Tamai *et al.*, 1988). Finally, when point (ii) is irrelevant, the optimum choice may be represented by instruments working in the frequency-domain with modulated light sources. Depending on the time resolution desired, either amplitude-modulated continuous-wave lamps and lasers, or intrinsically modulated sources, e.g., mode-locked lasers, may be adopted. Time-resolutions as short as a few tens of ps can be achieved with commercial systems utilizing xenon arc lamps and Pockels cells (Gratton and Barbieri, 1986). The reliability of such systems, together with their wavelength flexibility has certainly favored their recent popularity for time-resolved fluorescence spectroscopy in both photochemistry and photobiology. The discussion is still open on the performance of phase-modulation systems for the single-photon timing approach when mode-locked lasers are employed as a source, particularly when multi-exponential fluorescence decays have to be measured. This point has been discussed in more detail by Lakowicz *et al.* (1988).

From a practical point of view, the photobiological applications of time-resolved fluorescence spectroscopy present problems and limitations that are mainly related to the need for high sensitivity in cases where this requirement is concomitant with the occurrence of undesired light-induced effects. An examination of the ways by which these difficulties can be overcome is not only relevant to the attainment of reliable measurements, but may also provide useful guidelines on the choice of the most versatile system available.

Basic Principles of Time- and Frequency-Domain Measurements

The most natural approach to measure the time-evolution of the S_1 state population of a chromophore is to detect the time-resolved fluorescence pulsed signal emitted upon excitation with an extremely short light pulse. Ideally, the energy of the excitation pulse, E, should be totally emitted at time t=0. Thus the power, W(t), should be a δ-function of time, $W(t) = E\,\delta(0)$ and hence $E = \int W(t)\,dt = \int E\,\delta(0)\,dt$. This approach is convenient when the energy and pulse repetition rate of the source allow the use of a technique with sufficient time resolution. Nanosecond pulsed lamps coupled to a monochromator seldom allow excitation with high peak intensity. Thus these sources are preferably used in conjunction with single-photon timing detection techniques. Typically, the number of photons emitted in a fluorescence pulse are so few that a fast photomultiplier tube (PMT) would detect them as separated in time (Fig. 1, curve a). In this case, the PMT output in response to each excitation pulse would not correlate well with the fluorescence decay. However, when the output from many pulses is summed, the distribution of times at which the fluorescence photons are detected, yields the fluorescence time-decay. In comparison, ns-pulsed dye lasers, owing to their wavelength tunability and high peak powers, are ideal excitation sources when used in conjunction with detection techniques that utilize fast PMT's and signal averaging. In such a case, the fluorescence pulse contains a sufficiently high number of photons so that the time-resolved PMT current output (Fig. 1, curve b) is a reasonable noise replica of the fluorescence decay. Thus, for a given fluorescent sample, it is the number of photons contained in the fluorescence pulse (determined by the power of the excitation pulse) that dictates which time-resolved detection technique is most suitable.

In the study of easily photodamageable samples, one is forced to adopt the single-photon timing technique. The limited availability of high-intensity UV lasers also often limits one to the use of conventional ns flash lamps. It is worth mentioning that the number of photons one has to detect to measure a fluorescence decay with given accuracy is independent of the time-resolved technique adopted. Thus, in principle, when the fluorescence response to a single pulse emitted by the source operating at full power contains a reasonable number of photons, the fast detection of the fluorescence pulse should be preferred to single-photon timing. Exceptions to this apply when the source has a high enough repetition rate to compensate for the fact that only one fluorescent photon is detected per excitation pulse. PMT's available at present allow one to achieve shorter time resolutions with single-photon timing than with fast detection of high photon-flux pulses. This is true only if comparably short excitation pulses are used (Yamazaki *et al.*, 1985). Unfortunately,

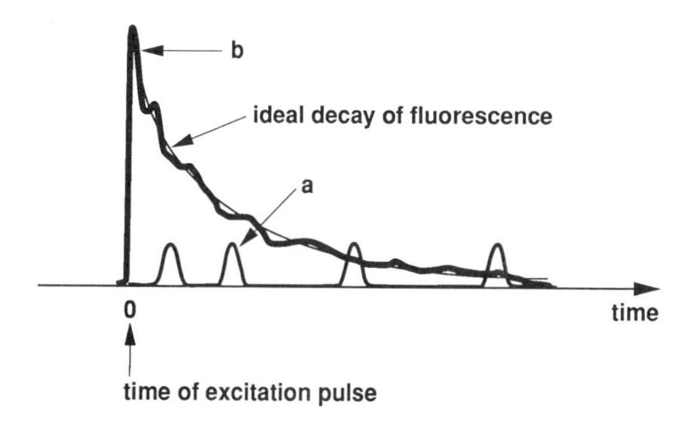

Figure 1. Typical time-resolved current output of a fast PMT detecting a fluorescence pulse excited, at time 0, by a δ-function light pulse. Either few photons, a, or many photons, b, may be detected depending on the power of the excitation pulse.

a cumbersome excitation apparatus, such as a picosecond dye laser, is needed in this case instead of the simpler short-pulse flash lamp.

In the time-domain mode, the fluorescence decay is determined by measuring the amplitude of the fluorescence pulse, from the time of the exciting pulse to times long after complete relaxation to the ground state. In the frequency domain mode, the fluorescence decay is determined by the phase delay and the modulation ratio of the emission with respect to excitation measured as a function of frequency. Weber (1977) has shown the equivalence of the two types of information which arises, for fluorescent systems responding linearly to the exciting light. This equivalence stems from the fact that time- and frequency-responses are Fourier transforms of each other. Consider a fluorescent system excited by a light pulse which is an ideal δ-function of time, t, with a multi-exponential decay given by,

$$I(t) = \Sigma_n [A_n e^{-t/\tau_n}]. \quad [1]$$

Each component of I(t) has a time-constant τ_n and an initial amplitude A_n. Define the normalized real and imaginary parts, G(f) and S(f), respectively, of the Fourier transform as,

$$G(f) = \int_o^\infty I(t) \cos(2\pi ft)\, dt \,/ \int_o^\infty I(t)\, dt \quad [2]$$

and

$$S(f) = \int_o^\infty I(t) \sin(2\pi ft)\, dt \,/ \int_o^\infty I(t)\, dt \,. \quad [3]$$

If the system is excited with light whose power is modulated at frequency f, the resulting fluorescence will be modulated at the same frequency, but the modulation amplitude will be decreased and its phase will be shifted with respect to the modulation and phase of the excitation. The phase, Φ, and the modulation, Γ, of the emission (i.e., the ratio of the modulation of the fluorescence to that of the exciting light) are given by,

$$\Phi = \tan^{-1} [S(f)/G(f)] \quad [4]$$

and

$$\Gamma = [S^2(f) + G^2(f)]^{1/2}. \quad [5]$$

Thus the measurement of Φ and Γ as a function of frequency provides the functions S(f) and G(f) and, hence, becomes equivalent to determining the time evolution of fluorescence, I(t).

Detection and Averaging of High Photon-Flux Fluorescence Pulses

When chromophores with particularly long S_1 lifetimes (e.g., pyrene derivatives) are investigated, the required instrumentation is relatively simple. The fluorescence signal can be detected by a photomultiplier (PMT) and its output can be sent directly to a digital oscilloscope that both displays and digitizes the signal. A similar approach is commonly used to detect the transient triplet state absorption in flash photolysis experiments. Once acquired, the data can be stored in the internal memory of the digitizing instrument itself, or in a computer. To increase the signal-to-noise ratio, many decay curves are accumulated and averaged. As shown in Fig. 2, the successive signals are synchronized by diverting part of the excitation pulse to a fast photodiode. The output from this fast diode is also recorded allowing the data from succesive pulses to be synchronized. Although some digitizers can also operate as averagers, averaging is usually carried out by a computer, since it is normally necessary to perform numerical analysis of the experimental fluorescence decay and fitting to theoretical models.

When sub-ns time resolution of the fluorescence decay is needed, different electronics are required. Earlier instruments, capable of sub-ns time resolution, employed boxcar averagers, i.e., gated integrators (Brahma *et al.*, 1985). Such averagers (Fig. 2), triggered by a signal obtained from the excitation pulse via a fast photodiode, perform the averaging of the output of the PMT for many repetitions of the exciting laser pulse. The boxcar averager measures the total amount of light from the fluorescence signal for only a brief period of time called a window. The window is much shorter than the decay time, and so measures only a part of the decay during any single excitation pulse. The full decay spectrum is built up by slowly moving the begining of the window relative to the excitation pulse. After many repititions, the entire decay curve has been obtained.

A method that allows virtually the same operations without some of the technical difficulties associated with the boxcar integration technique was developed by the author and colleagues (Bottiroli *et al.*, 1979). This system utilizes a very fast sampling oscilloscope (risetime = 25 ps) whose vertical output is digitized and averaged. The use of a sampling oscilloscope makes this method suitable for repetition rates of the exciting pulses of, at least, a few tens of Hz, permitting the use of sub-ns nitrogen-pumped dye lasers. Alternatively, either fast transient digitizers (Andreoni and Bücher, 1976) or the last generation digital signal analyzers such as the Tektronix DSA 602, may be used. These modern instruments have frequency bandwidths of the order of GHz.

The Single-Photon Timing Technique

In the single-photon timing technique, fluorescence decay is measured as the distribution of the times at which the PMT detects single fluorescence photons, relative to the

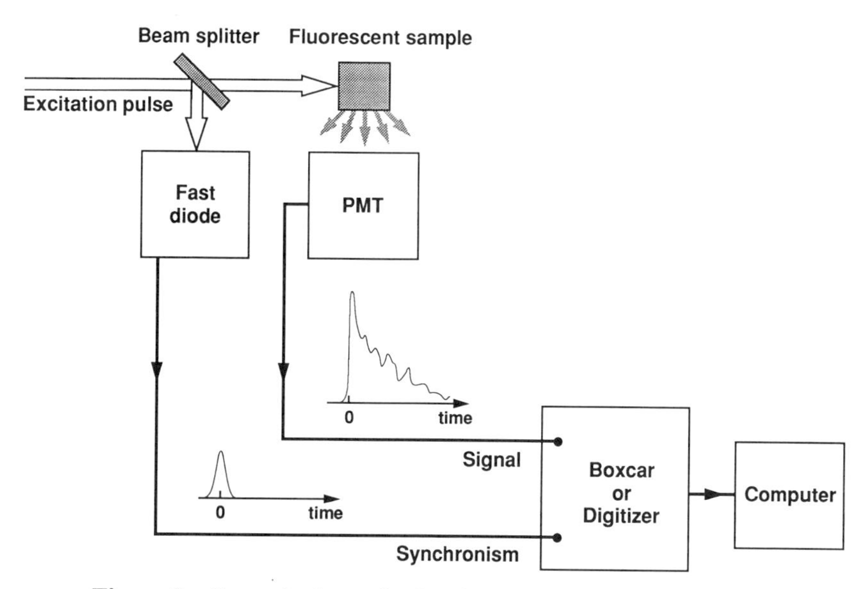

Figure 2. General scheme for fast detection and signal averaging.

time at which the exciting pulse is detected. This requires a suitably high number of repetitions of the excitation pulse.

The principles of operation of this technique are illustrated in Fig. 3. The figure shows a pulse of light exciting a fluorescent sample. Part of the excitation pulse is reflected by a beam splitter to a fast diode, while the remainder continues to the sample. The photomultiplier (PMT) detects the resulting fluorescence pulse from the sample. The goal of the technique is to determine the time interval between the excitation pulse and the fluorescence pulse, and this is accomplished by the clock in the figure. The signal from the fast diode is used to start the clock. The clock is stopped by the signal from the PMT which detects the fluorescence photons. Because the intensity of the excitation pulse is very low in this technique, the PMT typically detects only a single photon.

There are several restrictions on the equipment that can be used in such a set-up. For example, the time that it takes the fluorescence signal to traverse the PMT must be very reproducible. The fast diode can be a fast PMT, a vacuum diode, or a solid state photodiode. The clock is actually a system of electronics, which is shown as a block diagram in Fig. 4.

The clock, or single photon timing apparatus, has 3 major segments that will concern us. The heart of the system is the time to pulse-height converter. When it receives an appropriate signal, a start, it begins to generate a constantly increasing voltage pulse, i.e. a ramp of voltage. When it receives an appropriate stop signal, it stops generating the voltage ramp. Since the ramp has been increasing constantly from the start signal, the time interval between start and stop is proportional to the final voltage. The pulse shaping circuitry and constant-fraction discriminator ensure that the excitation pulse has the proper waveform to trigger the time to pulse-height converter. The same function is accomplished for the fluorescence photon signal by the amplifiers and discriminator shown at the bottom of the figure. The multichannel analyzer determines the height that the voltage ramp has reached, and passes this information to the computer which can be used to plot and analyze the complete distribution of delay times. This complete distribution is the fluorescence decay spectrum.

To ensure that the measured distribution represents an undistorted replica of the fluorescence decay, each exciting pulse should cause only a single photon to be detected by

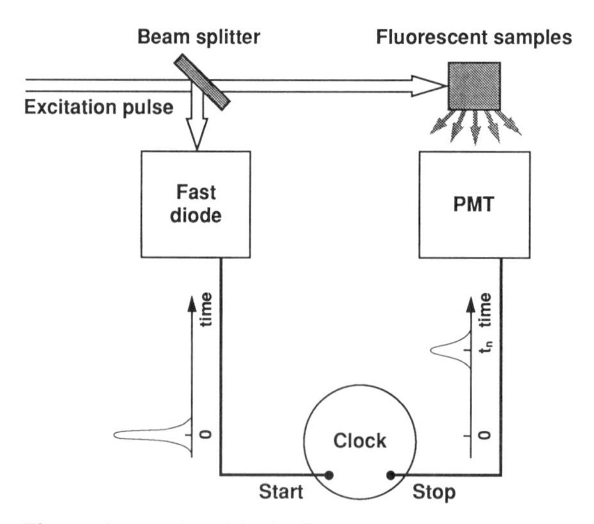

Figure 3. Typical block diagram of a single-photon timing apparatus.

the PMT. If this was not the case, the clock would stop too early for some pulses, and the tail of the distribution would never be recorded. (In fact, when more than one photon is emitted from a single excitation pulse on the sample, these photons are emitted within a time of the order of the S_1 lifetime.) Most PMTs can resolve single photons, as indicated in Curve a, Fig. 1, and hence detect them as time-separated current pulses. However, both constant-fraction discriminators and time to pulse-height converters have fairly long dead times. That is, transmission of one signal leaves them incapable of transmitting a second signal for a finite period of time. As a result, photons emitted at early times in the fluorescence decay would have a better chance of being measured than those emitted in the tail of the decay. To avoid this kind of artificial shortening of the measured decay, it is customary to use two scalers (Fig. 4). The scalers monitor the number of stop and start pulses, and check that their ratio is so low that the probability of multiple photon detection (per excitation pulse) is much less than the probability of detection of a single fluorescence photon. In practice, this means that the excitation pulse must be attenuated enough that less than 5% of the excitation pulses produce a detectable fluorescence signal. These considerations set an upper limit on the data acquisition rate that is feasible.

Phase-Modulation Fluorimetry

Phase modulation fluorimetry takes advantage of the time delay between excitation and fluorescence emission in a unique way. Because of fluorescence decay time, the fluorescence signal can persist long (~ns) after excitation. If an excitation source is used whose intensity oscillates (e.g. varies sinusoidally) at some frequency, f, then the resulting fluorescence signal would also vary with frequency, f, but its peaks and troughs would be delayed relative to the exciting light. That is, it would be phase shifted, Φ. In addition the amplitude of the fluorescence signal will depend on f. It is customary to measure this

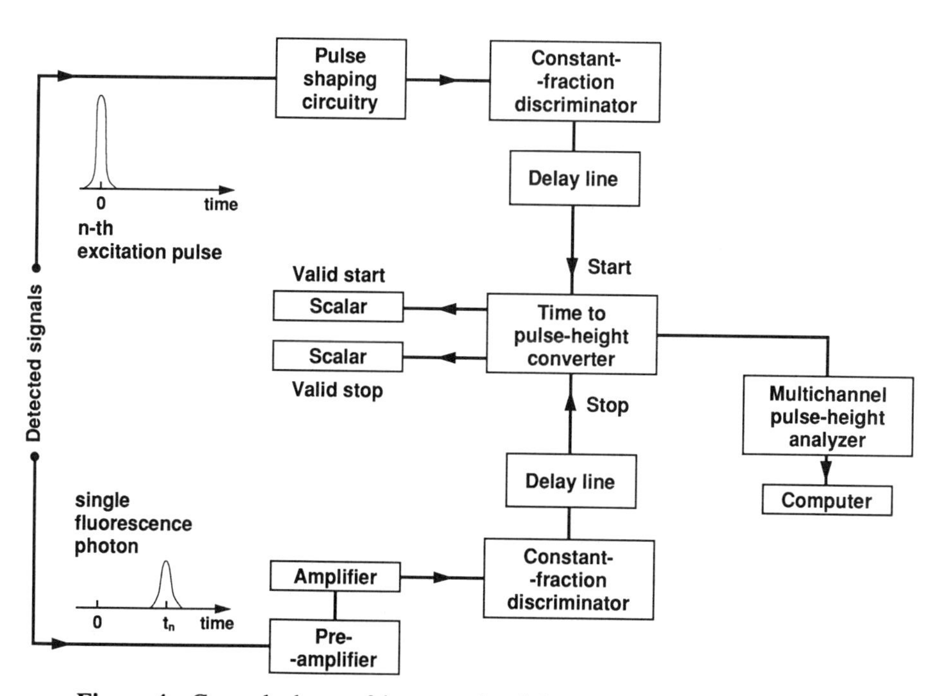

Figure 4. General scheme of time-correlated single-photon timing apparatus.

amplitude change as a modulation ratio. The modulation ratio is simply the ratio of the amplitude of the fluorescence signal to that of the exciting light.

Lifetime measurement by the phase-modulation technique involves the collection of phase shifts, Φ, and modulation ratios, Γ, of the fluorescence with respect to the exciting light, these variables being collected at several modulation frequencies, f. In the simplest case of a chromophore exhibiting a single-exponential fluorescence decay with lifetime τ, Eqs. 4 and 5 provide two independent evaluations of τ:

$$\Phi = \tan^{-1} (2\pi f \tau_{\Phi}) \quad [6]$$

and

$$\Gamma = [\, 1 + (2\pi f)^2 \tau_{\Gamma}^{\,2}]^{-1/2} \quad [7]$$

Therefore, τ can be determined either as $\tau_{\Phi} = (\tan\Phi)/(2\pi f)$ or as $\tau_{\Gamma} = [1/\Gamma^2 - 1]^{1/2}/(2\pi f)$. According to Eqs. 6 and 7, in order to measure lifetimes on the order of ns, the frequency f must be in the range of 10 to 100 MHz. For lifetime detections of the order of a few ps, Φ and Γ should be measured with a precision of 1/1000, which is not practical at such high frequencies. Spencer and Weber (1969) showed that this precision could be greatly improved by detecting the cross-correlation product* of the signal at frequency f by that at frequency f+δf, δf being a fixed frequency offset of the order of several Hz. As indicated in Fig. 5, the cross-correlation product is accomplished by modulating the gain of the photomultiplier that

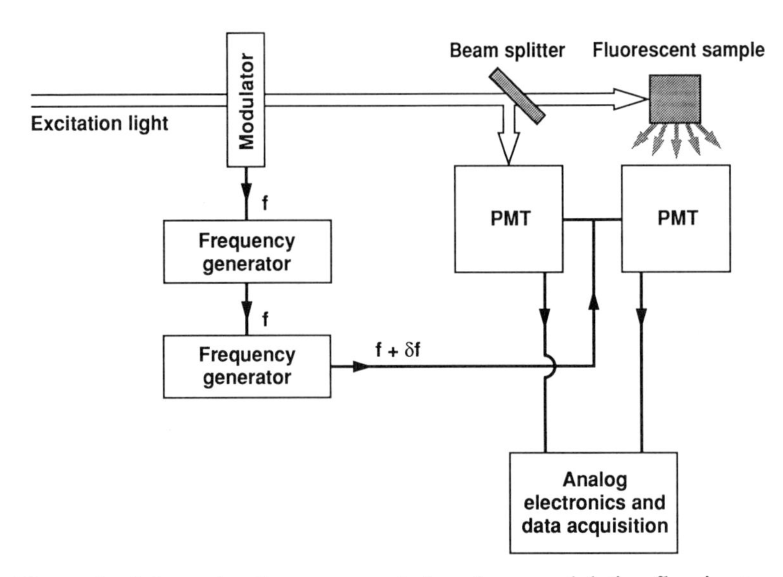

Figure 5. Schematic of a cross-correlation phase-modulation fluorimeter.

*A cross-correlation product is a clever way of determining that there is a relationship between two time-varying signals. If there's a blip in one, is there a blip in the other, and if so does it occur at the same time, or at some later time? In simple terms, the trick that's used is to multiply the two signals and to determine average amplitudes of the result. If the two signals are random, they will tend to cancel one another, that is one will be high when the other is low and vice-versa. If they are correlated, they will vary together and the average amplitude will be higher. If, on the other hand, there is a time delay between the correlated signals, we would expect the greatest average amplitude of the product if the second signal were translated back in time by an amount equal to the time delay, and this is just what is done. The average amplitude of the product of the signals is determined for different time displacements of the second signal.

detects the fluorescence signal at frequency f+δf, while the power of the exciting beam, and hence the fluorescent light, is modulated at frequency f. In order to obtain phase coherency between the signals of the two frequency generators, the one generating f+δf is driven by the oscillator, i.e., a quartz crystal of the one generating f. An identical PMT, whose gain is also modualted at frequency f+δf, detects a portion of the excitation light deviated by a beam splitter. The outputs of the PMTs thus contain components at frequency δf that are filtered by the electronics and these components are used to measure Φ and Γ as a function of f (Gratton and Limkeman, 1983). Since the use of cross-correlation detection results in the rejection of harmonics and other sources of noise, this approach has the further advantage that it does not require the excitation to be purely sinusoidal. Therefore, any time profile containing a frequency component that can be synchronized with the gain modulation of the PMT can be used for the excitation light. For instance, the source can be either a synchrotron (Gratton *et al.*, 1984; Conti *et al.*, 1984; Gratton and Barbieri, 1986), whose storage ring radio-frequency cavity is driven by the generator of frequency f (Fig. 5), or a mode-locked ion laser, in which case the frequency f drives the acousto-optic crystal modulating the laser losses, or any dye laser synchronously pumped by a mode-locked ion laser (Gratton and Barbieri, 1986; Lakowicz *et al.*, 1988; Valat *et al.*, 1988). As has been pointed out by Lakowicz et al. (1988), the improved frequency range of the responses of microchannel plate PMTs, such as the R2566U-6 (1% of maximum response at 10 GHz), might allow measurements at f ~ 10 GHz to become feasible in the near future. This is extremely interesting since, at such high frequencies, even ps decay time components would give noticeable Φ (see Eq. 6). The high accuracy that can be achieved by the cross-correlation technique may eventually show that fluorescence can be detected in the frequency domain with a time-resolution comparable to that by which it can be excited with the already available femtosecond laser sources.

EXPERIMENT 2: SELF AGGREGATION IN HEMATOPORPHYRIN AS MEASURED BY ABSORBANCE SPECTROSCOPY

OBJECTIVE

This laboratory exercise is designed to illustrate how aggregation phenomena can place severe limitations on the Beer-Lambert absorbance law.

LIST OF MATERIALS

Hematoporphyrin IX dihydrochloride (HP) (Formula weight = 671.62)
Phosphate buffer, pH 7.4
10 ml calibrated pipettes (3)
10 ml pipettes (3)
5 ml pipette
1.00 cm path length absorbance cuvettes, quartz (2)
Beakers, 50 ml (6)
Volumetric flask, 50 ml
Spectrophotometer

PREPARATIONS

Hematoporphyrin IX Dihydrochloride Stock Solution: Dissolve 3.4 mg HP in ~ 25 ml phosphate buffer in a 50 ml volumetric flask. Dilute to mark. This is the stock solution (100 μM) also called C_{max} for maximum contentration.

Series Dilutions: Prepare a solution of 50 μM HP ($C_{max}/2$) by adding 8 ml stock solution (C_{max}) with 8 ml buffer solution in a 50 ml beaker. Add 8 ml of the 50 μM solution to 12 ml buffer in a second 50 ml beaker to prepare the 20 μM ($C_{max}/5$) solution. Continue the series dilution in order to prepare 20 ml of 10 μM HP (10 ml of the 20 μM solution plus 10 ml buffer), 20 ml of 5 μM HP (10 ml of the 10 μM solution plus 10 ml buffer), 20 ml of 2 μM HP (8 ml of the 5 μM solution plus 12 ml buffer), 20 ml of 1 μM HP (10 ml of the 2 μM solution plus 10 ml buffer). You should be left with 8 ml of 50 μM ($C_{max}/2$), 10 ml of 20 μM ($C_{max}/5$), 10 ml of 10 μM ($C_{max}/10$), 12 ml of 5 μM ($C_{max}/20$), 10 ml of 2 μM ($C_{max}/50$) and 20 ml of 1 μM ($C_{max}/100$).

Spectrophotometer: Any conventional double beam, ratio recording visible-UV spectrophotometer, with a linear absorbance range of 0-2.5 will suffice. This experiment was successfully tested with the use of a Shimadzu UV-160.

Quartz cuvettes: The 1.00 cm path length quartz cuvettes should be cleaned thoroughly prior to the start of the experiment. Soaking the cuvettes overnight in 3 M nitric acid, followed by thorough rinsing with distilled water, is a good cleaning procedure.

EXPERIMENTAL PROCEDURE

1. Verify the aborbance baseline by recording the absorbance spectrum between 350 and 410 nm with phosphate buffer in both the analytical and reference cuvettes.
2. Empty the analytical cuvette of its buffer solution and fill it with the 1 μM HP solution ($C_{max}/100$). Replace the cuvette in the analytical compartment of the spectrophotome-

ter and record the absorbance spectrum using a vertical scale that allows easy reading of the absorbance values in the range 350-410 nm. Determine the absorbance peak wavelength accurately and call it λ_p.

3. Empty the measuring cuvette by pouring the contents into the beaker containing the 1 μM solution. Simply dry the cuvette, without washing it, and fill it with the 2 μM HP solution ($C_{max}/50$). Record the absorbance spectrum and take note of the absorbance value at λ_p.
4. Repeat these operations for all solutions, ending with that at concentration 100 μM (stock solution, C_{max}).
5. The wavelength of maximum absorbance will not be the same for all solutions. On all recorded spectra, determine the absorbance values, A, at both λ_p and at the wavelength of maximum absorbance (λ') of the most concentrated solution.
6. Using linear graph paper, plot the absorbance values at both λ_p and λ' against HP concentration.
7. Keep all solutions, since they are required for Experiment 3 (Fluorescence Lifetimes of Monomeric and Aggregated Hematoporphyrin).

DISCUSSION

An examination of the Beer-Lambert plots (step number 6 above) reveals that the linear relationship between absorbance and HP concentration is only observed for dilute solutions. Thus absorbance measurements could not be used to reliably determine the concentration of concentrated solutions from the linear component of the Beer-Lambert plot.

Since absorbance can be expressed as $A = \varepsilon c l$ where ε is the molar extinction coefficient, and l is the path length, it is revealing to plot the data as ε vs c. Conversion of A into ε is easily made by dividing each absorbance value by its corresponding concentration. Such a plot should yield an ε value of ~200 000 $M^{-1} \cdot cm^{-1}$ (Andreoni *et al.*, 1983) at low Hp concentration values. The plot of $\varepsilon(\lambda_p)$ vs. c should curve downwards, whereas the plot of $\varepsilon(\lambda')$ vs. c should curve upwards. At low HP concentrations where both plots meet, the ε value is representative of the monomer. At higher concentrations, HP self-associates to form aggregates whose absorption spectrum is blue-shifted compared to that of monomeric HP. At high concentrations, the ε value reflects that of the aggregated HP. Since the molar extinction coefficient is an intrinsic constant for a particular molecular species, any deviation from this value at higher concentration indicates a change in the molecular structure of the solute.

EXPERIMENT 3: FLUORESCENCE LIFETIMES OF MONOMERIC AND AGGREGATED HEMATOPORPHYRIN

OBJECTIVE

This laboratory exercise results in the quantitative evaluations of the fluorescence lifetimes of different fluorescent species in solutions, as well as in the determination of the fractions of each fluorescent species found in the solutions.

LIST OF MATERIALS

Fluorimetric cuvette, 1 × 1 × 4.5 cm, quartz
Hematoporphyrin IX dihydrochloride (HP) (Formula weight = 671.62)
Phosphate buffer, pH 7.4
5 ml pipette
10 ml calibrated pipettes (3)
10 ml beakers (6)
50 ml beakers (6)
Fluorescence lifetime measurement system

PREPARATIONS

Hematoporphyrin IX Dihydrochloride Stock Solution: Dissolve 3.4 mg HP in ~ 25 ml phosphate buffer in a 50 ml volumetric flask. Dilute to mark. This is the stock solution (100 μM) also called C_{max} for maximum contentration.

Series Dilutions: Prepare a solution of 50 μM HP ($C_{max}/2$) by adding 8 ml stock solution (C_{max}) with 8 ml buffer solution in a 50 ml beaker. Add 8 ml of the 50 μM solution to 12 ml buffer in a second 50 ml beaker to prepare the 20 μM ($C_{max}/5$) solution. Continue the series dilution in order to prepare 20 ml of 10 μM HP (10 ml of the 20 μM solution plus 10 ml buffer), 20 ml of 5 μM HP (10 ml of the 10 μM solution plus 10 ml buffer), 20 ml of 2 μM HP (8 ml of the 5 μM solution plus 12 ml buffer), 20 ml of 1 μM HP (10 ml of the 2 μM solution plus 10 ml buffer). You should be left with 8 ml of 50 μM ($C_{max}/2$), 10 ml of 20 μM ($C_{max}/5$), 10 ml of 10 μM ($C_{max}/10$), 12 ml of 5 μM ($C_{max}/20$), 10 ml of 2 μM ($C_{max}/50$) and 20 ml of 1 μM ($C_{max}/100$).

Fluorescence Lifetime Measurement System: Several commercial systems are available and the relative merits of these are discussed in the Introduction. The fluorescence detection system should preferably be red enhanced, since hematoporphyrin IX-dihydrochloride emits in the 600-700 nm region. This experiment was successfully tested using a Photon Technology Inc. LS-100 system.

EXPERIMENTAL PROCEDURE

Quartz cuvette: Very weak fluorescence can easily be measured by the high sensitivity of modern detectors. Thus it is very important that the quartz cuvette be scrupulously cleaned prior to the start of the experiment. Soaking the cuvette overnight in 3M nitric acid, followed by thorough rinsing with distilled water, is a good cleaning procedure.

Fluorescence Lifetime Measurement System: The operational procedure will of course depend on the type of instrument available. You should thus familiarize yourself with the instrument at hand, and make the corresponding adaptations. Since a PTI LS-100 instrument was used to test this experiment, relavent comments are included. The PTI LS-100 utilizes a nitrogen lamp as the excitation source and performs averaging on the fluorescence decay via a boxcar integrator. The integrator gate has a fixed duration and this is slowly swept across the decay. The N_2 lamp emission is filtered via an excitation monochromator, tuned at 337 nm (the most intense line emitted by the lamp in the spectral region of wavelengths absorbed by HP). If another excitation source is used, it would be preferable to excite at 400 nm. The emission is detected through a second monochromator, set at 610 nm. All settings, including the gate sweep velocity and the number of scans one wants to average, are performed by the operator via the computer keyboard connected to the instrument.

1. Optimize the excitation wavelength (preferably 400 nm) and emission wavelength (610 nm) on the system.
2. Fill the quartz cuvette with the 1 μM HP solution (C_{max}/100). Optimize the gate sweep velocity and the number of scans in order to obtain a relatively strong fluorescence signal. If the instrument permits, record the full fluorescence spectrum from 550 to 750 nm. Record the fluorescencce decay curve.
3. If the excitation pulse is only slightly shorter than the fluoresccence lifetime to be measured, it is necessary to also record the excitation time profile and deconvolute it from the measured fluorescence decay. To this end, a cuvette containing a scattering suspension (such as Intralipid) is substituted for the fluorescence sample. The scattered excitation pulse is measured and averaged using the same procedure as for measuring the HP fluorescence decay, except that the emission monochromator is set equal to the excitation monochromator. The averaged fluorescence decay can then be obtained with the computer calculated convolution of the average time-profile of the scattered excitation with a suitable multi-exponential decay law.
4. Steps 2 and 3 are repeated for all the other HP concentrations. The best-fit decay time constants and relative pre-exponential factors are determined for all solutions.

DISCUSSION

The fluorescence lifetime of a molecular species is an important parameter to evaluate in any photochemical/photobiological study. Its value can be related to changes in the molecular structure of the emitting species, as well as to changes in the environment. If a solution contains only one fluorescent species, a single exponential decay should be observed. Very dilute solutions of HP (<2 μM) contain essentially only monomeric HP. When the HP concentration is increased, self-association occurs, and aggregates or dimers are formed. Aggregates often exhibit increased radiationless deactivation, and thus have a shorter fluorescence lifetime. Dilute aqueous solutions containing the monomeric molecular species of HP have been reported to have fluorescence lifetimes of ~14 ns, whereas concentrated solutions of HP have lifetimes of ~4 ns, reflecting the aggregated species (Andreoni *et al.*, 1983). The results obtained in this study should be compared to those reported by Andreoni *et al.* (1983).

REVIEW QUESTIONS

1. A sample with absorbance 0.02 at $\lambda = 337$ nm has a fluorescence quantum yield value of 0.3 and a decay time of $\tau = 5$ ns.
   (i) How many fluorescence photons does the sample emit upon excitation with a nitrogen laser pulse of energy E = 100 μJ?

(ii) Does the duration of the excitation pulse affect the result?

Hint: Calculate the number of photons absorbed from that of the photons in the excitation pulse and multiply it by the fluorescence quantum yield.

2. Which technique among those described in this chapter would you adopt to measure the fluorescence decay of the sample of Question 1?

   Hint: Consider the number of photons emitted by the sample per exciting pulse calculated above, the value of τ expected and the typical specifications of nitrogen lasers as to pulse duration.

3. With reference to the setup in Fig. 3, consider an excitation source with 20 kHz repetition rate and suppose that the multichannel analyzer (MCA) is set at 1024 channels and that, due to both the time to pulse-height converter and the MCA settings, each MCA channel corresponds to 30 ps. How long does it take to accumulate 10 000 counts in peak channel of a detected fluorescence decay with a time constant of 8 ns?

   Note: The times at which fluorescence photons are detected, once converted to voltage pulses, cannot fully cover the dynamic range of the MCA because the background must also be measured. Thus, leave 100 channels corresponding to times earlier than the fluorescence photon detection times for the background.

   Hint: Consider that, for attaining the single photon regime, no more than 5% of the excitation pulses can produce detected photons. The channels in which the detected photons are counted are (1024-100) through which the counts decay from the maximum value, 10 000, with the decay law exp[-(channel number) (30 ps)/8 ns].

4. The fluorescence of a sample starts fading upon irradiation with 10 mJ light at the wavelength at which it is excited for measuring the fluorescence decay. For how long can it be safely exposed to the modulated excitation of a phase-modulation fluorimeter whose Pockels cell output beam has 50 μW average power?

5. Figure 5 shows how the cross-correlation product of the signal at frequency f by that at frequency f+δf is accomplished. Are the requirements of frequency response (risetime) of the two PMTs dictated by the value of f or by that of δf?

6. The monomer/dimer equilibrium of a chromophore at concentration C in a given solvent is such that the monomer and dimer concentrations are C_M and C_D, respectively, with $C_M + 2\,C_D = C$. If monomers and dimers have molar extinction coefficients ε_M and ε_D at a wavelength λ, what is the absorbance value at λ, $A(\lambda,C)$, of the solution at concentration C?

7. What is the simplest expected fluorescence decay law for the solution of Question 6 upon excitation with a δ-function light pulse at wavelength λ?

8. The microchannel plate photomultiplier Hamamatsu R1564U has a typical gain of 5×10^5 and maximum anode dark current of 5 nA. Assume that the quantum efficiency of the cathode at 420 nm is 15%. Calculate the rate at which photons (at 420 nm) would need to strike the cathode to generate an anode photcurrent equal to the dark current.

   Hint: Calculate the rate of electrons equivalent to 5 nA current at the anode and then the corresponding rate of electrons at the cathode.

9. A fluorescence light pulse at 420 nm is detected by the PMT in Question 8 and displayed by a suitably fast oscilloscope with 50 Ω input impedance as a voltage pulse of 100 mV peak. Since the fluorescence pulse is shorter than the risetime of the PMT (0.22 ns), the voltage pulse has a full width at half maximum (FWHM) equal to 0.22 ns. Assuming a Gaussian shape for the voltage pulse, calculate the number of photons in the

fluorescence pulse. It will result that only one of the techniques described might be used to measure the fluorescence decay time in this case. Assess which one, explaining why.

Hint: Convert voltage into current and integrate the Gaussian current pulse to find the integral charge in the PMT output pulse. Remember that the integral of a Gaussian is equal to the product of maximum amplitude by FWHM. Then use the data of Question 8 for gain and quantum efficiency.

ANSWERS TO REVIEW QUESTIONS

1. (i) The number of photons in the exciting pulse is the ratio of the pulse energy E to the energy of each photon: $N_{exc} = E/(hc\lambda_{exc}) = 1.87 \times 10^{14}$. The number of photons absorbed is then $N_{abs} = N_{exc}(1-10^{-0.02})$ and that of the photons emitted is obtained by multiplying N_{abs} by the fluorescence quantum yield: $0.3N_{abs} = 2.52 \times 10^{12}$.

   (ii) No, because the answer to point (i) depends only upon the energy of the excitation pulse, whereas the pulse duration affects the peak power of the excitation.

2. Small nitrogen lasers typically emit 100 μJ pulses (See Question 1) with durations of the order of magnitude of 1 ns (peak power $W = 10^{-4}$ J/10^{-9} s $= 100$ kW). If one supposes that 1% of the fluorescence photons calculated above can be collected onto the photocathode of a PMT with 5% quantum efficiency (average over the fluorescence spectral region), and 10^5 gain, one would obtain a PMT output containing the following charge value: $Q = 2.52 \times 10^{12} \times 0.01 \times 0.05 \times 105 \times 1.6 \times 10^{-19} = 2 \times 10^{-5}$ C. With a 1 ns PMT risetime, the output current pulse, i(t), would be an exponentially decaying pulse with a time constant virtually identical to the fluorescence decay time $\tau = 5$ ns. Therefore, its time integral would be $Q = i(t{=}0)\tau$. Note that a peak current value such as $i(t{=}0) = Q/\tau = 2 \times 10^{-5}$ C/ 5×10^{-9} s is enormous. The excitation must be attenuated by orders of magnitude to keep i(t=0) below the maximum peak current allowed to have a linear response from the PMT (say, 10 mA). We attenuate, for instance,by a factor of 10^6 and obtain i(t=0) = 4 mA as the peak value of the current output which, displayed by any instrument with 50 Ω input impedance would produce a voltage pulse $V(t) = 50\, i(t{=}0) \exp(-t/\tau)$, i.e., a peak voltage signal of 200 mV. The signal can be averaged to increase the signal-to-noise ratio. Note that, with a PMT having a more extended linear range, the excitation power could be increased to the advantage of the signal-to-noise ratio of the detected single fluorescence pulse.

3. The counts corresponding to the times at which the fluorescence photons are detected are distributed over 924 channels, i.e., over 924 × 30 ps = 27.72 ns. We must calculate the total number of counts $\Sigma_i C(i)$ for i=0 to 923, C(i) being the number of counts in the i-th channel. Since C(0) = 10 000, the fluorescence decay time is 8 ns and the time-width of the channels is 30 ps, the decay law of the counts throughout the MCA channels is 10 000 exp[- i (30 ps)/8 ns] whose integral equals $\Sigma_i C(i)$ and is easy to calculate. In fact, the integral of a decaying exponential is equal to the product of the initial amplitude, or pre-exponential factor, by the time constant measured in units of the independent variable. In our case, i is the independent variable, hence 8 000/30 ~ 267 channels, which is the time constant in units of channels, and 10 000 is the pre-exponential factor. Actually, we should calculate the definite integral over the interval i = 0 to 923. However, as C(i) decays by a factor of e every 267 channels, C(i) has already vanished, in practice, at i = 923 so our approximation is essentially correct. Thus $\Sigma_i C(i) = 10\,000 \times 267 = 2.67 \times 10^6$. It must be taken into account that, for attaining the single photon regime, no more than 5% of the excitation pulses can produce detected photons.

Thus the acquisition rate of the counts cannot exceed 5% of 10 kHz = 500 Hz (or counts per second). To accumulate 2.67×10^6 counts requires a time equal to $2.67\times10^6/500$ = 5340 s = 1.48 h = 1 h 29 min.

4. A beam of 50 μW power delivers 50 μJ energy every second. Hence, the maximum exposure time can be calculated as the ratio 10 mJ/50 μW = 200 s.

5. The frequency responses of the PMTs must be high at frequency f. See text and Lakowicz et al. (1988).

6. $A(\lambda,C) = \varepsilon_M C_M l + \varepsilon_D C_D l$ where l is the optical path length of the spectrophotometer cuvette.

7. If no interactions occur in the excited state between monomers and dimers, the fluorescence decay of the solution at concentration C is the sum of the fluorescence decays of the monomers and of dimers. Let us call τ_M and τ_D the corresponding decay times and remember that the number of monomers excited at time 0 by the light pulse is proportional to $\varepsilon_M C_M$ and that of dimers to $\varepsilon_D C_D$. Therefore, according to Eq. 1, the fluorescence decay law should be:

$$I(t) = \varepsilon_M C_M \exp(-t/\tau_M) + \varepsilon_D C_D \exp(-t/\tau_D)$$

For a discussion of the effects of excited state interactions on the fluorescence decay see Andreoni (1988) Fluorescence lifetimes of chromophores interacting with biomolecules, in *Photosensitization. Molecular, Cellular and Medical Aspects* NATO-ASI, Series H: Cell Biology, Vol. 15 (Edited by G. Moreno, R.H. Pottier and T.G. Truscott), pp. 29-38. Springer-Verlag, Berlin.

8. A 5 nA current corresponds to a rate of $5\times10^{-9}/1.6\times10^{-19} = 3.125\times10^{10}$ electron/s at the anode and to a rate of $3.125\times10^{10}/5\times10^5 = 6.25\times10^4$ electron/s leaving the cathode. Since 15% photons incident onto the cathode produce a photoelectron, such cathodic current would be generated by a rate of $6.25\times10^4/0.15 = 4.17\times10^5$ incident photons.

9. A 100 mV voltage display by the 50 Ω oscilloscope corresponds to 100 mV/50 Ω = 2 mA current peak. The integral of the Gaussian current pulse is equal to $(2\times10^{-3}\times 0.22\times10^{-9}) = 0.44\times10^{-12}$. $C = (0.44\times10^{-12}/1.6\times10^{-19})$ electrons = 2.75×10^6 electrons. Such a number of electrons in the anode pulse corresponds to a number of electrons at the cathode (prior to multiplication by the gain) equal to $2.75\times10^6/5\times10^5 = 5.5$ which, to be generated as photoelectrons by the cathode, require 5.5/0.15 = 36.7 photons in the fluorescence pulse. Apparently, the pulsed excitation source gives a relatively low number of fluorescence photons and the decay time is faster than the PMT risetime. For these reasons, averaging the PMT output is time-consuming and useless. Both single-photon time correlation and phase modulation techniques offer better time resolutions. Presumably, however, no continuous wave excitation source would allow excitation power higher than the peak power of the pulsed source used in the example. Therefore, phase modulation does not seem to be suitable, while the measurement might be feasible with single-photon timing, also because the PMT has a very low transit-time jitter (70 ps).

SUPPLEMENTARY READING

Grossweiner, L.I. (1989) Photophysics. In *The Science of Photobiology*, 2nd edition (Edited by K.C. Smith), pp. 1-45. Plenum Press, New York.

Birks, J.B. (Editor) (1975) *Organic Molecular Photophysics*, Vol. 1 & 2. John Wiley & Sons, London.

Birks, J.B. (1970) *Photophysics of Aromatic Molecules*. Wiley-Interscience, New York.

Cundall, R.B. and R.E. Dale (Editors) (1983) *Time Resolved Fluorescence Spectroscopy in Bio-chemistry and Biology*. Plenum Press, New York

Wehry, E.L. (Editor) (1976) *Modern Fluorescence Spectroscopy*, Vol. 2. Plenum Press, New York.

REFERENCES

Andreoni, A. and H. Bücher (1976) Fluorescence of uranium hexafluoride in the gas phase. *Chem. Phys. Letters* **40**, 237-240.

Andreoni, A., R. Cubeddu, S. De Silvestri, G. Jori, P. Laporta and E. Reddi (1983) Time-resolved fluorescence studies of hematoporphyrin in different solvent systems. *Z. Naturforsch* **38c**, 83-89.

Bottiroli, G., G. Prenna, A. Andreoni, C.A. Sacchi and O. Svelto (1979) Fluorescence of complexes of quinacrine mustard with DNA. I. Influence of the DNA base composition on the decay time in bacteria. *Photochem. Photobiol.* **29**, 23-28.

Brahma, S.K., M.P. Baek, D. Gaskill, R.K. Force, W.H. Nelson and J. Sperry (1985) The rapid identification of bacteria using time-resolved fluorescence and fluorescence excitation spectral methods. *Appl. Spectrosc.* **39**, 869-872.

Conti, F., T. Parasassi, N. Rosato, O. Sapora and E. Gratton (1984) Fluorescence studies using synchrotron radiation on normal and differentiated cells labeled with parinaric acid. *Biochim. Biophys. Acta* **805**, 117-122.

Gratton, E. and B. Barbieri (1986) Multifrequency phase fluorometry using pulsed sources. *Spectroscopy* **1**, 28-36.

Gratton, E., D.M. Jameson, N. Rosato and G. Webber (1984) Multifrequency cross-correlation phase fluorometry using synchrotron radiation. *Rev. Sci. Instrum.* **55**, 486-494.

Gratton, E. and M. Limkeman (1983) A continuously variable frequency cross-correlation phase fluorometer with picosecond resolution. *Biophys. J.* **44**, 315-324

Lakowicz, J.R., G. Laczko, I. Gryczynski, H. Szmacinski and W. Wiczk (1988) Gigahertz frequency-domain fluorometry: Resolution of complex decays, picosecond processes and future developments. *J. Photochem. Photobiol. B:Biol.* **2**, 295-311.

Pesek, J.E., H. Abpikar and J.F. Becker (1988) Fluorescence lifetime measurements of mercury/protein complexes. *Appl. Spectrosc.* **42**, 473-477.

Spencer, R.D. and G. Weber (1969) Measurement of subnanosecond fluorescence lifetimes with a cross-correlation phase fluorometer. *Ann. NY Acad. Sci.* **158**, 361-375.

Tamai, N., T. Yamazaki and I. Yamazaki (1988) Excitation energy relaxation of rhodamine B in Langmuir-Blodgett monolayer films: picosecond time-resolved fluorescence studies. *Chem. Phys. Letters* **147**, 25-29.

Valat, P., G.D. Reinhart and D.M. Jameson (1988) Application of time-resolved fluorometry to the resolution of porphyrin-photoproduct mixtures. *Photochem. Photobiol.* **47**, 787-790.

Weber, G. (1977) Theory of differential phase fluorometry: detection of anisotropic molecular rotations. *J. Chem. Phys.* **66**, 4081-4091.

Yamazaki, I., H. Kume, N. Tamai, H. Tsuchiya and K. Oba (1985) Microchannel-plate photomultiplier applicability to the time-correlated photon-counting method. *Rev. Sci. Instrum.* **56**, 1187-1194.

A Non-Invasive Technique for *in Vivo* Pharmacokinetic Measurements Using Optical Fiber Spectrofluorimetry

David A. RUSSELL
The Royal Military College of Canada
Canada

INTRODUCTION

Pharmacokinetics

The uptake and subsequent elimination of an exogenous drug from tissues and organs is important pharmalogical data which should be evaluated for any potential drug. Such data is invariably obtained by the use of *in vitro* pharmacokinetic studies. *In vitro* pharmacokinetic studies involve the injection of the drug of interest into numerous animals, at various time intervals the animals are sacrificed and several tissues are harvested. The drug is extracted from the tissues and its concentration determined usually by the use of a spectroscopic technique. Possibly the most sensitive of the spectroscopic techniques available is that of fluorescence spectroscopy. The combination of fluorescence spectroscopy with *in vitro* pharmacokinetic studies has, for example, proved invaluable for the evaluation of pharmalogical data of new photosensitizer drugs for the cancer treatment known as photodynamic therapy (PDT; Reddi *et al.*., 1987).

Unfortunately *in vitro* pharmacokinetic studies are time consuming, tedious and require a large number of animals for only a limited amount of acquired information. These problems have been overcome, to a large extent, with the development of a non-invasive *in vivo* pharmacokinetic technique (Pottier *et al.*., 1986; Pottier and Kennedy, 1989), whereby a single animal is used to obtain the complete pharmalogical data. This technique has been shown to produce results similar to those obtained using the *in vitro* studies (Biolo *et al.*, 1991). The *in vivo* pharmacokinetic technique utilises a nitrogen laser pumped dye laser as an excitation source and an intensified photodiode array for the detection system. The main drawback with this technique is that the equipment utilised may not be found in all photobiological laboratories. Therefore, for this laboratory experiment, a new instrumental design has been devised based on the use of a xenon arc-lamp as an excitation source and a photomultiplier tube as a detector. Results obtained with such a modified system compare favourably with the nitrogen laser/dye laser combination.

DAVID A. RUSSELL, Department of Chemistry and Chemical Engineering
The Royal Military College of Canada, Kingston, Ontario, K7K 5L0, CANADA

Photobiological Techniques, Edited by D.P. Valenzeno *et al.*
Plenum Press, New York, 1991

Selective Biodistribution of Photosensitizing Compounds – PDT

Numerous compounds, both natural and synthetic, have been proposed as drugs for use in the treatment of neoplastic tissue by PDT, see for example (van Lier, 1988). The predominant interest in these compounds is their ability to transfer energy, via the triplet state, to molecular oxygen ($^3O_2$) to form the oxidizing singlet oxygen ($^1O_2$) species which is thought to be responsible for the induced cytotoxic effect. However, all of these compounds also exhibit varying degrees of emission by fluorescence after excitation. This fluorescent energy can be used to locate the malignant tissue, and secondly to assess the bioselectivity of the drug for cancerous as apposed to healthy tissue. Thus, the use of the *in vivo* pharmacokinetic technique is a powerful tool in the assessment of the bioselectivity of new photosensitizer drugs.

The drug which is to be used in these studies is Uroporphyrin I which is structurally related to the drug currently used in PDT, hematoporphyrin derivative (HpD). It should be stressed however that this technique is not restricted to monitor drugs used in PDT alone, as it can be applied to any drug which has luminescent properties.

EXPERIMENT 4: NON-INVASIVE *IN VIVO* PHARMACOKINETICS

OBJECTIVE

This laboratory exercise will demonstrate the use of a simple optical fiber based spectrofluorimetric technique which may be used to obtain *in vivo* pharmacokinetic data non-invasively.

LIST OF MATERIALS

Xenon arc-lamp (75 W) source or other high intensity, stable light source
400 nm interference filter, 10 nm bandpass
10 cm focal length lens (2)
Optical fiber probe* (approximately 10 mm in diameter)
650 nm interference filter, 10 nm bandpass
Photomultiplier tube, including associated power supply and current amplifier (Photon Technology International, 01-612)
Light-tight box
Voltmeter or display device such as a chart recorder
Mouse (Skh:HR-1 hairless mouse, aged between 6-10 weeks)
Uroporphyrin I, 1.1 mg
Disposable syringe (1 ml) with a 27 gauge (or smaller) needle
HCl, 1 M, 5.0 ml
NaOH, 0.1 M, (A few drops)
NaCl, 0.1 g
Na_2HPO_4, 0.05 g
KH_2PO_4, 0.02 g
Analytical balance

PREPARATIONS

Phosphate Buffered Saline: A saline solution buffered to the physiological pH of 7.4 is required. Phosphate buffered saline (PBS) solution is ideal. Dissolve 0.085 g NaCl, 0.043 g Na_2HPO_4 and 0.012 g KH_2PO_4 in approximately 9 ml of distilled water in a 10 ml volumetric flask. When dissolved, add distilled water to a final volume of 10 ml.

Uroporphyrin Standard: This uroporphyrin I (UP) solution will be used as an optical alignment standard. Dissolve 2.5 μg of UP in 5.0 ml of 1 M HCl to produce a 0.5 μg/ml solution. Fill a 1 × 1 cm fluorescence cuvette with this standard.

Injectable Uroporphyrin: This UP solution is to be used for the *in vivo* pharmacokinetic measurements. Dissolve 1.0 mg of UP in the absolute minimum amount of 0.1 M NaOH (in practice this will be approximately 2-3 drops). Add 2.0 ml of the PBS solution. This will give a final concentration of 0.5 mg/ml. Inject 0.1 ml of this solution for every 10 g of body weight, i.e., 5.0 mg/kg body weight.

---

* The optical fiber probe, shown in Fig. 1, is an in-house built probe consisting of 7 silica optical fibers. This probe configuration may be replaced by a simpler design consisting of 2 optical fibers, one for excitation and the other for collection of the emitted fluorescence. However, with this less complicated instrumental design there are inherent problems with regards reproducibilty of the *in vivo* fluorescence signal (Russell *et al.*, 1991).

Intraperitoneal (IP) Injection: This type of injection is the simplest method for introducing an exogenous drug into an animal subject. The standard veterinary technique, which complies with the country and regional laws and standards governing such work should be adopted by the experimenter.

Optical Configuration: The arrangement of the spectroscopic instrumentation is shown in Fig. 1. The light from the xenon arc-lamp is focused onto a single excitation optical fiber by a 10 cm focal length lens. The wavelength of the exciting light is selected using the 400 nm band pass interference filter. The fluorescence emission collected by the 6 optical fibers surrounding the excitation fiber of the probe is focused by means of a second lens onto a photomultiplier tube. The emission spectrum of UP, in the acidic solvent has two fluorescence maxima, at 597 nm and 657 nm. The peak at 657 nm is the larger of the two maxima and therefore the variation of fluorescence intensity of the UP *in vivo* is monitored at approximately this wavelength by inserting a 650 nm band pass interference filter in front of the photomultiplier tube. The ends of the 6 optical fibers, the lens and the 650 nm interference filter are all contained in a light tight box to which the photomultiplier is attached. The signal from the photomultiplier can then be simply displayed by means of a voltmeter or a chart recorder.

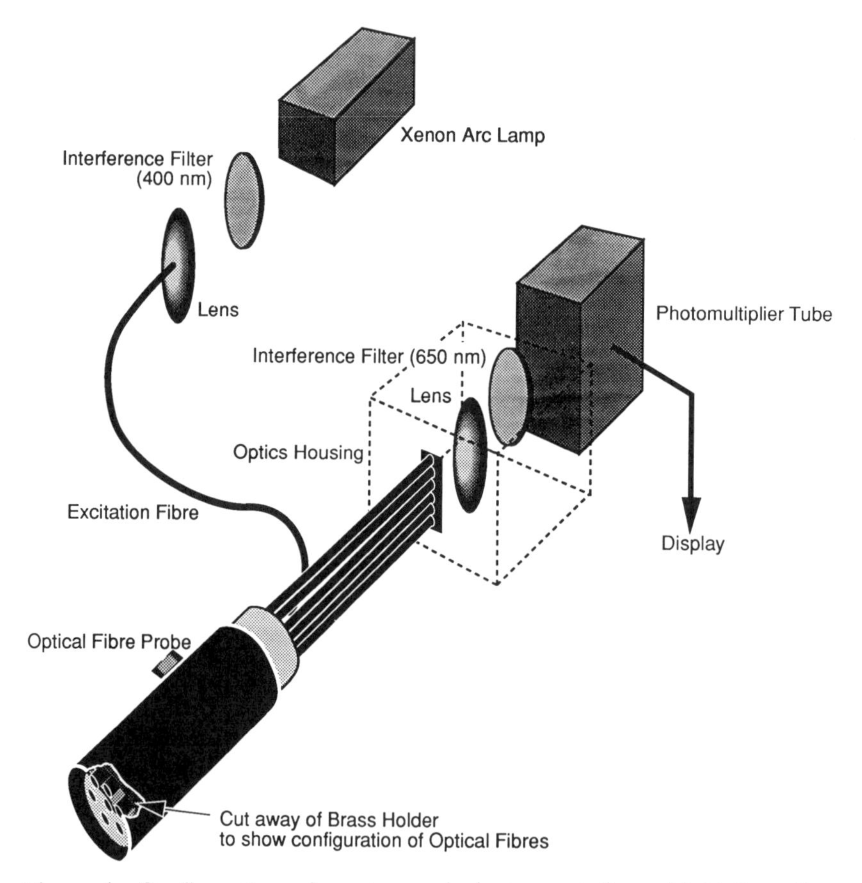

Figure 1. Configuration of spectroscopic instrumentation. Light from the xenon arc lamp is conducted by an optical fiber to the probe head. Six optical fibers, concentric with that from the lamp, collect the emitted fluorescence from the subject and conduct it to the photomultiplier.

Alignment Procedure: Once the instrumentation has been arranged as in Fig. 1, it is important to optimize the alignment of the optical components. This is best achieved by using the UP standard solution. Place the UP standard against the end of the optical fiber probe. *Be sure the lab lights are turned off.* Position the arc-lamp and lens so that the excitation light is focused directly onto the excitation fiber. With fluorescence studies, the greater the intensity of light incident on the sample (in this work the UP standard or the mouse) the larger the fluorescence signal obtained. Thus, as the light from the arc-lamp is brought into focus on the excitation optical fiber, the fluorescence obtained from the UP standard will reach its maximum. Similarly, the fluorescence detected by the six optical fibers should be focused directly onto the photomultiplier, via a lens, such that the maximum fluorescence intensity is obtained on the display unit.

EXPERIMENTAL PROCEDURE

In this experiment measurements will be taken from the abdomen and skinfold (behind the neck) areas of the mouse. Background measurements from the mouse will be made on these two sites, so that any background fluorescence signal can be subtracted from the fluorescence signal obtained from the injected drug. The experimenter should practice holding the mouse firmly to ensure reasonable reproducibility of *in vivo* signal and to prevent the animal from biting. For the abdomen this can be achieved by holding the skinfold at the back of the neck with one hand and one of the hind legs with the other hand. The skinfold measurement may be obtained by holding one end of the skinfold in each hand.

BE SURE THAT THE LAB LIGHTS ARE DIMMED.

1. Determine the mass of the mouse.
2. Hold the abdomen of the mouse against the head of the optical probe. Read and record the intensity of the background fluorescence from the photomultiplier output.
3. Hold the skinfold against the optical probe and record the fluorescence intensity as in step 2.
4. Repeat steps 2 and 3 until three measurements have been made at each site. Try to complete these 6 readings within 2 min.
5. Place the cuvette of UP standard in front of the optical probe. It should be flush with the surface of the probe head. Record the fluorescence intensity of the standard.
6. Calculate the volume of injectable UP solution (0.5 mg/ml) which is needed to inject 5.0 mg/kg of body weight.
7. Inject the volume of UP solution, determined in step 6, intraperitoneally (IP). Note the time.
8. At 5 min intervals repeat the measurements in steps 2-5. Continue for 1.5 h after the time of injection.
9. Calculate the average fluorescence intensity for each anatomical site at each time interval.
10. The fluorescence intensities obtained from the UP Standard should be constant (within approximately 5%) throughout the 1.5 h of measurement. Otherwise, normalize the data by dividing each averaged fluorescence intensity by the intensity obtained from the respective UP standard.

11. Plot the averaged fluorescence intensity for each data point as a function of time after injection to obtain pharmacokinetic curves for UP at both the abdomenal and skinfold positions.

DISCUSSION

The pharmacokinetic results should show a sharp rise of fluorescence intensity for both anatomical sites as the UP drug accumulates in the tissues of the animal. The fluorescence intensity from the abdomenal area of the mouse reflects the accumulation of the UP in the skin as well as internal organs such as the liver, which tends to concentrate such drugs. The time taken for the maximum fluorescence intensity to be obtained from the abdomen is approximately 20 min. The skinfold site however, takes slightly longer for its fluorescence intensity maximum to occur at approximately 30 min. The fluorescence from the skinfold is due to the presence of the UP drug in the skin only, suggesting that the drug accumulates in the internal organs faster than in the skin.

The time by which the fluorescence intensity maximum is reached in the various tissues can be affected by the injection mode. In this work an IP injection was used, which is the simplest method for the introduction of the UP into the mouse. However if an intravenous (IV) injection had been used to inject the UP directly into the animal's circulatory system, the intensity maxima would have occurred more promptly.

This non-invasive *in vivo* pharmacokinetic technique may be used for any fluorescing drug. However, the hydrophobicity of any such drug should be considered prior to any pharmacokinetic study. UP is a hydrophilic porphyrin and as such will be eliminated from the mouse quite rapidly. However as the hydrophobicity of the drug increases, so does the time of residency in the mouse, and therefore the analysis time and the patience of the experimenter must be increased accordingly.

REVIEW QUESTIONS

1. In this experiment a 400 nm interference filter was used to select the excitation wavelength. Could other interference filters be used for this purpose? If yes, what would be the possible wavelength(s) used?

2. Why is the 650 nm interference filter necessary? Could it be replaced by another filter?

3. If a drug was developed which had an excitation maximum in the UV region of the spectrum, *e.g.*, 220 nm, which part of the current instrumentation would cause concern with regards to its suitability for use for such an *in vivo* pharmacokinetic study?

4. The degree of hydrophobicity of an exogenous drug determines, to some extent, its pharmacokinetic behaviour. Suggest other characteristics which might affect such a drug's pharmacokinetic behavior.

ANSWERS TO REVIEW QUESTIONS

1. The appropriate use of interference filters for the selection of the excitation wavelengths is dependent on the excitation spectrum of the drug of interest. Uroporphyrin I is a typical porphyrin in that it has an excitation spectrum containing a Soret band

and 4 smaller Q bands. The Soret band at approximately 400 nm is the preferential choice for excitation, however the smaller Q bands may be used. Therefore, the wavelength of any Q band maximum could be used for selection of other possible interference filters, *i.e.*, approximately 550, 595, 615 or 630 nm.

2. One of the inherent problems with obtaining non-invasive *in vivo* pharmacokinetic data using fluorescence spectroscopy is that background fluorescence and reflected excitation light can easily give erroneous results. This can largely be prevented by the suitable selection of a filter which will only pass a wavelength of detected fluorescence which is known to come from the exogenous drug. Hence the selection of the 650 nm interference filter, as it is known that UP has a fluorescence maxima at approximately this wavelength. Other filters could be chosen, for instance at 600 nm, providing that the experimenter is confident that this wavelength is not prone to interferences from other sources.

3. The present instrumentation is designed to monitor drugs that fluoresce in the visible, particularly the red region of the spectrum, as photosensitizer drugs for PDT are optimized for this region for light transmission properties through skin. However, if pharmacokinetic behavior of a UV excited drug was to be investigated it would be necessary to ensure that all optical components were capable of transmitting such radiation. Particular attention should be paid to the optical fiber as often the transmission characteristics degenerate considerably towards the UV region of the spectrum. It would also be advisable to ensure that quartz lenses are used. Further, since the chromophores found in the skin absorb strongly at 220 nm, the *in vivo* fluorescence technique would largely be limited to the skin surface.

4. a. Chemical Structure
   b. Hydrophobicity/hydrophilicity
   c. State of ionization and aggregation
   d. Chemical interactions after injection, *e.g.*, removal of side chain groups making the drug more hydrophobic.

SUPPLEMENTARY READING

Grossweiner, L.I. 1989. Photophysics. In *The Science of Photobiology*, 2nd edition (Edited by K.C. Smith), pp. 1-45. Plenum Press, New York.

REFERENCES

Biolo, R., G. Jori, J.C. Kennedy, P. Nadeau, R. Pottier, E. Reddi and G.Weagle (1991) A comparison of fluorescence methods used in the pharmacokinetic studies of Zn(II)phthalocyanine in mice. *Photochem. Photobiol.*, **53**, 113-118.

Pottier, R.H., Y.F.A. Chow, J.-P. LaPlante, T.G. Truscott, J.C. Kennedy and L.A. Beiner (1986) Non-invasive technique for obtaining fluorescence excitation and emission spectra *in vivo*. *Photochem. Photobiol.*, **44**, 679-687.

Pottier, R.H. and J.C. Kennedy (1989) Photodiode array fluorescence technique for measuring drug clearance rates *in vivo*. *J. Photochem. Photobiol. B(Biol.)*, **3**, 135-137.

Reddi, E., G. Lo Castro, R. Biolo and G. Jori (1987) Pharmacokinetic studies with zinc(II)-phthalocyanine in tumour-bearing mice. *Br. J. Cancer*, **56**, 597-600.

Russell, D.A., P. Nadeau, R.H. Pottier, G. Jori and E. Reddi (1991) A comparison of laser and arc-lamp spectroscopic systems for *in-vivo* pharmacokinetic measurements of photosensitizers used in photodynamic therapy. In *Laser Systems for Photobiology and Photomedicine*, (Edited by S. Martellucci and A. Chester), pp. 193-199. Plenum Press, New York.

Van Lier, J. (1989) New sensitizers for photodynamic therapyof cancer. In *Light in Biology and Medicine*,.Vol. 1, (Edited by R.H. Douglas, J. Moan and F. Dall'Acqua), pp. 133-141. Plenum Press, New York.

Quantum Yield of a Photochemical Reaction

Roy H. POTTIER and David A. RUSSELL
The Royal Military College of Canada
Canada

INTRODUCTION

Photochemistry – The Foundation of Photobiology

It can be said that the foundation of photobiology is photochemistry. Photochemical reactions involve "bimolecular" interactions between a light quantum and a molecule, as well as the subsequent chemical and physical changes which result from this interaction. Light is *always* one of the initial reactants in a photochemical process. Photophysical processes are transitions which interconnect excited states with each other or excited states with the ground state, via radiative and/or radiationless processes. Photochemical processes, on the other hand, are transitions from an electronically excited state that yield structures of different configurations than the reactant ground state.

Basic Laws of Photochemistry

There are two basic laws of photochemistry, and these laws apply to all photochemical systems, be they simple gaseous atoms or complex biological molecules found in living systems. The first law of photochemistry, also called the Grotthus-Draper law, can be stated as: *only the light which is absorbed by a molecule can be effective in producing photochemical changes in the molecule.* From the first law, it is obvious that there must be some overlap between the wavelengths of light entering the reaction cell (or biological system) and those absorbed by the system under investigation, if a photochemical reaction is to take place. For quantitative work, this law must be rigorously applied to *all* aspects of a given photochemical system. This implies that, in addition to knowing the absorbance spectrum of the molecule being photolyzed, one should also know the spectral characteristics of the light source, filter(s) (or band width of the light emerging from the excitation monochromator), reaction cell windows and solvent, as well as the spectral characteristics of the photolysis products.

The second law of photochemistry, sometimes referred to as the Stark-Einstein law, is related to the quantization of light absorption by matter, and can be stated as: *the absorption of light by a molecule is a one quantum process, so that the sum of the primary process quantum yields must be unity.* The quantum yield (Φ_B) of photoconversion of reactant A into photoproduct B is defined as:

ROY H. POTTIER AND DAVID A. RUSSELL, Department of Chemistry and Chemical Engineering
The Royal Military College of Canada, Kingston, Ontario, K7K 5L0, Canada

Photobiological Techniques, Edited by D.P. Valenzeno *et al.*
Plenum Press, New York, 1991

$$\Phi_B = \frac{\text{Molecules of B formed per unit volume per unit time}}{\text{Quanta of light absorbed by A per unit volume per unit time}} \qquad [1]$$

The denominator is related to the absorbance characteristics of the reactant. The numerator, on the other hand, is related to the fraction of excited molecules that lead to the product B in question, with respect to all other competing processes which lead to the destruction or deactivation of the excited molecule. It is the sum of all the primary process quantum yields that must be unity. As such, one must always specify the process of a quantum yield (e.g. dissociation, isomerization, fluorescence, phosphorescence, radiationless transitions). The Stark-Einstein law also holds true for absorbance between a metastable state and higher electronically excited states (such as in triplet-triplet flash photolysis), and for biphotonic processes (successive absorption of a molecule to two electronic states via two photons).

The first step in any photochemical reaction is the absorption process, sometimes referred to as "photon-annihilation" (Rodgers, 1988). Of special significance to photochemical/photobiological reactions is the long lived triplet state (10^{-5} s to several seconds). Molecules in this energetic metastable state, at room temperature in liquid solutions, may undergo on the order of 10^9 collisions during their lifetimes.

Energy Surface Pathways

A convenient way to view a photochemical reaction is via an energy surface pathway, as illustrated in Fig. 1. This figure can be used to illustrate three main distinctions between photochemical and thermal reactions. First, the energy of activation of a photochemical reaction is provided by the absorption of light, as compared to the absorption of heat in a thermal reaction. Second, a photochemically activated molecule may differ significantly

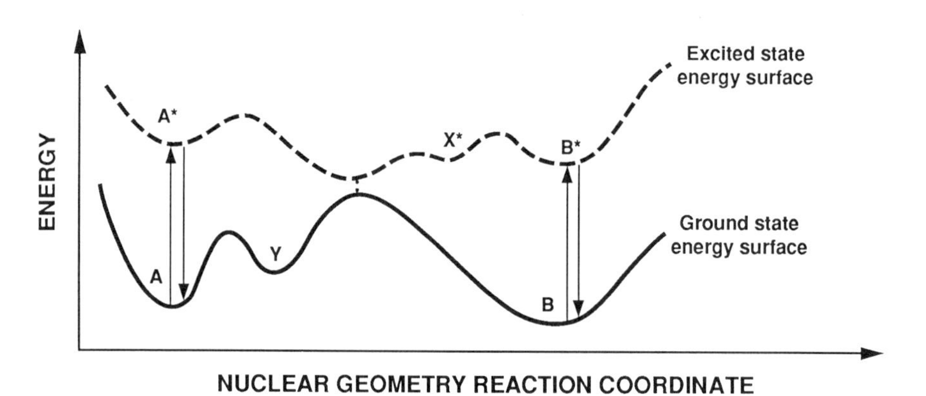

Figure 1. Comparison of thermal and photochemical reactions for reactant (A) converted into product (B). Star (*) indicates excited species. X and Y are metastable intermediate species. Vertical lines pointing up indicate absorption processes, those pointing down indicate emission. In a heat induced reaction the path is $A \xrightarrow{\Delta H} Y \rightarrow B$, where ΔH is the enthalpy of reaction. In a photochemical reaction, the path is $A \xrightarrow{h\upsilon} A^* \rightarrow B$, where hυ is the absorbed photon energy. Spin-orbit coupling induces the passage from the excited state surface to the ground state surface, indicated by the dotted line where the two surfaces are in close proximity.

from that of a thermally activated molecule. The excited molecule is in fact an *electronic isomer* of the corresponding ground state molecule. Third, product formation is far more thermodynamically accessible from a photoexcited state than from a ground state. In fact, certain photoreactions can occur at temperatures near absolute zero.

Energetic Requirements of Photoreactions

In order for a photochemical reaction to proceed, its energetic requirements must be met. Since it is the photon that contains the energy required to produce the excited state, an examination of typical photon energies that are transferred in the absorption process would be instructive. This energy is given by $\Delta E = h\upsilon$ where ΔE is the energy difference between the excited and ground state, h is Planck's constant (6.63×10^{-34} J·s) and υ is the frequency of the photon in s^{-1}. Since $c = \lambda \cdot \upsilon$, where c is the speed of light in a vacuum (3.00×10^{8} m·s^{-1}) and λ is the wavelength of the photon:

$$\Delta E = h\,c / \lambda \qquad [2]$$

The amount of energy transferred through the absorption of one mole of photons (an Einstein) of 600 nm light is thus:

$$\Delta E = (6.63 \times 10^{-34} \times 3.00 \times 10^{8} \times 6.02 \times 10^{23})/(600 \times 10^{-9}) = 200 \text{ kJ·mol}^{-1}.$$

Table 1 shows the energy per Einstein available from photons of different wavelengths.

Table 1

Available energy per Einstein at various wavelengths

λ (nm)	200	300	400	500	600	700	800	900	1000	1100	1200	1300
ΔE (kJ·mol^{-1})	599	399	299	240	200	171	150	133	120	109	99.8	92.0

The energy required to break chemical bonds is well within the range of bond energies found in many organic biomolecules. Some of the weakest bonds, such as 0-0 bonds, have typical bond energies of ~160 kJ·mol^{-1}, while the stronger bonds, such as 0-H bonds, require ~440 kJ·mol^{-1}. Since 250 nm radiation corresponds to 479 kJ·mol^{-1}, one might conclude that such photons could be used to photoreact virtually all common organic compounds. A reexamination of Fig. 1, however, reveals that the excited state must also cross energy barriers (activation energy) in order to proceed to product formation. Further, the rate of the reaction must also be considered, as well as the rates of all other competing processes. The rate of a reaction can be expressed as a function of the activation energy:

$$\text{Rate of reaction} = Ae^{-E_a/RT} = A \times 10^{-E_a/(0.0192T)} \qquad [3]$$

where A (s^{-1}) is a measure of the probability of reaction from a state which possesses the minimum energy for reaction and E_a ($kJ \cdot mol^{-1}$) is a measure of the activation energy required for the reaction to proceed. For unimolecular reactions, A is typically 10^{13} s^{-1}, and for bimolecular reactions, it is ~10^7 s^{-1}. E_a values range from a few $kJ \cdot mol^{-1}$ to ~170 $kJ \cdot mol^{-1}$. For long lived triplet states of the order of 1 s, a unimolecular rate constant at room temperature (298 K), having an activation energy of 25 $kJ \cdot mol^{-1}$, can be expected to have a value in the vicinity of 10^8 s^{-1}. Such a rate is favorable for product formation, but if the activation energy is of the order of 80 $kJ \cdot mol^{-1}$, the rate drops to ~10^{-1} s^{-1}, indicating virtually no reaction. For bimolecular photoreactions, where A~10^8 s^{-1}, an activation energy of 25 $kJ \cdot mol^{-1}$ would have a rate of only ~10^3 s^{-1}. Since the excited state lifetimes are usually much shorter than one second, photoreactions are only favorable if the activation energy is less than ~80 $kJ \cdot mol^{-1}$ for triplet states and less than ~20 $kJ \cdot mol^{-1}$ for singlet states.

Some important photoreactions in photobiology are: 1) Linear addition to an unsaturated molecule (e.g. cysteine + thymine), 2) Cycloaddition of an unsaturated molecule (e.g., thymine dimer formation in DNA), 3) Photofragmentation (e.g., riboflavin → lumiflavin), 4) Photooxidation (e.g., photooxidation of cholesterol into 3 β-hydroxy - 5α - hydroperoxy - Δ^6 -cholestene), 5) Photohydration (e.g., UV hydration of uracil), 6) Cis-trans isomerization (e.g., all-trans retinal→11-cis retinal) and 7) Photorearrangement (e.g., 7-dehydrocholesterol→pre-vitamin D_3). For further details on basic molecular photochemistry, in depth coverage is provided by Turro (1978). Photobiological applications are discussed by Grossweiner and Smith (1989).

Quantum Yields

One of the most useful quantities that can be experimentally evaluated for any photochemical process is the quantum yield of reaction, also called the quantum efficiency. The quantum yield is the probability that the absorption of one photon at a given wavelength will lead to a single observed event, such as the photolysis of one molecule, or the emission of one photon. Although conceptually simple, as indicated in Eq. 1, quantum yield determinations can be quite difficult, especially if the fraction of incident light absorbed is measured directly. Multireflections inside the photolysis cell can lead to more light being absorbed by the system than that which is measured. These difficulties have been covered in detail by Calvert and Pitts (1966). A more accurate method of quantum yield determination relies on the use of a secondary actinometer solution with an accurately known quantum yield. This method requires that the fractions of light absorbed by reactant and actinometer solution be the same. Regardless of the true value of the fraction of light absorbed by the reactant or the actinometer solution, it must be the same, since any reflection, absorption, or scattering of the light by the cell windows is identical in the two cases, and the number of quanta absorbed by each solution must be identical if the incident light intensity is kept constant in the two experiments. If the reactant is irradiated for t_B seconds and forms n_B molecules of product B, and the chemical actinometer solution forms n_A molecules of its product A when irradiated for t_A seconds, the quantum yield of product B must be given by

$$\Phi_B = (n_B t_A \Phi_A)/(n_A t_B). \qquad [4]$$

In this experiment the quantum yield of 7-dehydrocholesterol (a steroid, also known as ergosterol or provitamin D) photoconversion into previtamin D_3 will be determined. This reaction is an important reaction that takes place in mammalian skin, and is photoactivated by sunlight. Previtamin D_3 is thermally converted to vitamin D_3 by body heat. Vitamin D_3

in turn binds to a plasma vitamin D-binding protein for transport into the circulation. Vitamin D_3 prevents rickets, which is characterized by defective ossifications and disturbance of calcium uptake, resulting in weakness and deformation of the bones, especially in growing children who do not get adequate sunshine. The photochemical mechanism of this reaction has been extensively studied [Holick, *et al.*, 1980]. The primary photoproduct (previtamin D_3) undergoes two side reactions to produce the isomers lumisterol and tachysterol.

EXPERIMENT 5: DETERMINATION OF THE QUANTUM YIELD OF FORMATION OF PREVITAMIN D_3 FROM 7-DEHYDROCHOLESTEROL

OBJECTIVE

This laboratory exercise is designed to illustrate the use of actinometry in determining the quantum yield of a photochemical reaction.

LIST OF MATERIALS

7-Dehydrocholesterol (7-DHC), 2 mg
Previtamin D_3 (p-D_3), 1 mg
Potassium Ferrioxalate, 3 g
1,10-phenanthroline, 0.1 g
Methanol (spectroscopic grade), 33 ml
Sulfuric acid, 0.5 M, 460 ml
Sodium acetate, 1 M, 600 ml
295 nm narrow bandpass filter or excitation monochromator set at 295 nm
10 cm focal length, convex quartz lens (2)
1 cm × 1 cm × 4.5 cm quartz cuvettes (2)
100 ml red volumetric flasks (3)
1000 ml red volumetric flask
1 and 5 ml graduated pipettes
2 ml pipette
Pasteur pipettes
Stopclock
UV light source, such as a stabilized 75 W high brilliance Xe arc lamp. (A 250, or 500 W Xe arc will also do.)
Double beam ratio recording spectrophotometer capable of multicomponent analysis.
Dark room (or facilities to significantly reduce room lighting)

PREPARATIONS

PREPARATIONS AND THE EXPERIMENT MUST BE CARRIED OUT IN A DARK ROOM, USING A RED PHOTOGRAPHIC SAFELIGHT, OR UNDER SUBDUED LIGHT.

Potassium Ferrioxalate Actinometer Stock Solution (0.006 M): Dissolve 2.95 g of solid potassium ferrioxalate, $K_3Fe(C_2O_4)_3 \cdot 3H_2O$, in 800 ml of distilled water in a 1 liter red volumetric flask. Add 100 ml of 0.5 M H_2SO_4, dilute to 1 liter with distilled water and mix well.

NOTE: The $K_3Fe(C_2O_4)_3 \cdot 3H_2O$ crystals should be green in color. Light exposure will cause the pure green crystals to decompose, with a color change to a yellowish brown. Pure solid $K_3Fe(C_2O_4)_3 \cdot 3H_2O$ may be prepared by mixing 3 volumes of 1.5 M $K_2C_2O_4$ solution with 1 volume of 1.5 M $FeCl_3$ solution (reagent grade chemicals). The precipitated $K_3Fe(C_2O_4)_3 \cdot 3H_2O$ is then recrystallized 3 times from warm water and air dried at 45°C. The solid can be stored in the dark for extended periods of time without change.

1,10-Phenanthroline Solution: Dissolve 0.1 g 1,10-phenanthroline in 50 ml distilled water, dilute to 100 ml. This yields a 0.1% (mass/volume) phenanthroline solution.

Buffer Solution: Mix 600 ml of 1 M sodium acetate (CH_3COONa) with 360 ml of 0.5 M H_2SO_4. Dilute with distilled water to 1 liter.

7-Dehydrocholesterol Solution: Dissolve 7.4×10^{-4} g of 7-dehydrocholesterol in 5 ml methanol, dilute to 10 ml with methanol in a red volumetric flask. In a 1 cm pathlength absorbance cuvette, this solution should have an absorbance value of near unity.

Preparation of Mixed Standard Solutions: Dissolve 9.0×10^{-4} g of previtamin D_3 in 5 ml methanol, dilute with methanol to 10 ml in a red volumetric flask. Dissolve 9.0×10^{-4} g of 7-dehydrocholesterol in 5 ml of methanol, dilute with methanol to 10 ml in a red volumetric flask. The following amounts of these solutions are then taken to prepare the mixed standards in a 5 ml volumetric (red) flask. These standard solutions are then diluted to the 5 ml mark using methanol. From the exact masses of p-D_3 and 7-DHC, determine the respective concentrations in each mixed standard.

	Volume of pre-D_3 (ml)	Volume of 7-DHC (ml)
Standard 1	1	3
Standard 2	2	2
Standard 3	3	1

Photolysis System: The photolysis system used to successfully test this experiment is illustrated in Fig. 2.

Spectrophotometer: In this experiment, the analysis of product formation is made by spectral analysis of the irradiated solution. As such, the spectrophotometer used must be capable of multicomponent analysis. Low priced commercial instruments are now readily available with this capability. This experiment was successfully tested using a Shimadzu UV-160 spectrophotometer.

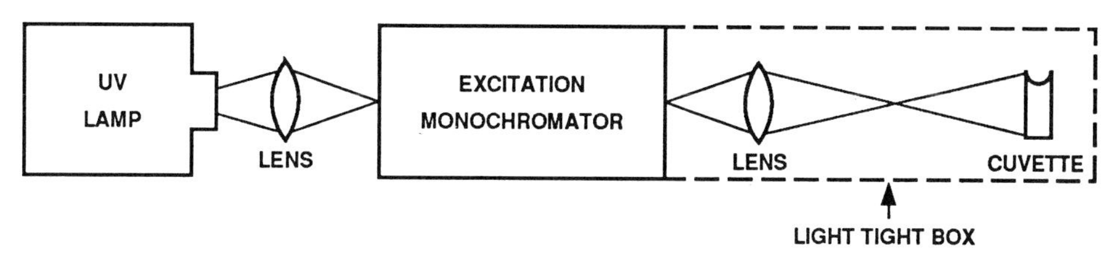

Figure 2. Schematic of photolysis system. The UV lamp was a 75 W Xe, high brilliance arc, stabilized by the use of an optical feedback system (PTI LOS200X). The excitation monochromator was a 0.2 meter, PTI, set at 295 nm. The monochromator may be replaced by a 295 nm narrow bandpass filter, in combination with a heat reflecting mirror. It is important that the entire volume content of the cuvette be irradiated.

EXPERIMENTAL PROCEDURE

Actinometer photolysis

1. Pipette 5 ml of the actinometer stock solution into a 100 ml red volumetric flask. Dilute to 100 ml with distilled water. Pipette 3.5 ml (V_1) of this diluted actinometer solution into a 1 × 1 × 4.5 cm quartz cuvette. Place this cuvette at the photolysis site, and start the stopclock. Irradiate for 5 min ($t_{Fe^{+2}}$).

2. After photolysis, pipette 2 ml (V_2) of the irradiated actinometer solution into a 5 ml (V_3) red volumetric flask. Add 0.4 ml of 0.1% 1,10-phenanthroline solution and 1 ml of buffer solution. Dilute to the 5 ml mark with distilled water. Mix well and let stand in the dark for 1 h. Label this solution as "irradiated".

3. While the irradiated solution is standing in the dark, prepare a similar parallel solution (2 ml of the diluted, unirradiated actinometer solution + 0.4 ml of the 0.1% 1,10-phenanthroline solution + 1 ml buffer solution, diluted to 5 ml with water) and allow it to stand in the dark for 1 h. Label this solution as "non-irradiated".

Photolysis of 7-Dehydrocholesterol

4. Pipette 3.5 ml of the 7-DHC solution into a 1 × 1× 4.5 cm quartz cuvette and irradiate for 15 min ($t_{P\text{-}D_3}$).

Calibration of Spectrophotometer for Multicomponent Analysis

5. While the 7-DHC solution is being irradiated, the spectrophotometer should be calibrated using the three mixed standard solutions. Record a base line between 200 and 500 nm, with both reference and analytical cuvettes filled with methanol.

6. Record the absorbance spectra of the three mixed standards between 200 and 500 nm, and enter them into the memory of the spectrophotometer, along with the concentrations of 7-DHC and pre-vitamin D_3 in each standard solution. The exact order in which these solutions and their concentration values are entered into the instrument will depend on the type of spectrophotometer used.

Analysis of Irradiated 7-DHC Solution

7. At the end of the 15 min irradiation of the 7-DHC solution, place the cuvette containing the irradiated 7-DHC into the analytical compartment of the spectrophotometer and record its spectrum from 200 to 500 nm. Enter this spectrum into the spectrophotometer memory and determine the amount of previtamin D_3 formed during irradiation.

Analysis of the Actinometer Solution

8. The number of moles of reduced iron ($n_{Fe^{+2}}$) formed during the photolysis of the ferrioxalate actinometer can easily be determined by spectrophotometry. At the end of the 1 h standing period, place the two actinometer solutions (irradiated and non-irradiated) into the analytical and reference positions (respectively) of the spectrophotometer. Measure the absorbance (A) at 510 nm. Since $A=\varepsilon cl$, where ε is the molar extinction coefficient of the Fe^{+2}-phenanthroline complex (1.11×10^4 l·mol^{-1}·cm^{-1}), c is the concentration in mol·l^{-1}, and l is the pathlength (1.00 cm), the number of moles of the Fe^{+2} complex can easily be calculated from:

$$n_{Fe^{+2}} = (V_1 V_3 A)/(V_2 \varepsilon l)$$

Quantum Yield of Formation of Previtamin D_3

9. The quantum yield of photoconversion of 7-dehydrocholesterol into pre-vitamin D_3 may be calculated via:

$$\Phi_{p\text{-}D_3} = (\Phi_{Fe^{+2}}\, n_{p\text{-}D_3}\, t_{Fe^{+2}}) \,/\, (n_{Fe^{+2}}\, t_{p\text{-}D_3})$$

The quantum yield of the actinometer, $\Phi_{Fe^{+2}}$, is equal to 1.24 (Calvert and Pitts, 1966) at 295 nm.

DISCUSSION

In this experiment, you have determined the quantum yield of formation of previtamin D_3 from 7-dehydrocholesterol, at a wavelength of 295 nm. The quantum yield of reaction of previtamin D_3 ($\Phi_{p\text{-}D_3}$) has been reported to be 0.23 (MacLaughlin, *et al.*, 1982). This could have been done by accurately measuring the number of photons absorbed by the reactant, using a photometric approach. If such a method is to be used, care must be taken to assure that the measured fraction of light transmitted by the reactant solution can be used to calculate the actual light absorbed by the reactant molecules. For example, scattering of incident light by turbid samples and reflection at the reaction cell surfaces can easily lead to errors in the determination of the number of photons absorbed. For precise quantum yield determinations, correction factors have been determined for these effects (Calvert and Pitts, 1966).

Although quantum yield determinations by the use of chemical actinometers are generally easier and can produce very accurate results, a proper choice of actinometer must be made. A list of currently known chemical actinometers has recently been published (Braslavsky and Kuhn, 1987) and should be consulted for their advantages and applicabilities. When chemical actinometers are used, it is important to assure that the same number of photons are absorbed by both the actinometer and the sample under study. This is easily done for the initial reaction time, by adjusting the concentrations of the actinometer and sample such that they both have the same absorbance. However, once photolysis begins, the rate of change in absorbance of both solutions may differ significantly. Depending on the photoproduct(s) formed from each solution, this could lead to a different number of photons absorbed by the solutions. In order to minimize this possible difference, it is best to limit the extent of both reactions to less than 10% of the total reaction.

Subsequent photoreactions or dark reactions from the photoproduct being studied can also lead to errors in the quantum yield determined. In the example studied, it is well established that previtamin D_3 forms three isomers (lumisterol, tachysterol and vitamin D_3) (Jacobs *et al*, 1977). In such a case two approaches can be taken to assure correct quantum yield determination. One approach is to separate the various products formed, say by HPLC, then to calculate the quantum yield based on the specific product of interest (previtamin D_3 in our case). The other approach is to determine the fraction of the product of interest by spectral analysis. This can only be done accurately if the spectral details of all the products formed are known, and that spectral separation can be easily achieved. HPLC analysis is normally more time consuming, but is a more reliable approach.

Competitive or consecutive reactions can sometimes be eliminated, or partially reduced, by a proper choice of experimental conditions. For example, the successive thermal reaction of previtamin D_3 into vitamin D_3 can be essentially eliminated by carrying out the 7-dehydrocholesterol photolysis at low temperature (Holick *et al.*, 1980). A proper choice of excitation wavelengths can also be used to favor the desired photoreaction and reduce the extent of side or successive reactions. Thus although previtamin D_3 (and subsequently vitamin D_3) is normally prepared by irradiating 7-dehydrocholesterol at 254 nm (Hg lamps), the formation of the two isomers (lumisterol and tachysterol) is greatly reduced by using 295 nm radiation (MacLaughlin *et al.*, 1982).

For a rigorous determination of the quantum yield of photoconversion of 7-dehydrocholesterol into previtamin D_3, the amount of lumisterol and tachysterol should have been determined. In this experiment, these two secondary reactions were neglected. The experiment could be made more rigorous by the preparation of mixed standards containing known fractions of lumisterol and tachysterol, as well as the 7-DHC and previtamin D_3. Since spectrophotometer multicomponent analysis normally requires n + 1 mixed standard solutions, where n is the number of components to be analyzed, a minimum of five mixed standard solutions would be required. While testing this experiment, $\Phi_{p\text{-}D_3}$ values on the order of 0.20-0.25 were obtained, well within the range of the published value of MacLaughlin *et al.*, (1982).

The accurate determination of quantum yields is an important step in most photochemical/photobiological studies. This experiment has been somewhat simplified in order to illustrate the principle involved during a relatively short laboratory period. Accurately determined quantum yields may require considerably more time.

REVIEW QUESTIONS

1. 7-dehydrocholesterol, dissolved in ethanol, has a molar extinction coefficient of 15 000 $l\cdot mol^{-1}\cdot cm^{-1}$ at 280 nm. Chlorophyll *a*, dissolved in carbon tetrachloride, has a molar extinction coefficient of 80 000 $l\cdot mol^{-1}\cdot cm^{-1}$ at 665 nm. Calculate the photon energy, in $kJ\cdot mol^{-1}$, that is absorbed by these molecules, at the given wavelengths. Per photon absorbed, which molecule gains more energy? Per incident photon, which molecule has the higher capacity to absorb photons?

2. Assume that you have money to invest, and you wish to support the development of new drugs. You are approached by a photochemist who needs capital to set up an industry to produce a new anti-tumor drug by a photochemical reaction. The raw material is readily available at low cost and dissolves readily in water. You are shown an absorbance spectrum of the raw material (reactant). The compound absorbs at 600 nm with an extinction coefficient of 100 000 $l\cdot mol^{-1}\cdot cm^{-1}$. Sunlight is to be used to photoexcite the reactant, in order to produce the new drug. In order to check out your investment, you consult the chemical literature on reaction rate data. You find that the proposed reaction is a unimolecular reaction with a preexponential factor (A) of 10^{13} s^{-1} and an activation energy (E_a) of 170 $kJ\cdot mol^{-1}$. Calculate the rate of the proposed reaction at 25°C. Would it be wise to invest in such a venture?

3. Explain why most C-H bonds (bond energy ~420 $kJ\cdot mol^{-1}$) in organic biomolecules are not quickly broken when irradiated with 250 nm radiation.

4. The concentrations of the diluted actinometer solution and that of the 7-dehydrocholesterol were chosen such that at 295 nm, both solutions had equal absorbance values (1.0). If a mistake is made in the preparation of the 7-dehydrocholesterol

solution, and an absorbance value of 1.2 is obtained, how would this affect the validity of the actinometry method of quantum yield determination? Would this lead to a lower or higher (or equal) value of $\Phi_{p\text{-}D_3}$?

5. 7-dehydrocholesterol has its first vibronic absorbance band at 290 nm, and has its corresponding phosphorescence at 390 nm. Tetracycline absorbs at 400 nm and phosphoresces at 510 nm. Could tetracycline photosensitize 7-dehydrocholesterol? Could 7-dehydrocholesterol photosensitize tetracycline? The drawing of an approximate energy level (singlets and triplets) diagram will aid you in answering this question. Remember that wavelength is not directly proportional to energy (but frequency is).

ANSWERS TO REVIEW QUESTIONS

1. 7-DHC. Photon energy:

   $\Delta E = (hc)/(\lambda) = (6.63 \times 10^{-34} \times 3.00 \times 10^{8} \times 6.02 \times 10^{23})/(280 \times 10^{-9}) = 428\ \text{kJ}\cdot\text{mol}^{-1}$

   Chlorophyll *a*. Photon energy:

   $\Delta E = (6.63 \times 10^{-34} \times 3.00 \times 10^{8} \times 6.02 \times 10^{23})/(665 \times 10^{-9}) = 180\ \text{kJ}\cdot\text{mol}^{-1}$

   7 DHC - gains more energy per photon absorbed, since it absorbs higher energy photons (UV).

   Chl *a* - has a higher capacity to absorb photons (higher molar extinction coefficient).

2. Rate $= A \times 10^{-E_a/(0.0912T)}$

   $= 10^{13} \times 10^{-170/(0.0192 \times 298)}$

   $= 2 \times 10^{-17}\ \text{s}^{-1}$ (no reaction, not wise to invest).

3. 250 nm radiation corresponds to 479 kJ·mol^{-1} ($\Delta E = hc/\lambda$). This is greater than the bond energy of the C-H bond. However, activation energy must also be provided. If this excitation energy is greater than 59 kJ·mol^{-1}, reaction cannot proceed. Further, deactivation processes will compete with photochemical process.

4. The chemical actinometry method of quantum yield determination is based on both solutions (actinometer and reactant under investigation) absorbing the same number of photons. If this is not the case, one solution will absorb more photons than the other, and a direct comparison cannot be made. If the 7-DHC concentration was higher than that required for equal initial absorbance, it will absorb more photons than the actinometer solution, and thus convert a greater amount of p-D_3. This would produce a higher value of $n_{p\text{-}D_3}$, yielding an artificially high value of $\Phi_{p\text{-}D_3}$.

5. Convert wavelengths to frequency or kJ·mol^{-1} as shown in the table below.

(nm)	ΔE (kJ·mol^{-1})	Transition
290	413	$S_0 \rightarrow S_1$
390	307	$T_1 \rightarrow S_0$
400	299	$S_0 \rightarrow S_1$
510	235	$T_1 \rightarrow S_0$

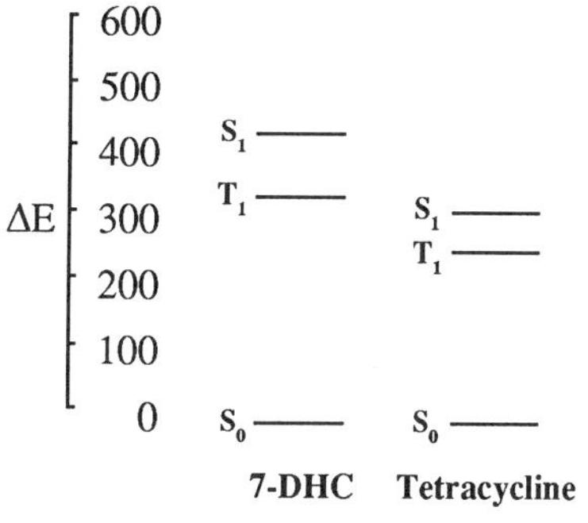

ΔE
600
500
400
300
200
100
0

7-DHC: S_1, T_1, S_0
Tetracycline: S_1, T_1, S_0

Photosensitization normally occurs via triplet-triplet energy transfer. Tetracycline cannot photosensitize 7-DHC, since the triplet energy of tetracycline is lower than that of 7-DHC. 7-DHC could photosensitize tetracycline, by exciting 7-DHC at 290 nm (413 $kJ \cdot mol^{-1}$), having singlet-triplet intersystem crossing to the 7-DHC triplet (307 $kJ \cdot mol^{-1}$), then energy transfer to the lower lying triplet state of tetracycline (235 $kJ \cdot mol^{-1}$). However, the 290 nm radiation would most probably be absorbed directly by the tetracycline via higher excited singlet states, followed by internal conversion to its S_1 state and possibly intersystem crossing to its triplet state. In order to prevent this from occurring, a long pass filter removing all radiation below 290 nm would have to be used, which would also prevent the 7-DHC from absorbing any radiation. The relative energy levels of typical photosensitizer/acceptor molecules is shown below.

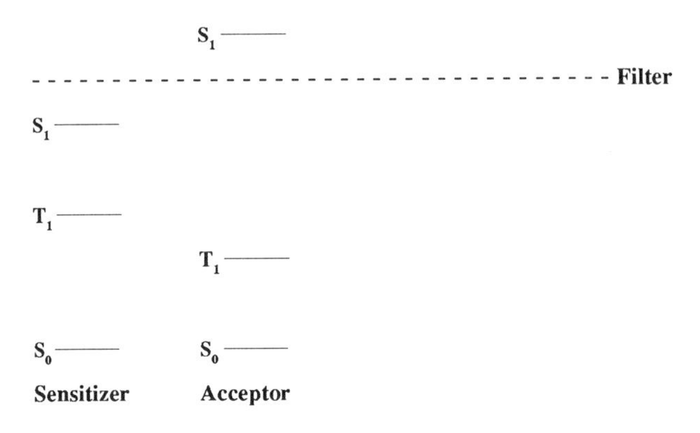

ACKNOWLEDGEMENTS

Special thanks are expressed to Dr. Michael F. Holick, Boston University School of Medicine, for supplying the 7-dehydrocholesterol and the previtamin D_3 that were used in the development of this experiment, as well as his helpful comments on the optimization of the experimental design.

SUPPLEMENTARY READING

Grossweiner, L.I. and K.C. Smith (1989) Photochemistry, In *The Science of Photobiology*, 2nd edition (Edited by K.C. Smith), pp. 47-78. Plenum Press, New York.

Simon, J.P. (1971). *Photochemistry and Spectroscopy*. Wiley-Interscience, New York. Covers the essentials of photochemistry and spectroscopy of organic compounds.

Schulman, S.G. (1977). *Fluorescence and Phosphorescence Spectroscopy: Physicochemical Principles and Practice*. Pergamon Press, Toronto. Contains many applications on biological systems.

Jacobs, H.J.C. and E. Hevinga (1979) Photochemistry of vitamin D and its isomers and of simple trienes. In *Advances in Photochemistry*, Vol. 11 (Edited by J.N. Pitts, Jr., G.S. Hammond, K. Gollnick and D. Grosjean), pp. 305-373.

REFERENCES

Braslavsky, S.E. and H.J Kuhn (1987) IUPAC Commission on photochemistry: Project chemical actinometers. *IAPS Newsletter*, May, pp. 27-45.

Calvert, J.G. and J.N. Pitts Jr., (1966) Experimental methods in photochemistry, *Photochemistry*, Chapter 7, pp. 686-814. John Wiley & Sons, Inc., New York.

Grossweiner, L.I. and K.C. Smith (1989) Photochemistry, In *The Science of Photobiology*, 2nd edition, (Edited by K.C. Smith), pp. 47-78. Plenum Press, New York.

Holick, M.F., J.A. MacLaughlin, M.B. Clark, S.A. Holick, J.T. Potts Jr., R.R. Anderson, I.H. Blank, J.A. Parrish and P. Elias (1980) Photosynthesis of previtamin D_3 in human skin and the Physiologic Consequences. *Science*, **210**, 203-205.

Jacobs, H.J.C., F. Boomsma, E. Havinga and A. van der Gen (1977) The photochemistry of previtamin D and tachysterol. *Rec. J. Roy. Netherlands Chem. Soc.* **94**, 113-118.

MacLaughlin, J.A., R.R. Anderson and M.F. Holick (1982) Spectral Character of Sunlight modulates photosynthesis of previtamin D_3 and its photoisomers in human skin. *Science*, **216**, 1001-1003.

Rodgers, M.A.J. (1988) Primary photochemical processes. In *Photosensitisation: Molecular, Cellular and Medical Aspects.* (Edited by G. Moreno, R.H. Pottier and T.G. Truscott), pp. 11-18. Springer-Verlag, Heidelberg.

Turro, N.J. (1978) *Modern Molecular Photochemistry.* The Benjamin/Cummings Publishing Co., Inc. Don Mills, Ontario.

Photochemistry and Photobiology of Tetrapyrrole Pigments

Antony F. McDONAGH
University of California, San Francisco
U.S.A.

INTRODUCTION

Photobiological Requirements

For light to produce a photobiological effect in a living organism at least two conditions must be fulfilled. First, light must be absorbed by a photoreceptor molecule within the organism. Second, absorption and excitation of the photoreceptor must cause a significant biological response, generally via some photochemical reaction of the primary photoreceptor.

Most of the molecules that make up living organisms, such as carbohydrates, lipids, proteins, and nucleic acids, do not absorb light in the visible or near-UV region of the spectrum and do not undergo photochemical reactions that have photobiological significance. However, life is never completely colorless, and all organisms contain some molecules that do absorb light and can, therefore, potentially lead to a photobiological response.

Tetrapyrrole Pigments

One important class of light-absorbing compounds that occurs in all living organisms is comprised of tetrapyrrole pigments – so-called because they contain four, five-membered C_4N rings formally related to pyrrole (Fig. 1). Two types of tetrapyrrole occur in nature, cyclic and open-chain. In the latter, the four rings are linked by carbon bridges in an open linear array. Pyrrole, itself does not absorb visible light, but many naturally-occurring tetrapyrrole compounds do and are photobiologically active. Naturally-occurring photobiologically active tetrapyrroles include the cyclic porphyrins and chlorophylls and the open-chain phycobilins and phytochrome. In this laboratory exercise you will study some photochemical reactions of bilirubin (Fig. 1), a yellow naturally-occurring open-chain tetrapyrrole, and octaethylporphyrin (Fig. 1), a synthetic cyclic tetrapyrrole. Bilirubin has

ANTONY F. McDONAGH, Gastrointestinal Unit and The Liver Center, Box 0538
University of California, San Francisco, San Francisco, CA 94143, U.S.A.

Photobiological Techniques, Edited by D.P. Valenzeno *et al.*
Plenum Press, New York, 1991

been chosen because it is a readily available photolabile compound with a strong absorption band in the visible region of the spectrum whose photochemistry is clinically important in the treatment of jaundice in babies. Octaethylporphyrin was chosen because it is a stable commercially-available porphyrin that can be used to demonstrate the phenomenon of photosensitization, a property that is currently being explored for the treatment of cancer. The experiments are designed to demonstrate the use of absorbance and absorbance difference spectroscopy for detecting and distinguishing different types of photochemical reactions and the use of high performance liquid chromatography for the separation and detection of small quantities of photoproducts.

Photoisomerization of Bilirubin *In Vitro* and *In Vivo*

Bilirubin is a yellow compound, produced *in vivo* from hemoglobin and other heme-containing pigments, which is present in blood in millimolar concentrations. In people with liver disease bilirubin and its metabolites often accumulate in skin, causing the yellow pigmentation known as jaundice. Bilirubin has a broad absorption band with a peak at about 450 nm. When irradiated with light of wavelengths near 450 nm, bilirubin can undergo several photochemical reactions, depending on the experimental conditions. Some of these are isomerization reactions, reactions in which the structure and shape of the molecule is changed while leaving its chemical constitution intact. The first experiment in this section demonstrates one type of isomerization reaction, reversible isomerization about a double bond, that occurs rapidly when bilirubin in solution is exposed to light. A similar type of reaction is of fundamental importance in the mechanism of vision and in the control of growth and movement in plants by the tetrapyrrolic plant pigment phytochrome. The second experiment shows that complexation of bilirubin with protein can have a marked influence on its photochemistry, as is frequently true for photoreceptor pigments. This experiment also reveals a second type of photoisomerization reaction that bilirubin can undergo. The third experiment shows that visible light can penetrate tissue and cause photoisomerization of bilirubin *in vivo* when people sunbathe.

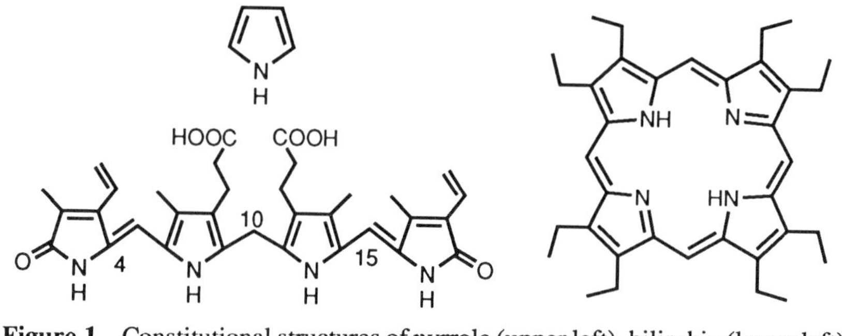

Figure 1. Constitutional structures of pyrrole (upper left), bilirubin (lower left) and octaethylporphyrin (right).

EXPERIMENT 6: CONFIGURATIONAL ISOMERIZATION OF BILIRUBIN IN AN ORGANIC SOLVENT

OBJECTIVE

The purpose of this experiment is to demonstrate photochemically induced configurational (*cis-trans*) isomerization and the phenomenon of a photostationary state. The experiment also shows that absorbance difference spectrophotometry can be a more sensitive tool than absorbance spectrophotometry for following photochemical reactions and it illustrates the use of HPLC for detecting photoproduct formation.

Bilirubin contains two unsymmetrically substituted exocyclic double bonds (Fig. 2). These have a *Z* configuration [For an explanation of the *Z*, *E* nomenclature for double bond configurational isomers see Orchin et al. (1980)]. When bilirubin, in solution, is exposed to blue light the exocyclic double bonds undergo rapid configurational isomerization leading to formation of 4*E*,15*Z* and 4*Z*,15*E* isomers from the 4*Z*,15*Z* starting material. (Small amounts of a fourth 4*E*,15*E* isomer are formed also, but this will be neglected in this discussion). Because the reaction is reversible and the photoisomers and starting material have overlapping absorption spectra, irradiation of bilirubin with a broad-spectrum light-source eventually leads to formation of a photostationary mixture of isomers. The composition of this photostationary mixture is dependent on the wavelength of irradiation. Isomer formation can be readily followed by absorbance difference spectroscopy, and the individual isomers can be separated by high performance liquid chromatography (HPLC). The 4*E*,15*Z* and 4*Z*,15*E* photoisomers are less extensively hydrogen bonded than the parent 4*Z*,15*Z* isomer and are therefore more polar. This is reflected in their lower retention times on reversed phase HPLC. They are also thermally and photochemically unstable, and revert readily to the starting material, particularly in the presence of acids.

This experiment is based on the work of Lightner, et al. (1979), McDonagh, et al. (1979), and McDonagh, et al. (1982a).

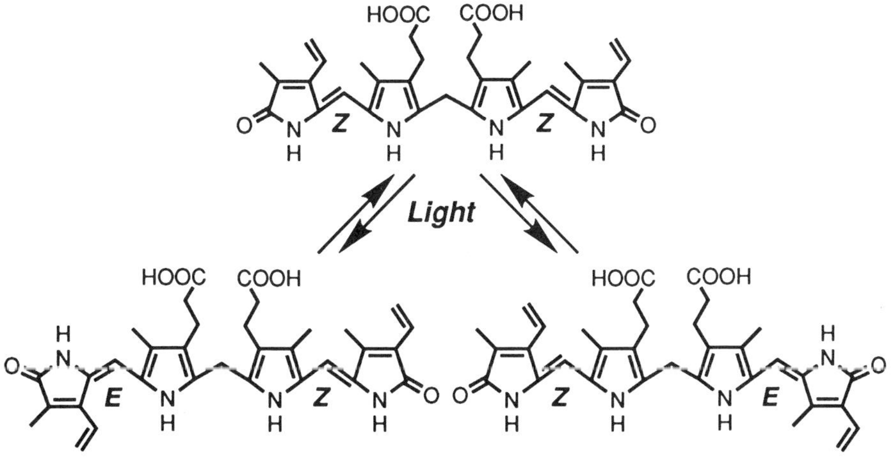

Figure 2. Configurational photoisomerization of bilirubin.

LIST OF MATERIALS

Chloroform (containing ~1% ethanol stabilizer)
Triethylamine (should be colorless)
Bilirubin (Porphyrin Products, Logan, Utah)
Trifluoroacetic acid
Di-*n*-octylamine (Aldrich Chem. Co., Cat. No. D20,114-6)
HPLC-grade methanol
Glacial acetic acid
HPLC-grade water
Acetone wash bottle for rinsing cuvettes
Spectrophotometer cuvettes with stoppers, glass or quartz, 1 cm pathlength (2)
Round bottom flasks, 5 ml, with standard taper ground glass necks suitable for attachment to a rotary evaporator (3)
Disposable glass micropipettes suitable for measuring 2 μl (Drummond Sci. Co.)
Beaker (~500 ml) or bucket containing crushed ice
Erlenmeyer flask, 25 ml (1)
Volumetric flask, 1 liter (1)
Volumetric flask, 500 ml (1)
Volumetric flasks, 10 ml (2)
Volumetric and Pasteur pipettes and bulbs
Measuring cylinders
Stopwatch
50 μl syringe for injecting solutions into HPLC (1)
200 μl syringe for rinsing out the HPLC loop between injections (1)
Lighting fixture equipped with a 20 Watt Westinghouse Special Blue (F20T12/BB) fluorescent bulb. The emission spectrum of this bulb has a large overlap with the absorbance spectrum of bilirubin.
Dual beam spectrophotometer
Rotary evaporator
Microbalance
HPLC analyses require an isocratic pump system with an injector fitted with a 20-μl loop, an absorbance detector capable of monitoring near 450 nm, a C18 reversed phase column (4.6 mm × 25 cm) and preferably a precolumn (4.6 mm × 4.5 cm) containing the same stationary phase.

PREPARATIONS

EXCEPT WHERE NOTED, THESE EXPERIMENTS SHOULD BE DONE IN A DARKROOM OR UNDER SUBDUED LIGHTING, AND SOLUTIONS CONTAINING BILIRUBIN SHOULD BE CAREFULLY PROTECTED FROM LIGHT OR MANIPULATED UNDER AN ORANGE OR RED SAFELIGHT.

CAUTION: Chloroform is toxic to the liver. Operations involving the preparation and transfer of chloroform solutions should be done in a fume hood or well-ventilated area.

Solvent for HPLC: To a 250 ml Erlenmeyer (conical) flask add about 100 ml of HPLC-grade methanol (MeOH) and tare on a a top-loading balance. With a Pasteur pipette add 24.15 g of di-*n*-octylamine, mix, and then add 6.005 g of glacial acetic acid. (When pure, di-*n*-octylamine crystallizes near room temperature. If necessary, it can be liquified by warming the bottle with your hands.) Transfer the solution quantitatively to a 1 liter volumetric flask with

MeOH and make up to the mark with MeOH. For HPLC dilute this solution with 4% water (*e.g.* add 20 ml HPLC-grade water to a 500 ml volumetric flask and make up the volume of the solution to 500 ml with the di-*n*-octylamine acetate solution). The solvent should preferably be filtered through a 0.45 micron filter before use. The solvent should be used at a flow rate of ~0.75 ml/min and the column should be equilibrated with the solvent for about 20-30 min before use.

Chloroform/triethylamine solvent: Mix 25 ml of chloroform and 25 ml of triethylamine. This mixture, which should be stored in a stoppered bottle, will be referred to as C/T.

Bilirubin solution: Weigh 1.8 mg of bilirubin and transfer quantitatively to a 10 ml volumetric flask. Add about 7 ml of C/T and swirl gently until the bilirubin is completely dissolved. Then make the solution volume up to 10 ml with C/T. Dilute 1 ml of this solution to 10 ml with C/T in a volumetric flask and wrap the flask with aluminum foil. Use this stock solution (BR solution) for the experiments.

EXPERIMENTAL PROCEDURE

Photostationary State

1. Transfer ~3.5 ml of C/T to each of two 1-cm pathlength cuvettes and place these in the sample and reference holders of a dual beam spectrophotometer.
2. Run a baseline absorbance spectrum from 350-650 nm (Absorbance range 0-2.0).
3. Rinse out the sample cuvette with acetone and dry it in a light stream of dry air.
4. Transfer ~3.5 ml of the BR solution to the cuvette.
5. Run the absorbance spectrum of the BR solution, and then transfer the BR solution back to the orginal stock solution with a Pasteur pipette.
6. Wash (acetone) and dry the sample and reference cuvettes.
7. Transfer to each a portion (~3.5 ml) of the BR solution and replace in the spectrophotometer.
8. Set the absorbance scale from –0.2 to +0.1 (full scale).
9. Run an absorbance difference spectrum baseline (this should be flat).
10. Remove the sample cuvette and place it directly in front of the Special Blue light for 10 s.
11. Gently invert the cuvette to mix the contents.
12. Replace the cuvette in the spectrophotometer and run a difference spectrum (t = 10 s). The difference spectrum should show a marked positive peak near 490-500 nm, and two loss peaks between about 400 nm and 475 nm, with an isosbestic point at about 478 nm.
13. Irradiate the sample solution for a further 10 s, mix and run the difference spectrum again (t = 20 s).
14. Repeat this procedure for total irradiation times of t = 30, 40, 50 and 60 s or until there is no further large change in the intensity of the difference spectrum (photostationary state). The tight isosbestic point near 478 nm should be maintained.
15. Once the photostationary state has been reached, transfer the unirradiated solution in the reference cuvette back to the volumetric flask containing the original BR stock solution.
16. Rinse and dry the cuvette.

17. Fill the cuvette with solvent (C/T), replace it in the reference beam of the spectrophotometer, switch to the 0-2.0 absorbance range, and run the absorbance spectrum of the irradiated BR solution.

18. Compare the spectrum with that of the unirradiated soluton run previously. Plot the change in absorbance at ~500 nm from the difference absorbance specta versus time to demonstrate the formation of a photostationary state.

HPLC Analysis

19. Remove two 1 ml samples from the irradiated cuvette and one 1 ml sample from the orginial unirradiated solution and transfer each to a 5 ml round bottom flask.

20. Evaporate the solutions to dryness at room temperature on a rotary evaporator (or, alternatively, in a stream of air or nitrogen in a fume hood) until the residue no longer has a fishy smell.

21. Redissolve the residue from one of the irradiated samples in 1 ml of $CHCl_3$. Add 2 μl trifluoroacetic acid (TFA) to this solution. Swirl the flask to mix the contents and immediately evaporate the solution to dryness. You will thus obtain three residues derived, respectively, from the unirradiated bilirubin solution, the irradiated bilirubin solution, and the irradiated bilirubin solution treated with TFA. Keep these on ice in the dark until they can be analysed by HPLC.

22. Dissolve each residue in 0.2 ml of ice cold HPLC solvent and keep the solutions on ice.

23. Remove 50 μl of the first residue solution with a syringe and inject into the 20-μl loop of the HPLC injector.

24. Repeat step 23 for the other two residue solutions.

DISCUSSION

The overall change in absorbance of the bilirubin solution during the irradiation is small, yet HPLC analysis clearly reveals that substantial formation of photoisomers has occurred. Only small changes in the total absorbance are seen because the photoisomers have absorbance spectra that are very similar to that of the starting material. Using absorbance difference spectroscopy magnifies the changes in the absorption spectrum during the reaction and makes it easier to detect when photoequilibrium has been reached. The appearance of a positive peak in the difference spectrum proves that at least one new chromophoric substance is formed on irradiation.

The HPLC analysis should indicate the formation of two major isomers (4*E*,15*Z* and 4*Z*,15*E*). Photoisomers of bilirubin will be eluted before the unchanged 4*Z*,15*Z*-bilirubin. The 4*E*,15*Z* photoisomer elutes slightly faster than the 4*Z*,15*E* photoisomer. [The bilirubin used as starting material will probably contain a small proportion (~10%) of two isomeric impurities and these may appear in the chromatogram as two bands or shoulders flanking the main bilirubin band. These impurities are structural isomers of bilirubin in which the order of the side-chains is different from that in natural bilirubin (McDonagh and Assisi, 1971, 1972). They are formed artifactually during the commercial isolation of bilirubin from pig bile and gallstones.] Despite the formation of at least two main photoproducts, isosbestic points are seen in the difference spectrum. This demonstrates that, contrary to what is often taught, an isosbestic point does not necessarily indicate stoichiometric conversion of one chromophoric substance to another. Note that treatment of the photoequilibrium reaction mixture with trifluoroacetic acid results in rapid and complete disappearnce of the photoproducts because of acid-catalyzed thermal reversion of the photoisomers to starting material. This is evidence, but not proof, that the photoproducts seen on HPLC are in fact isomers of the starting material.

Bilirubin is prone to oxidative degradation and other photochemical reactions (see Experiments 7 and 8 below). Because of this a true photostationary state cannot be achieved with air-saturated solutions and on prolonged irradiation the shape of the difference spectrum changes and there is loss of the initial sharp isosbestic points.

As an alternative to the C/T solvent, MeOH containing 1% (by volume) of concentrated ammonium hydroxide solution may be used (Lightner *et al.*, 1979). If a Westinghouse Special Blue fluorescent light is not available, an ordinary broad-spectrum white light fluorescent tube may be substituted. With such a light source, the formation of the photostationary state may be slower and the amplitude of the difference spectrum and the concentration of photoisomers produced may not be so large as with the Special Blue bulb.

REVIEW QUESTIONS

1. How may chromophores are present in the bilirubin molecule?
2. What do the terms *Z* and *E* mean?
3. What is an isosbestic point and what does it signify?
4. What factors determine the proportion of photoisomers that are present at the photostationary state?
5. What would be the effect of using a red emitting light instead of a blue or a broad-spectrum white light in this experiment?

ANSWERS TO REVIEW QUESTIONS

1. Two.
2. *Z* and *E* are proscripts used to unambiguously designate the stereochemical arrangement of atoms (or groups) around a double bond (or ring). They are derived from the German words *zusammen*, meaning together, and *entgegen*, meaning opposite. Unsymmetrically-substituted double bonds that are not contained within a ring system can exist in two possible configurations, which are referred to as *Z* and *E*, respectively.
3. An isosbestic point is the wavelength where the absorptions (or molar extinction coefficients) of two light-absorbing species are equal. The adjective isosbestic is derived from the Greek *iso*, meaning equal, and *sbestos*, meaning extinguished. The appearance of an isosbestic point during a reaction is generally assumed to indicate formation of a single species from another or equilibrium between two species.
4. The wavelength of thc light and the relative molar extinction coefficients of the photoisomers at that wavelength.
5. No reaction would occur because bilirubin does not absorb red light.

SUPPLEMENTARY READING

Grossweiner, L.I. and K.C. Smith. (1989) Photochemistry. In *The Science of Photobiology*, 2nd edition (Edited by K.C. Smith), pp. 47-78. Plenum Press, New York.

Lightner, D.A., T.A. Wooldridge and A.F. McDonagh (1979) Photobilirubin: An early bilirubin photoproduct detected by absorbance difference spectroscopy. *Proc. Nat. Acad. Sci.* USA **76**, 29-32.

McDonagh, A.F., D.A. Lightner and L.A. Palma (1979) Geometric isomerisation of bilirubin IXα and its dimethyl ester. *J. Chem. Soc. Chem. Commun.* 110-112.

McDonagh, A.F., L.A. Palma, F.R. Trull and D.A. Lightner (1982) Phototherapy for neonatal jaundice. Configurational isomers of bilirubin. *J. Am. Chem. Soc.* **104**, 6865-6867.

Lightner, D.A. and A.F. McDonagh (1984) Molecular mechanisms of phototherapy for neonatal jaundice. *Accounts Chem. Res.* **17**, 417-424.

McDonagh, A.F. and D.A. Lightner (1985) 'Like a shrivelled blood orange'—bilirubin, jaundice, and phototherapy. *Pediatrics* **75**, 443-455.

McDonagh, A.F. and D.A. Lightner (1988) Phototherapy and the photobiology of bilirubin. *Seminars in Liver Disease* **8**, 272-283.

EXPERIMENT 7: PHOTOISOMERIZATION OF BILIRUBIN BOUND TO HUMAN SERUM ALBUMIN

Bilirubin is produced continuously in the body by metabolism of hemoglobin and is transported in human blood as an association complex with serum albumin. Surprisingly, perhaps, complexation with protein does not prevent configurational ($Z \rightarrow E$) photoisomerization of bilirubin. However, in the presence of human serum albumin the reaction is highly stereoselective and isomerization occurs almost exclusively at the C-15 double bond, with little occurring at the C-4 double bond. Thus, complexation of the tetrapyrrole with albumin has a pronounced influence on the stereochemistry of the reaction. Binding to albumin also facilitates a second type of photoisomerization (Fig. 3) in which the bilirubin molecule undergoes intramolecular cyclization. This reaction has a low quantum yield, but, unlike the configurational photoisomerization, is irreversible. Consequently, when bilirubin is irradiated in the presence of excess albumin, configurational isomers first appear, followed by the gradual accumulation of the structural isomer. These two types of isomerization (configurational and structural) are important in the treatment of non-infectious jaundice in newborn babies. In this treatment the babies are irradiated with blue or broad spectrum white light (*not* UV light). Irradiation converts the bilirubin that has accumulated in the babies' bodies to photoisomers. These are more polar than bilirubin and can be excreted more readily. In this way, the light treatment (phototherapy) helps to stimulate the excretion of bilirubin from the body and prevent its accumulation.

OBJECTIVE

The purpose of this experiment is to show that proteins can influence the photochemical reactions of bound chromophores, and to illustrate the effect of competing reversible and irreversible photochemical reactions of the same substrate.

This experiment is based on the work of McDonagh, et al. (1982b).

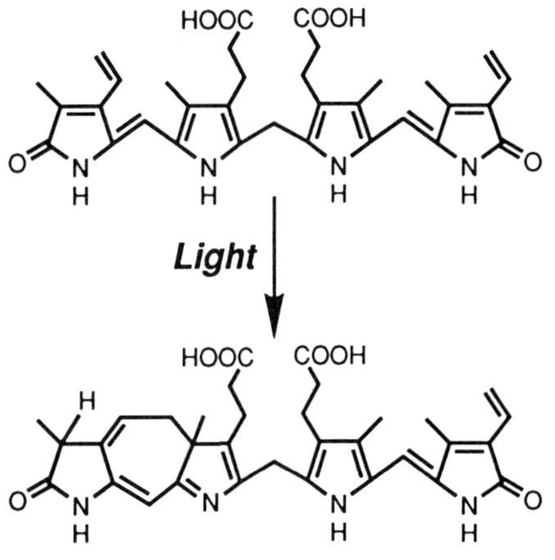

HOOC COOH

O N N N N O
H H H H

Light

HOOC COOH

H

O N N N N O
H H H

Figure 3. Structural photoisomerization of bilirubin (upper structure) to lumirubin (lower structure) which is intramolecularly cyclized.

LIST OF MATERIALS

Human serum albumin (HSA, Cohn Fraction V, Sigma Chemical) [Note: Bovine serum albumin should not be substituted for human serum albumin in this experiment]
0.1 M sodium hydroxide solution (freshly prepared)
Bilirubin (Porphyrin Products, Logan, Utah)
Di-*n*-octylamine (Aldrich Chemical Co., Catalog No. D20,114-6)
HPLC-grade methanol
Glacial acetic acid
HPLC-grade water
Spectrophotometer cuvettes, glass or quartz, 1-cm pathlength (2)
Volumetric flask, 10 ml (1)
Ice bucket with crushed ice
Pasteur and volumetric pipettes and bulbs
Small glass centrifuge tubes (disposable glass culture tubes, 6 × 50mm)
Stopwatch
20 μl disposable glass micropipettes
50 μl syringe for injecting solutions into the HPLC column (1)
200 μl syringe for rinsing out the HPLC loop between injections and for measuring out microliter quantities of HPLC solvent (1)
Lighting fixture equipped with a 20 watt Westinghouse Special Blue (F20T12/BB) fluorescent bulb. The emission spectrum of this bulb has a large overlap with the absorbance spectrum of bilirubin
Dual beam spectrophotometer
Microbalance
Centrifuge (or microfuge)
Vortex mixer
HPLC equipment and solvent as in Experiment 6

PREPARATIONS

EXCEPT WHERE NOTED, THESE EXPERIMENTS SHOULD BE DONE IN A DARKROOM OR UNDER SUBDUED LIGHTING AND SOLUTIONS CONTAINING BILIRUBIN SHOULD BE CAREFULLY PROTECTED FROM LIGHT OR MANIPULATED UNDER AN ORANGE OR RED SAFELIGHT.

Bilirubin/HSA solution: Weigh out 20 mg of HSA into a 10 ml volumetric flask. Add ~6 ml distilled water and mix gently to dissolve the protein (don't shake, because frothing will occur). Weigh out 1.8 mg of bilirubin and transfer to a small test tube. Add 2.0 ml of 0.1M NaOH and mix gently until the bilirubin is *completely* dissolved (clear, orange solution). Add 0.2 ml of this solution immediately to the HSA solution with a Pasteur pipette, mix gently, and finally make up the BR/HSA solution to 10 ml with water. Use this stock bilirubin/HSA solution. The solution is stable in the dark and may be stored for a few days without deterioration if refrigerated.

EXPERIMENTAL PROCEDURE

Photoirradiation

1. Fill two 1-cm pathlength spectrophotometer cuvettes (sample and reference) with the bilirubin/HSA solution and use difference spectrophotometry to measure the time required to reach a photostationary state (as in Experiment 6) as the sample is irradiated with blue fluorescent light.

2. When the sample is at or near photoequilibrium, remove 0.5 ml from the irradiated solution, transfer to a small stoppered vial or test-tube and put on ice in the dark. Continue irradiation of the remaining sample for twenty times the time it took to reach photoequilibrium.
3. Remove a second 0.5 ml from the irradiated solution and 0.5 ml from the unirradiated reference sample and store these samples on ice in the dark. You will now have three 0.5 ml samples representing respectively, bilirubin/HSA solution that has not been irradiated, bilirubin/HSA solution that has been irradiated to near photoequilibrium, and bilirubin/HSA solution that has been irradiated beyond photoequilibrium.
4. Next analyze these solutions by HPLC. If the analyses cannot be run on the same day freeze the samples rapidly by immersing them in a dry-ice acetone slush and store them in the frozen state.

HPLC Analysis

5. Use the 200 μl syringe to add 80 μl of HPLC solvent to a microfuge tube. To this add 20 μl of sample solution with a disposable micropipette.
6. Mix the contents of the tube thoroughly with a Vortex mixer.
7. Centrifuge to pack the precipitated protein at the bottom of the tube, and carefully aspirate 50 μl of the supernate into the HPLC injection syringe.
8. Inject this solution into the HPLC injector and chromatograph with conditions as in Experiment 6.

DISCUSSION

HPLC of the irradiated samples shows that complexation of bilirubin with albumin does not inhibit $Z \rightarrow E$ photochemical isomerization of the pigment, but it does make the reaction more stereospecific. By comparing the reaction products from Experiment 6 with those from this experiment it can be seen that formation of one of the *E,Z* photoisomers (the 4*E*,15*Z* isomer) is almost completely suppressed in the presence of serum albumin. Complexation with albumin also facilitates a second type of isomerization in which one end of the bilirubin molecule undergoes intramolecular cyclization (Fig. 3). This reaction is much slower than the $Z \rightarrow E$ isomerization, but it is irreversible and does not lead to a photoequilibrium mixture. Thus continued irradiation of the bilirubin/HSA solution after photoequilibrium has been reached does not lead to increased formation of *Z,E* isomers, but it does lead to slow appearance of a new peak, lumirubin, in the chromatogram. Formation of lumirubin *in vivo* is important in the phototherapy of jaundiced babies with visible light.

REVIEW QUESTIONS

1. Why might binding of bilirubin to albumin influence the stereoselectivity of isomerization?
2. Why does only one of the two chromophores in bilirubin undergo intramolecular photocyclization?
3. Which of the following two reactions seems to have the highest quantum yield: photoisomerization of bilirubin to 4*Z*,15*E* bilirubin or photocyclization of bilirubin to lumirubin?

ANSWERS TO REVIEW QUESTIONS

1. Because steric interactions between the bilirubin molecule and bulky groups of the protein might inhibit rotation and isomerization at one of the *Z*-configuration double bonds.
2. Because only one chromophore bears a suitably placed vinyl substituent.
3. Photoisomerization to 4*Z*, 15*E* bilirubin.

SUPPLEMENTARY READING

McDonagh, A.F., L.A. Palma, and D.A. Lightner (1982b) Phototherapy for neonatal jaundice. Sterospecific and regioselective photoisomerizaton of bilirubin bound to human serum albumin and NMR characterization of intramolecularly cyclized photoproducts. *J. Am. Chem. Soc.* **104**, 6867-6869.

Lightner, D.A. and A.F. McDonagh (1984) Molecular mechanisms of phototherapy for neonatal jaundice. *Accounts Chem. Res.* **17**, 417-424.

McDonagh, A.F. and D.A. Lightner (1985) 'Like a shrivelled blood orange'—bilirubin, jaundice, and phototherapy. *Pediatrics* **75**, 443-455.

McDonagh, A.F. and D.A. Lightner (1988) Phototherapy and the photobiology of bilirubin. *Seminars in Liver Disease* **8**, 272-283.

EXPERIMENT 8: PHOTOISOMERIZATION OF BILIRUBIN *IN VIVO*

As shown in Experiment 7, bilirubin undergoes two types of photoisomerization in the presence of human serum albumin: configurational isomerization to 4*Z*,15*E* bilirubin and structural photoisomerization to lumirubin. Configurational photoisomerization of bilirubin has a high quantum yield and the 4*Z*,15*E* bilirubin photoisomer that is formed has a long half-life in plasma. Since visible light can penetrate skin, and since bilirubin is present in the blood it should be possible to photoisomerize bilirubin *in vivo* by exposure of the body to bright light. Photoisomerization of bilirubin in adults has no known physiological or clinical significance, but it provides a convenient example of a readily detectable photochemical reaction that can occur within the body.

OBJECTIVE

The objective of this experiment is to measure photoisomerization of bilirubin, which has occurred *in vivo*, by testing for the presence of bilirubin photoisomers in blood taken from a volunteer before and after sunbathing.

This experiment is based on the work of McDonagh (1986).

LIST OF MATERIALS

Small glass centrifuge tubes (*e.g*. disposable glass culture tubes, 6 × 50mm)
Hematocrit tubes (preferably non-heparinized)
Seal-Eze (plasticine type substance for sealing hematocrit tubes)
Alcohol swabs
Automatic lancet and disposable needles
Centrifuge or microfuge
Vortex mixer
HPLC equipment and solutions as in Experiment 6

PREPARATIONS

Blood sampling technique: Collect blood samples from a finger-stick wound in dim light as follows. Clean the end of a finger or thumb with an alcohol swab. Stab with an automatic lancet. Collect blood by capillary attraction into ~6 hematocrit tubes. Seal one end of each tube with the plastic sealant. After about 10 min place the hematocrit tubes inside a test-tube and centrifuge until there is a clear separation of the red cells and plasma (or clot and serum). Break the tubes at the red cell/plasma (or clot/serum) interface and blow the plasma (or serum) from all six tubes into a small tube.

Volunteer Preparation: For this experiment, the volunteer should avoid exposure to sunlight as much as possible for at least 12 h (preferably 24 h) before the start of the experiment. If possible the volunteer should fast for ~8 h before the start of the experiment.

EXPERIMENTAL PROCEDURE

Blood Sample

1. At the beginning of the experiment collect a blood sample and separate the serum (or plasma) as described above.

2. Store the serum sample on ice in the dark until it can be analyzed.

Photoexposure

3. Have the volunteer lie or exercise in sunlight for 60-120 min wearing only shorts or a brief two-piece bathing suit. Use of a sun protectant to block out UV and prevent sunburn is recommended in sunny weather.
4. At the end of the exposure collect a second blood sample and separate the serum or plasma.

HPLC Analysis

5. HPLC the two serum samples as in Experiment 7 above. Compare the chromatograms with those obtained in Experiment 7 and note the effect (if any) of sunbathing.

DISCUSSION

This experiment demonstrates that sunlight penetrates the skin sufficiently to cause endogenous bilirubin to undergo isomerization. As observed *in vitro* in Experiment 7, for bilirubin bound to HSA, the reaction *in vivo* is stereoselective for formation of the 4*Z*,15*E* isomer.

REVIEW QUESTIONS

1. If sunlight exposure were continued for a longer period, would the concentration of photoisomer relative to that of unisomerized bilirubin continue to increase indefinitely?
2. What factors influence the relative concentration of photoisomer and unisomerized bilirubin *in vivo*?
3. Would it make any difference to the experimental outcome if the subject were exposed to sunlight that had passed through window glass?
4. Would application of sun-tan oil prevent bilirubin from photoisomerizing *in vivo*?

ANSWERS TO REVIEW QUESTIONS

1. No. A photostationary ratio of photoisomer and unisomerized bilirubin would eventually be reached (probably within the first 2-3 h of sunlight exposure).
2. The wavelength composition of the sunlight and the relative rates of excretion of bilirubin and the photoisomer.

3. & 4. No. The effective wavelengths in this experiment are in the visible where bilirubin absorbs most strongly and where light penetrates most deeply into tissue. Light in the visible region of the spectrum is not absorbed by window glass or most sun-tan oils.

SUPPLEMENTARY READING

McDonagh, A.F. (1986) Sunlight-induced mutation of bilirubin in a long-distance runner. *New Eng. J. Med.* **314**, 121-122.

EXPERIMENT 9: PORPHYRIN SENSITIZED PHOTOOXIDATION OF BILIRUBIN

Porphyrins are intensely colored cyclic tetrapyrroles that absorb light strongly in the visible and near UV region of the spectrum. Many porphyrins have high photochemical quantum yields for formation of excited triplet states, and the excited triplet molecules thus formed often have relatively long lifetimes. In air-saturated solutions intermolecular energy transfer from an excited triplet state porphyrin molecule to a dissolved ground state molecular oxygen molecule can lead to formation of singlet oxygen, which is molecular oxygen in an excited singlet state. Thus porphyrins are effective photosensitizers for the generation of singlet oxygen. Singlet oxygen has a short lifetime, but is very reactive. If it is formed in tissues in living organisms it can cause severe toxic damage. Thus, irradiation of tissues containing porphyrins with visible light can be lethal to cells in the illuminated volume. This property is exploited in the experimental treatment of cancer with porphyrins and light (photodynamic therapy; Diamond *et al.*, 1971). It is also responsible for the painful and damaging photodermatoses that can occur in patients with rare genetic diseases called porphyrias in which porphyrins accumulate in the skin.

Bilirubin reacts avidly with singlet oxygen, breaking down to form products of lower molecular weight that are colorless. Consequently, bilirubin is bleached irreversibly in the presence of singlet oxygen. Bilirubin can, therefore, be used as a probe to detect singlet oxygen and to test the ability of sensitizers to catalyze singlet oxygen formation (Diamond *et al.*, 1977).

Another yellow pigment that interacts with singlet oxygen is β-carotene. β-carotene quenches singlet oxygen. That is, it deactivates singlet oxygen by removing and dissipating its excitation energy and converting the oxygen back to the ground state. In this way β-carotene can inhibit reactions in which singlet oxygen participates. Patients with the rare disease protoporphyria often take β-carotene to reduce their skin photosensitivity and allow them to spend more time outdoors.

OBJECTIVE

In this experiment, bleaching of bilirubin will be used to demonstrate porphyrin sensitized formation of singlet oxygen and β-carotene will be used to inhibit the reaction.

This experiment is based on the work of McDonagh (1971).

LIST OF MATERIALS

Bilirubin (Porphyrin Products, Logan, Utah)
Octaethylporphyrin (Porphyrin Products)
Beta-carotene (Sigma Chemical Co.)
Chloroform
Acetone wash bottle for rinsing cuvettes
Spectrophotometer cuvettes (2)
10 ml volumetric flasks (5)
1 ml volumetric pipettes and pipette bulb
Pasteur pipettes
Stopwatch

20 W fluorescent light fixture holding a broad spectrum white light tube (*e.g.*, Durotest 20 W Vitalight) (1)
Dual beam spectrophotometer

PREPARATIONS

EXCEPT WHERE NOTED, THESE EXPERIMENTS SHOULD BE DONE IN A DARKROOM OR UNDER SUBDUED LIGHTING, AND SOLUTIONS CONTAINING BILIRUBIN SHOULD BE CAREFULLY PROTECTED FROM LIGHT OR MANIPULATED UNDER AN ORANGE OR RED SAFELIGHT.

CAUTION: Chloroform is toxic to the liver. Operations involving the preparation and transfer of chloroform solutions should be done in a fume hood or well-ventilated area.

Bilirubin solution: Dissolve 0.88 mg of bilirubin in 10 ml of chloroform. This produces a 150 μM solution which will be referred to as solution A.

Octaethylporphyrin solution: Dissolve 0.08 mg of octaethylporphyrin in 10 ml of chloroform. This produces a 15 μM solution which will be referred to as solution B.

β-carotene solution: Dissolve 0.81 mg of β-carotene in 10 ml of chloroform. This produces a 150 μM solution which will be referred to as solution C.

EXPERIMENTAL PROCEDURE

Control Photooxidation

1. Dilute solution A 1:10 with chloroform in a volumetric flask and transfer 3.5 ml to a spectrophotometer cuvette.
2. Record the absorbance spectrum of this solution versus a chloroform blank from 350-650 nm (absorbance scale 0-1.2, full scale).
3. Place the cuvette containing the sample solution immediately in front of a Vitalite 20 W fluorescent bulb and expose the solution to light for 1 min.
4. Record the spectrum again.
5. Repeat this procedure with the same sample for total irradiation times of 3 and 6 min. This will give four absorbance spectra corresponding to total irradiation times of 0, 1, 3 and 6 min.
6. Discard the contents of the cuvettes. Rinse the cuvettes with acetone and blow them dry in a slow stream of dry air.

Porphyrin-Sensitized Photooxidation

7. Dilute solution B 1:10 with chloroform, and transfer 3.5 ml of this solution to the reference cuvette of the spectrophotometer.
8. Mix 1 ml of solution A and 1 ml of solution B in a 10 ml volumetric flask and make the volume of the solution up to 10 ml with chloroform.
9. Transfer 3.5 ml of this solution to the sample cuvette of the spectrophotometer and record the absorbance spectrum as in step 2 above.
10. Expose the sample solution to light from the Vitalite tube and record spectra at intervals as in steps 3-5, ensuring that the cuvette is placed at the same position relative to the light source as it was in the control studies.

11. Discard the solutions and rinse and dry the cuvettes.

Inhibition by β-Carotene

12. Mix 1 ml of solution B and 1 ml of solution A in a volumetric flask and make the volume of the solution up to 10 ml with chloroform. Place 3.5 ml of this solution in the reference cuvette.
13. Mix 1 ml of solution A, 1 ml of solution B and 1 ml of solution C in a volumetric flask and make up to 10 ml with chloroform.
14. Transfer 3.5 ml of this solution to the sample cuvette and record the absorbance spectrum.
15. Expose the solution to light and record spectra at intervals as in steps 3-5.
16. Discard the solutions and rinse and dry the cuvettes.
17. For each irradiation experiment (control, porphyrin, β-carotene) plot absorbance at 450 nm (or % decrease in absorbance at 450 nm) versus total time of irradiation.

DISCUSSION

Bilirubin undergoes slow photooxidation in addition to photoisomerization when exposed to light in solution. For this reason the chromophore is slowly bleached when exposed to light alone (control photooxidation). When exposed to light in the presence of a relatively low concentration of porphyrin (porphyrin-sensitized photooxidation), the bilirubin chromophore is bleached much more rapidly. This is evidence, but not proof, for the formation of singlet oxygen by porphyrin photosensitization. Further evidence for the intermediacy of singlet oxygen is the slowing down of the porphyrin-sensitized reaction by the presence of β-carotene (inhibition by β-carotene).

One flaw in the design of the experiment is that no correction is made for absorption of light by beta-carotene and the effect this might have on the reaction. Because dilute solutions are used, however, this inner filter effect is likely to be negligibly small.

REVIEW QUESTIONS

1. How many atoms are present in a molecule of singlet oxygen?
2. In the control photooxidation, why is the decrease in absorbance in the first 1 min of irradiation larger than that in the subsequent 2 min?
3. Bilirubin absorbs maximally at 453 nm in chloroform. Would this be the best wavelength to use for porphyrin-sensitized photodestruction of bilirubin?
4. Newborn babies excrete bilirubin in their feces. Bilirubin in the feces stains cloth diapers yellow. These stains are often difficult to remove completely by washing, but disappear if the washed diapers are hung outdoors to dry. Why?

ANSWERS TO REVIEW QUESTIONS

1. Two.
2. Because the decrease during the first minute reflects absorbance losses caused by both photooxidation and $Z \rightarrow E$ photoisomerization. However, photoisomerization,

to a photoequilibrium mixture of isomers, is complete within the first minute and thereafter the absorbance losses are almost entirely due to photooxidation.

3. No, it would not, because the absorbance of the porphyrin photosensitizer is relatively low at 453 nm.

4. Because the bilirubin is photooxidized to colorless compounds (photobleached).

SUPPLEMENTARY READING

McDonagh, A.F. (1971) The role of singlet oxygen in bilirubin photo-oxidation. *Biochem. Biophys. Res. Commun.* **44**, 1306-1311.

ACKNOWLEDGEMENTS

This work was supported by US Public Health Service Grant DK-26307 and NATO Grant 0149-85.

REFERENCES

Diamond, I., S.G. Granelli, A.F. McDonagh, S. Nielsen, C.B. Wilson and R. Jaenicke (1971) Photodynamic therapy of malignant tumors. *The Lancet* **ii**, 1175-1177.

Diamond, I., S.G. Granelli and A.F. McDonagh (1977) Photochemotherapy and photodynamic toxicity: simple methods for identifying potentially active agents. *Biochem. Med.* **17**, 121-127.

Lightner, D.A., T.A. Wooldridge and A.F. McDonagh (1979) Photobilirubin: An early bilirubin photoproduct detected by absorbance difference spectroscopy. *Proc. Nat. Acad. Sci.* USA **76**, 29-32.

Lightner, D.A. and A.F. McDonagh (1984) Molecular mechanisms of phototherapy for neonatal jaundice. *Accounts Chem. Res.* **17**, 417-424.

McDonagh, A.F. (1971) The role of singlet oxygen in bilirubin photo-oxidation. *Biochem. Biophys. Res. Commun.* **44**, 1306-1311.

McDonagh, A.F. (1986) Sunlight-induced mutation of bilirubin in a long-distance runner. *New Eng. J. Med.* **314**, 121-122.

McDonagh, A.F. and F. Assisi (1971) Commercial bilirubin: a trinity of isomers. FEBS *Lett.* **18**, 315-317.

McDonagh, A.F. and F. Assisi (1972) The ready isomerization of bilirubin IX-α in aqueous solution. *Biochem. J.* **129**, 797-800.

McDonagh, A.F. and D.A. Lightner (1985) 'Like a shrivelled blood orange'—bilirubin, jaundice, and phototherapy. *Pediatrics* **75**, 443-455.

McDonagh, A.F. and D.A. Lightner (1988) Phototherapy and the photobiology of bilirubin. *Seminars in Liver Disease* **8**, 272-283.

McDonagh, A.F., D.A. Lightner and L.A. Palma (1979) Geometric isomerisation of bilirubin IXα and its dimethyl ester. *J. Chem. Soc. Chem. Commun.* 110-112.

McDonagh, A.F., L.A. Palma, F.R. Trull and D.A. Lightner (1982a) Phototherapy for neonatal jaundice. Configurational isomers of bilirubin. *J. Am. Chem. Soc.* **104**, 6865-6867.

McDonagh, A.F., L.A. Palma, and D.A. Lightner (1982b) Phototherapy for neonatal jaundice. Sterospecific and regioselective photoisomerizaton of bilirubin bound to human serum albumin and NMR characterization of intramolecularly cyclized photoproducts. *J. Am. Chem. Soc.* **104**, 6867-6869.

Orchin, M., F. Kaplan, R.S. Macomber, R.M. Wilson, and H. Zimmer. (1980) *The Vocabulary of Organic Chemistry.* John Wiley & Sons, New York.

The Flavin-EDTA System in the Study of Photocatalytic Redox Reactions

Aurelio SERRANO
CSIC - Universidad de Sevilla
Spain

INTRODUCTION

Light-driven redox processes currently deserve much attention, not only because they are basic processes in photobiology but also because they have many technological applications, i.e. the production of several energy-rich compounds. Amongst others, flavins have been used as efficient photosensitizers in those model systems that use light energy to drive redox reactions of biological macromolecules.

This report describes a set of experiments, easily performed by students, which clearly demonstrate the possibility of constructing an artificial model system that reproduces a biological redox reaction, namely, the reduction of mitochondrial cytochrome *c*. The pigment used to capture light energy and drive photochemical reactions is flavin-mononucleotide (FMN), which is a well known biological flavin.

Flavins as Light/Redox Energy-Transducing Systems

The term "flavin" refers to the yellow chromophoric and redox-active prosthetic group of a class of respiratory enzymes occurring widely in animals and plants, namely the flavoproteins. Flavins are based upon the nitrogen heterocycle 7,8-dimethylisoalloxacine:

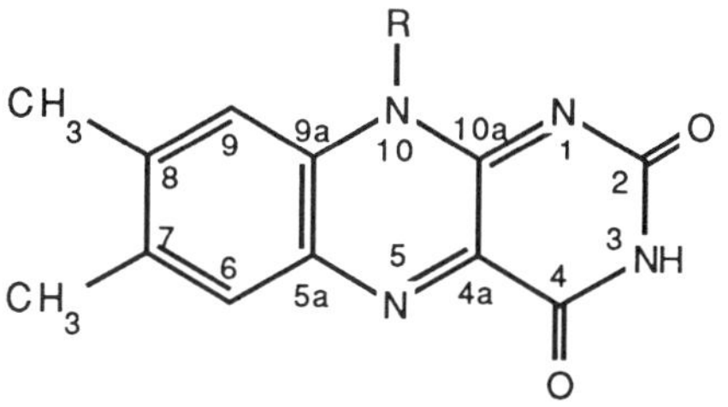

The particular side-chain, R, in FMN is a ribityl, phosphorylated in its C-5' hydroxy group.

AURELIO SERRANO, Instituto de Bioquimica Vegetal y Fotosintesis,
CSIC-Universidad de Sevilla, Apdo. 1113, 41080-Sevilla, Spain

Photobiological Techniques, Edited by D.P. Valenzeno *et al.*
Plenum Press, New York, 1991

Flavins exhibit a bright yellow colour in aqueous solutions, which is a consequence of its intense absorption of light (extintion coefficient, $1.2 \times 10^4\,M^{-1} \cdot cm^{-1}$) in the blue region of the visible spectrum ($\lambda_{max} \approx 450$ nm). Flavin electronic excitation involves the promotion of one electron in a π bonding orbital to the π^* antibonding orbitals. The unpaired electron spins, in the bonding and the antibonding orbitals, are antiparallel and lead to an unstable diamagnetic excited singlet state. Subsequent spin reversion quickly leads to a "relatively stable" paramagnetic triplet state with parallel electron spins. Thus, upon light absorption the flavin molecule becomes excited in a metastable state (lifetime, 10-100 μs) that is frequently called the lowest triplet state (^3F). It is from this state that most of the light-driven electron transfer reactions sensitized by flavins start.

The triplet state has an extra free-energy content (0-0 spectroscopic energy) of 2.07 eV relative to the ground state, which accounts for the shift of the midpoint redox potential value of flavins from -0.22 V at pH 7 in the dark, to $+1.85$ V in the light. This is the reason why flavins can operate as light/redox energy-transducing systems. The oxidized form in its ground state (F) is first excited by light to the ^3F state, which is then reduced at a high redox potential (low energy) to generate the corresponding flavosemiquinone ($FH^{\cdot}$). This is, in turn, oxidized in the dark at a low redox potential (high energy), thus completing the cycle.

EXPERIMENT 10: PHOTOCATALYTIC REDOX REACTION

OBJECTIVE

A light-driven system consisting of FMN as photosensitizer, EDTA (ethylenediaminetetraacetate) as electron donor, and horse heart cytochrome *c* as electron acceptor is presented as a simple model system to study biological reductive reactions. Flavin promotes the light-induced transfer of electrons from an appropriate electron donor, such as EDTA, to a suitable electron acceptor, such as the electron carrier protein cytochrome *c*, thus catalyzing the photoreduction of this hemoprotein by electrons from a sacrificial donor that is consumed throughout the process.

LIST OF MATERIALS

Ethylenediaminetetraacetate, disodium salt (Na_2EDTA)
Flavin-mononucleotide (FMN)
Potassium monohydrogen phosphate (K_2HPO_4)
Potassium dihydrogen phosphate (KH_2PO_4)
Horse heart cytochrome *c*
Potassium iodide (KI)
Sodium azide (Na_3N)
Bovine erythrocyte superoxide dismutase
Transparent (glass or plastic) cuvettes (volume, 3-70 ml) provided with plastic or rubber stoppers or 3 ml optical glass or quartz cuvettes (1 × 1 × 4.5 cm) with stoppers
White light source (e.g. slide projector with a 150 W halogen lamp).
Spectrophotometer. A PYE UNICAM SP8-100 UV-visible recording spectrophotometer or a similar model is suitable; however, high-speed microprocessor-controlled UV-visible spectrophotometers with graphic video display (i.e. Beckman DU-7 or Shimadzu UV160U) provide operational prompts.
Nitrogen cylinder and regulator
Tubing for nitrogen

EXPERIMENTAL PROCEDURE

A. Photoreduction of FMN with EDTA

1. To a 3 ml cuvette add:

   2.7 ml of 25 mM potassium phosphate buffer, pH 7

   0.3 ml of 0.2 M Na_2EDTA in phosphate buffer

   0.015 ml of 10 mM FMN in phosphate buffer (This stock solution should be stored at 4°C *in the dark.*)

2. Record the absorbance spectra (300-600 nm; 0-0.6 A) at time zero and after irradiation with white light (200 $W \cdot m^{-2}$) for 30 s, 1 min and 6 min. Flavin photoreduction is indicated by the dramatic decrease of absorbance and the eventual complete decoloration of the sample (see Fig. 1A).

3. Flavin reoxidation. After air exposure record the absorbance spectrum of the yellowish solution and compare it with the initial spectrum.

B. Control Irradiation without Electron Donor

4. Repeat the same procedure as in part A, but without EDTA in the solution. Increase the buffer volume to 3 ml. No photoreduction should be observed (see Fig. 1B).

C. Inhibition of FMN Photoreduction by Quenchers of Triplet States

5. Perform FMN photoreduction as described above (internal control for photoreduction).
6. Reoxidize the flavin and record the absorbance spectrum.
7. Add 0.1 ml of 2M Na_3N in phosphate buffer or 0.15 ml of 4 M KI in phosphate buffer.
8. Record the absorbance spectra and irradiate the sample as above. No absorbance changes should be observed in the FMN spectrum (see Fig. 1C).

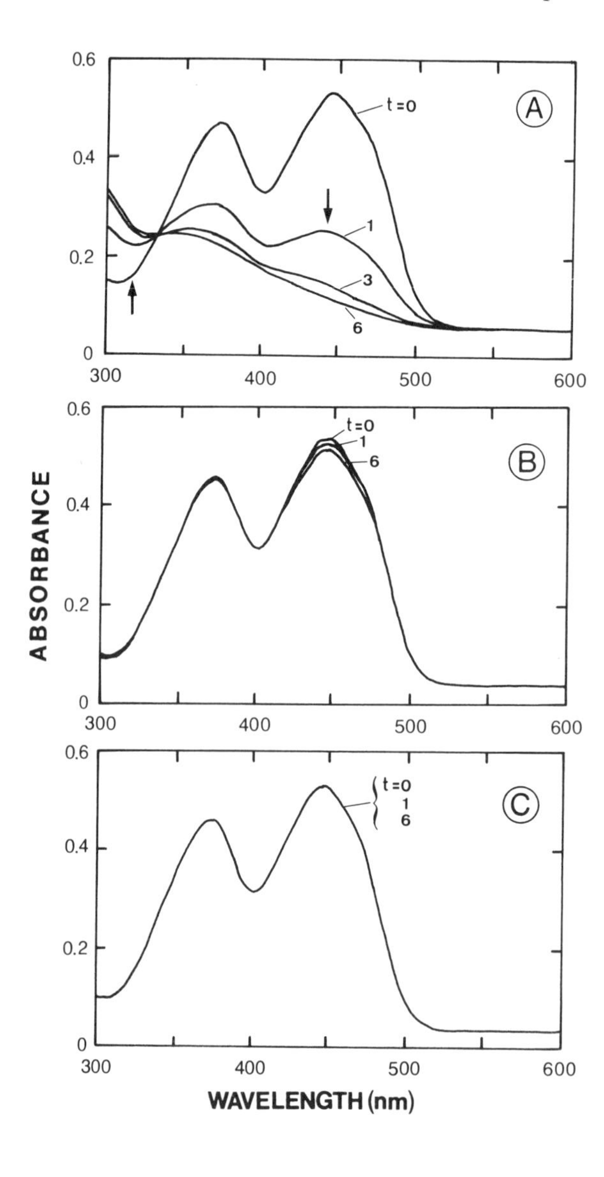

Figure 1. Absorption spectra of FMN recorded under different experimental conditions: A) Complete FMN/EDTA system, at time zero and after 1, 3 and 6 min of irradiation (the orientation of arrows indicates the direction of absorbance changes around the isosbestic point); B) Without EDTA, at time zero and after 1 and 6 min of irradiation; and C) CompleteFMN/EDTA system in the presence of Na_3N, at time zero and after 1 and 6 min of irradiation.

D. Reduction of Cytochrome *c* by the FMN/EDTA Photosystem

Irradiation procedure

9. To a 3 ml cuvette add:

   2.6 ml of 25 mM potassium phosphate buffer, pH 7

   0.3 ml of 0.2 M Na_2-EDTA in phosphate buffer

   0.015 ml of 1 mM FMN in phosphate buffer (This stock solution should be stored at 4°C *in the dark.*)

   0.1 ml of 1.5 mM cytochrome *c* in phosphate buffer (This stock solution should be stored at 4°C.)

10. Record the absorbance spectra (300-600 nm, 0-0.8 A) at time zero and after irradiation for 1 , 2 , 4 , 8 and 15 min. Note that microaerobic conditions are not required in this case. Cytochrome *c* reduction is characterized by the appearance of two new absorption peaks at 521 and 550 nm (see Fig. 2) and can be quantitated by the absorption increase at 550 nm (α-band).

DISCUSSION

Photoreduction of FMN with EDTA

A transparent 30 ml cuvette provided with a rubber or plastic stopper can be used as the reaction cell, which contains 0.01-0.1 mM FMN and 20-50 mM Na_2EDTA in 25 mM phosphate buffer, pH 7.0. The reaction mixture is then bubbled with nitrogen in order to evacuate dissolved oxygen; otherwise the very reactive flavin semiquinone will react with O_2, thus preventing total reduction of the flavin. The photochemical reaction starts upon irradiation of the reaction cell with a source of white light (e.g., a slide projector). In the course of the photochemical reaction, aliquots are removed at different time intervals using a gas tight syringe. Alternatively, and working in a lower scale, the experiment can be done in 3 ml optical glass or quartz cuvettes provided with plastic or rubber stoppers, so that the samples are not exposed to air. The flavin reduction is checked by recording the visible absorbance spectra in the range between 200 and 600 nm. The two characterstic peaks of oxidized flavin will eventually disappear so they are not present in the spectrum of the colorless, fully reduced form (Fig. 1A). The following reactions take place:

$$\mathrm{FMN} \xrightarrow{h\nu} {}^{1}\mathrm{FMN}$$

$${}^{1}\mathrm{FMN} \longrightarrow {}^{3}\mathrm{FMN}$$

$${}^{3}\mathrm{FMN} + \mathrm{EDTA} \longrightarrow \mathrm{FMNH}^{\cdot} + \text{Oxidized EDTA derivates}$$

$$\mathrm{FMNH}^{\cdot} + \mathrm{FMNH}^{\cdot} \longrightarrow \mathrm{FMNH_2} + \mathrm{FMN}$$

The process can be continuously monitored by the decrease of absorbance at 450 nm. Reoxidation of the flavin after irradiation can be easily obtained by air readmission and shaking the sample vigorously. A rapid increase in absorbance at 450 nm is then easily observed, the yellowish solution having the spectrum of oxidized flavin. When the pigment becomes oxidized, the active oxygen species superoxide, and eventually hydrogen peroxide, are generated:

$$FMNH_2 + O_2 \longrightarrow FMNH^{\cdot} + H^+ + O_2^{\cdot-}$$

$$O_2^{\cdot-} + O_2^{\cdot-} + 2\,H^+ \longrightarrow H_2O_2 + O_2$$

However, the initial absorption values of the peaks are not fully recovered even after long exposure to air (irreversible bleaching), thus indicating some photodestruction of pigment during irradiation. In agreement with this hypothesis, a very small, irreversible absorption decay is observed after irradiation in the absence of EDTA, where excitation, but no flavin photoreduction, is expected (Fig. 1B). On the other hand, it is also interesting to check the strong inhibitory effect that the presence of a quencher of the triplet state (i.e. Na_3N, 50-100 mM, or KI, 0.1-0.2 M) produces on the photoreduction process. Since there is no significant population of excited pigment molecules in this case, photodestruction should not occur, as is inferred from the complete absence of change in the absorbance spectrum during irradiation (Fig. 1C).

Reduction of Cytochrome *c* by the FMN/EDTA Photosystem

Horse heart cytochrome *c*, a well known electron carrier protein that participates in the respiratory electron transport chain, belongs to a class of water-soluble non-autooxidizable hemoproteins of low molecular weight which are present in all living systems. Many of these electron carrier proteins are reduced by flavoproteins.

The photoreduction of cytochrome *c* by the FMN/EDTA photosystem can be followed by the absorbance increase at 550 nm (extinction coefficient, reduced minus oxidized, $1.9 \times 10^{-4} M^{-1} \cdot cm^{-1}$). Cytochrome reduction is characterized by a peak splitting in the 500-570 nm region of the spectrum (Fig. 2). In this case the reaction mixture contains EDTA, FMN and 20-50 μM cytochrome *c*. It is interesting to note that in contrast to FMN photoreduction, cytochrome *c* photoreduction occurs with similar efficiency under anaerobic conditions as well as aerobic conditions. This is due to the fact that superoxide, generated after flavin oxidation by oxygen, also reduces the hemoprotein:

Anaerobic conditions

$$FMNH_2 + \text{cyt } c_{ox} \rightarrow FMNH^{\cdot} + \text{cyt } c_{red} + H^+$$

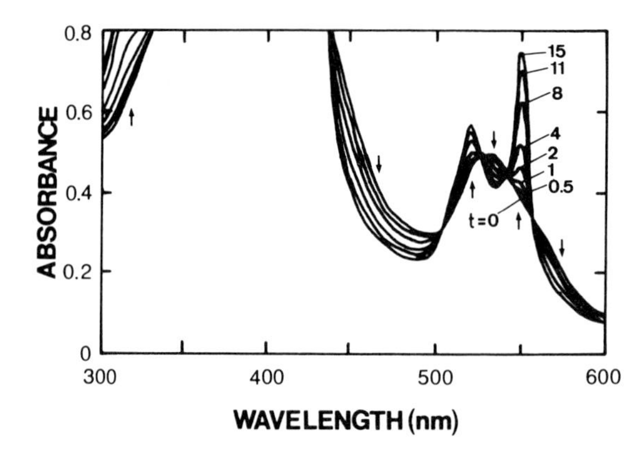

Figure 2. Horse heart cytochrome *c* photoreduction by the FMN/EDTA photosystem. Absorption spectra were recorded at time zero and after 0.5, 1, 2, 4, 8, 11 and 15 min of irradiation. The orientation of the arrows indicates the direction of the absorbance changes around the isosbestic points. The absorbance at wavelengths between 500 and 600 nm is due only to cytochrome *c*.

Aerobic conditions

$$FMNH_2 + O_2 \rightarrow FMNH^{\bullet} + H^+ + O_2^{\bar{\bullet}}$$

$$O_2^{\bar{\bullet}} + \text{cyt } c_{\text{ox}} \rightarrow O_2 + \text{cyt } c_{\text{red}}$$

These are simplified reactions because flavin semiquinone can also reduce the cytochrome, although with a lower efficiency.

The inhibitory effect of triplet quenchers on cytochrome *c* photoreduction by the FMN/EDTA system could also be tested. It is also interesting to check the effect of superoxide dismutase (about 1 μM in the reaction mixture), which inhibits cytochrome photoreduction under aerobic conditions but has no influence under anaerobic conditions.

REVIEW QUESTIONS

1. Taking into account the chemical structure of the following flavins: riboflavin, FMN, and flavin adenine mononucleotide (FAD), give a possible explanation to the fact that the efficiency of FAD as a photosensitizer in the flavin/EDTA photosystem is significantly lower (about 20%) than those of riboflavin and FMN.
2. Explain why quenchers of the triplet state completely inhibit the photoreduction reactions driven by the flavin/EDTA photosystem.
3. Imagine that you wish to construct a "photoreactor" for the production of the energy-rich compound hydrogen peroxide using the FMN/EDTA photosystem. Describe a simple method to recover the photosensitizer at the end of each photoproduction cycle, when all the electron donor has been consumed, so as to separate it from the photoproduct.

ANSWERS TO REVIEW QUESTIONS

1. In spite of its chemical analogy with FMN and riboflavin, FAD is a poor photosensitizer in flavin/EDTA photosystems because of the energy dissipation effect of the adenine group. This effect is probably a consequence of the formation of an intramolecular energy transfer complex between the flavin and the adenine rings.
2. In the presence of a quencher of the triplet state (i.e., I^-) the excited flavin (^3F) rapidly reacts with this chemical agent and is therefore deactivated before photochemical reactions take place.
3. Taking advantage of their different molecular sizes hydrogen peroxide (molecular mass, 34 g·mol^{-1}) can easily be removed from FMN (molecular mass, 478 g·mol^{-1}) by gel filtration chromatography on an appropriate gel matrix (e.g., Sephadex G-10).

ACKNOWLEDGEMENTS

The author thanks Profs. M.A. De la Rosa and M. Losada for helpful suggestions and Profs. D. Averbeck and W. Nultsch for their interest. Supported in part by grant PB87-401 of CICYT (Spain).

SUPPLEMENTARY READING

Grossweiner, L.I. and K.C. Smith (1989) Photochemistry. In *The Science of Photobiology*, 2nd edition (Edited by K.C. Smith), pp. 47-78. Plenum Press, New York.

REFERENCES

De la Rosa, F., M.A. De la Rosa, A.G. Fontes and C. Gomez-Moreno (1987) Photochemical production of hydrogen peroxide (in Spanish). *Investigacion y Ciencia* (Spanish edition of Scientific American) **125**, 8-15.

De la Rosa, M.A., P.F. Heelis, K.K. Rao and D.O. Hall (1987) Flavin-mediated hydrogen peroxide production by biological and chemical photosystems. *In Flavins and Flavoproteins* (Edited by D.E. Edmonson and D.B. McCormick), pp. 597-600, Walter de Gruyter, Berlin.

Grätzel, M. (Editor) (1983) *Energy Resources Through Photochemistry and Catalysis*. Academic Press, New York.

Losada, M., M. Hervas, M.A. De la Rosa and F. De la Rosa (1983) Energy transduction by bioelectrochemical systems. *Bioelectrochem. Bioenerg.* **11**, 193-230

Schmidt, W. and W.L. Butler (1976) Flavin-mediated photoreactions in artificial systems: a possible model for the blue-light photoreceptor pigment in living systems. *Photochem. Photobiol.* **24**,71-75.

Photosensitization: Reaction Pathways

Johan E. VAN LIER
University of Sherbrooke
Canada

INTRODUCTION

Thousands of naturally occurring and synthetic dyes can function as photosensitizers and inflict biological damage in the presence of light (Spikes, 1989 and references cited). The action is initiated by the absorption of a photon to yield an excited sensitizer and is followed by many competing dark reactions which ultimately result in the alteration of vital biomolecules. Reactions of the excited sensitizer can involve electron or hydrogen transfer, usually with a reducing substrate (Type I reaction) or interaction with oxygen (Type II reaction) (Foote, 1976 and references cited). The latter usually involves energy transfer to yield singlet molecular oxygen. Both Types I and II pathways may compete, with the predominant route being determined by such factors as oxygen and substrate concentrations, the proximity of the sensitizer to the substrate as well as the nature of the sensitizers and the substrate. Both pathways ultimately lead to the formation of oxidized products and radical chain reactions resulting in extensive biological damage. Living organisms contain many enzymic and nonenzymic antioxidant mechanisms to protect against reactive oxygen species and the inability to control the latter has been defined as oxidative stress (Sies, 1986).

In order to demonstrate the extent of the involvement of a Type I or Type II process in the photodynamic action of a sensitizer, a number of techniques may be employed. Most of them yield rather ambiguous answers since the complexity of the living cell or organism masks the initial event while the ample availability of scavengers usually strongly reduces the lifetime of the initially produced reactive species. Subsequent reactions likely involve different oxidation mechanisms, which further confuses the matter.

In addition to the direct detection of radical intermediates via electron spin resonance spectroscopy, or of singlet oxygen by its weak luminescence at 1270 nm representing the singlet to ground state decay, sensitization mechanisms can be elucidated indirectly from the effect of added quenchers. For example, β-carotene efficiently inactivates singlet oxygen via radiationless decay of the intermediate triplet β-carotene, while substrates such as polyunsaturated fatty acids give different hydroperoxy derivatives with singlet oxygen as compared to those obtained via radical mediated autoxidation processes. Varying substrate or oxygen

JOHAN E. VAN LIER, MRC Group in the Radiation Sciences , University of Sherbrooke
Faculty of Medicine, J1H 5N4, Sherbrooke, Quebec, Canada

Photobiological Techniques, Edited by D.P. Valenzeno *et al.*
Plenum Press, New York, 1991

concentrations affects the kinetics of Type I and Type II product formation differently, whereas added D_2O enhances the lifetime of singlet oxygen and thereby the yield of products derived from this pathway.

A number of endogenous substrates give characteristic products in Type I or II reactions. Among these substrates are membrane constituents such as cholesterol, DNA bases such as thymine and guanine, and certain amino acids including cysteine, tryptophan and histidine (Foote, 1976; Spikes, 1989). In this exercise we will examine the various products that can be obtained from the interaction of cholesterol with photosensitizers, light and oxygen. As already mentioned, the complexity of the chemical environment in the living cell often renders diagnostic tests to identify *in vivo* photosensitization mechanisms inconclusive and accordingly, the experiments will be conducted in homogeneous solutions. Different sensitizers and possible inhibitors of putative reactive intermediates will be employed using cholesterol as a substrate, and products will be characterized by their mobility upon silica gel thin layer chromatography and their color response to selective spray reagents.

Oxygen

Molecular oxygen is a key participant in the photosensitized degradation of biomolecules. In order to understand its reactivity a basic understanding of its electron configuration is required (Turro, 1978). The eight electrons of atomic oxygen are divided over the 1s, 2s and 2p orbitals as follows.

$$(1s)^2\,(2s)^2\,(2p)^4 \qquad [1]$$

In molecular oxygen, the combined 8 electrons of the p orbitals of the two oxygen atoms participate in σ and π bonding orbitals as follows.

$$(\sigma_{2p})^2\,(\pi_x)^2\,(\pi_y)^2\,(\pi_x^*)^1\,(\pi_y^*)^1 \qquad [2]$$

The 2 highest-energy electrons, i.e. the electrons in the π_x^* and π_y^* orbitals, can occupy one or both orbitals and they can exist in different spin states: a lowest energy triplet state ($^3\Sigma$) and two singlet states ($^1\Delta$ and $^1\Sigma$). In the $^3\Sigma$ ground state, the two electrons are unpaired and in separate orbitals, in the $^1\Delta$ excited state the two electrons are paired and in the same orbital while in the highest-excited state the electrons are paired but in different orbitals (Fig. 1).

The higher-excited $^1\Sigma$ singlet oxygen species, with an energy of 37.5 kcal·mol^{-1} above that of ground state molecular oxygen, has a very short lifetime in aqueous solution ($<10^{-11}$ sec) and if formed at all, is likely quenched to the $^1\Delta$ state before reacting (Foote, 1976). Thus, the longer lived $^1\Delta$ singlet state is believed to be the only singlet oxygen species involved in photosensitization reactions in solution. As can be seen from the orbital description (Fig. 1), ground state triplet oxygen behaves like a diradical while singlet oxygen reveals zwitterionic reactivity due to facile polarization of its electronic structure by π electrons of the substrate (Scheme 3).

$$:O-O \leftrightarrow O-O: \;\equiv\; {}^{(+)}O-O^{(-)} \leftrightarrow {}^{(-)}O-O^{(+)} \qquad [3]$$

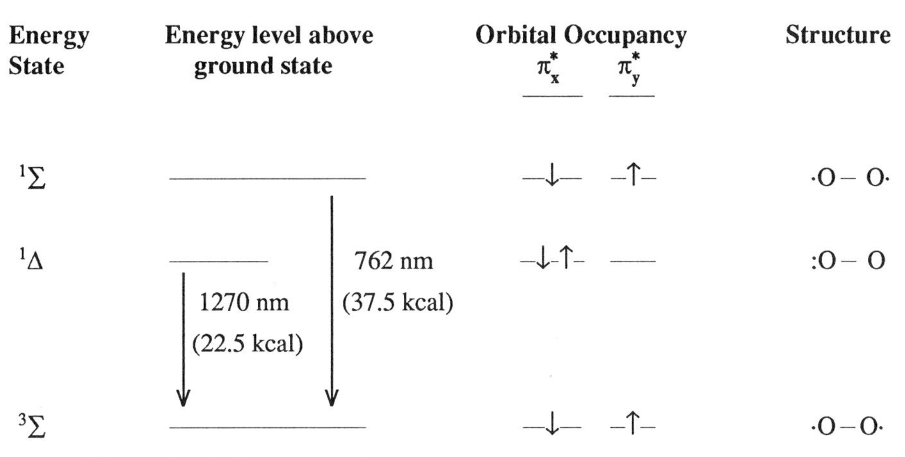

Figure 1. Orbital description and energy states of molecular oxygen.

Photosensitization

The photophysical processes between the ground state and various excited states of the sensitizer, including radiative and radiationless (dark) processes, can be depicted in an energy diagram. The first step usually involves a singlet-singlet absorption (Sens + hν → ^{1}Sens). The short lived ^{1}Sens species can be stabilized via a singlet-singlet emission, called fluorescence (^{1}Sens → Sens + hν), internal conversion (^{1}Sens → Sens + heat) or intersystem crossing to a triplet state (^{1}Sens → ^{3}Sens + heat). Lifetimes of ^{3}Sens states can be in the ms range while those of ^{1}Sens states usually are in the ns range. The triplet energy may be dissipated via a radiative triplet-singlet emission called phosphorescence (^{3}Sens → Sens + hν) or a reaction of the triplet with a substrate. The quenching mechanism of ^{3}Sens can be distinguished as Type I or II mechanisms (Schenk, 1963; Foote, 1976).

$$\text{Type I: } {}^3\text{Sens} + \text{SH} \rightarrow {}^\bullet\text{SH or S}^+ \qquad [4]$$

$$\text{Type II: } {}^3\text{Sens} + O_2({}^3\Sigma) \rightarrow O_2({}^1\Delta) \qquad [5]$$

A Type I reaction involves an electron or hydrogen transfer between the ^{3}Sens and a substrate (SH) to yield an ion radical or free radical species (Scheme 4), whereas a Type II reaction involves interaction between ^{3}Sens and ground state molecular oxygen. The latter usually involves energy transfer to yield singlet oxygen (Scheme 5). Electron transfer between ^{3}Sens and O_2 to yield the superoxide anion O_2^- is far less efficient (Lee and Rodgers, 1987). Both mechanisms ultimately lead to hydroperoxide formation and their subsequent rearrangements, as well as peroxidative chain reactions, yield a spate of secondary oxygenated products (Girotti, 1990).

Due to its short lifetime and high reactivity $O_2(^1\Delta)$ is difficult to detect directly in biological systems by its luminescence at 1270 nm resulting from its decay to $O_2(^3\Sigma)$ (Firey et al., 1988). A number of endogenous substrates give different products in the Type I and II pathways, while many other substrates only yield oxidation products with either one of the possible pathways. Product analysis thus provides an indirect means to probe photosensitization reaction mechanisms in complex media (Spikes, 1989; Foote, 1976).

Examples of Type I hydrogen-transfer processes include the reaction of triplet benzophenone with reducing substrates such as alcohols and amines (Spikes, 1989). Triplet menadione, or 2-methyl-1,4-naphthoquinone (MQ)†, efficiently reacts via a Type I electron-transfer process with thymine while the latter shows little reactivity towards $O_2(^1\Delta)$ (Wagner et al., 1984; 1990). In contrast, tryptophan gives characteristic tricyclic hydroperoxides with singlet oxygen while exhibiting little reactivity in photosensitized Type I pathways (Langlois et al., 1986). Cholesterol on the other hand gives characteristic oxidation products with a variety of activated oxygen species (Smith, 1981; 1987; Smith and Johnson, 1989) and its oxidation via Type I and II pathways will be examined more closely in this laboratory exercise.

Photosensitizers

The sensitizers selected for this experiment, i.e. the quinone derivative menadione (MQ) and tetrasulfonated aluminum phthalocyanine (AlOH-TSPc), are of current interest (Fig. 2). Menadione is used as a probe for the oxidative degradation of DNA components and phthalocyanines are extensively studied as possible second generation photosensitizers for the photodynamic therapy (PDT) of cancer (van Lier, 1990). Quinones play central roles in the electron transport chain of photosynthesis and aerobic respiration (Amesz, 1973), and synthetic quinones are used in chemotherapy of cancer and as antibiotics (Hartly et al., 1988). Quinones have a strong absorption band in the far-UV ($\varepsilon \approx 10^4$ $M^{-1}\cdot cm^{-1}$) as well as a moderate band in the near-UV or visible ($\varepsilon \approx 10^3$ $M^{-1}\cdot cm^{-1}$) (Bruce, 1967).

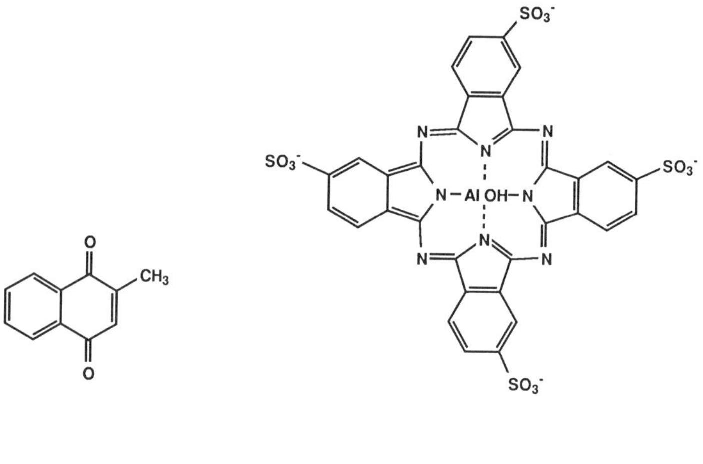

Figure 2. Molecular structure of photosensitizers. The molecular structure of menadione (MQ) and tetrasulfonated aluminum hydroxy phthalocyanine (AlOH-TSPc) are shown.

---

†**ABBREVIATIONS:** MQ, menadione or 2-methyl-1,4-naphthoquinone; 5α-OOH, 5α-hydroperoxy-3β-hydroxycholest-6-ene; 7α- or 7β-OOH, 7-hydroperoxy-3β-hydroxycholest-5-ene; AlOH-TSPc, tetrasulfonated aluminum hydroxy phthalocyanine; 5α-OH, 3β,5α-dihydroxycholest-6-ene; 7α- or 7β-OH, 3β,7-dihydroxycholest-5-ene; PDT, photodynamic therapy.

The MQ photosensitized photooxidation of nucleic acid derivatives involves electron transfer to triplet MQ to yield the MQ radical anion and the pyrimidine or purine radical cation which subsequently react with water and/or ground state oxygen to yield hydroperoxy derivatives as secondary oxidation products (Wagner et al., 1990). With cholesterol as a substrate, a similar electron or hydrogen transfer Type I process is envisaged, yielding the epimeric 7-hydroperoxides as characteristic products.

Sulfonated phthalocyanines in aqueous solution, on the other hand, are good sensitizers of singlet oxygen. Formation of $O_2(^1\Delta)$ is believed to be the principal cytotoxic species responsible for the tumor response during PDT of experimental animal tumors (van Lier, 1990).

The absorption spectra in methanol of MQ, AlOH-TSPc and β-carotene, which will be used in our experiment as a scavenger of $O_2(^1\Delta)$, are presented in Fig. 3. In this solvent the AlOH-TSPc is largely monomeric as suggested by the strong absorption maximum at 678 nm (ε = 190 000 $M^{-1} \cdot cm^{-1}$). Aggregation in aqueous media results in a blue shift and broadening of the absorption band together with a decrease in the ε (Bernauer and Fallab, 1961). Intramolecular interactions of the excited dye molecules render the aggregate inactive as a photosensitizer.

Substrate

The substrate for our photosensitization experiment, cholesterol, is a natural membrane constituent of mammalian cells and the precursor of the bile acid and steroid hormones. Cholesterol is present throughout nature. It is found in many human foodstuffs, and high blood levels of this sterol have been associated with the formation of plaques in the human aorta and the development of atherosclerosis. Cholesterol reacts with activated oxygen species and free radicals to yield various hydroperoxy and epoxy derivatives which may rearrange or initiate radical chain reactions resulting in a spate of secondary oxidized sterols (oxysterols). Among the latter are many as yet unidentified biologically active components

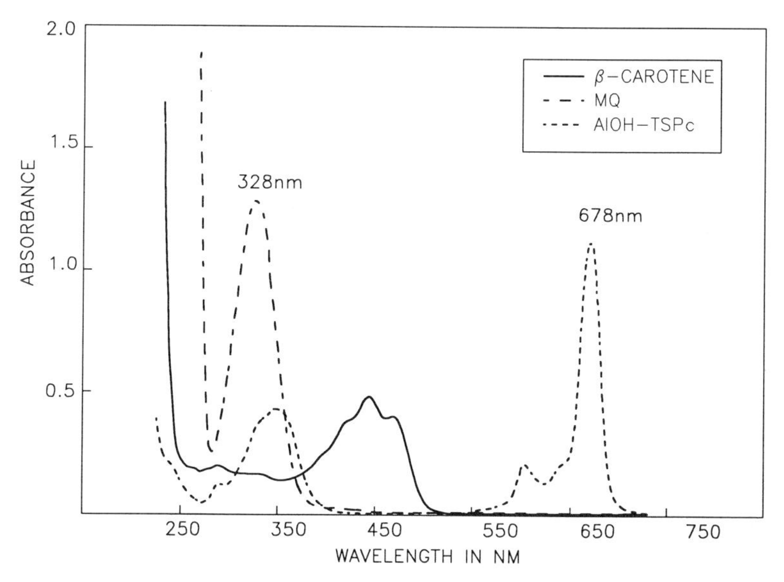

Figure 3. Absorption spectra in MeOH of menadione (MQ, 0.5 mM), aluminum tetrasulfophthalocyanine (AlOH-TSPc, 5.6 μM) and β-carotene (30 μM).

while others have been identified as effective enzyme inhibitors and cytotoxic as well as mutagenic agents (Smith, 1981; 1987; Smith and Johnson, 1989).

In a Type II process, cholesterol 1 reacts with singlet oxygen to give 5α-hydroperoxy-3β-cholest-6-ene (5α-OOH), 2, as a major product (Fig. 4). The superoxide anion (O_2^-) is unreactive towards cholesterol. In radical mediated Type I oxidation pathways the epimeric 7α- and 7β-hydroperoxycholesterols (7α-OOH and 7β-OOH), 3 and 4, are the main products. Upon borohydride or triphenylphosphine reduction the 5α-, 7α- and 7β-OOH, 2,

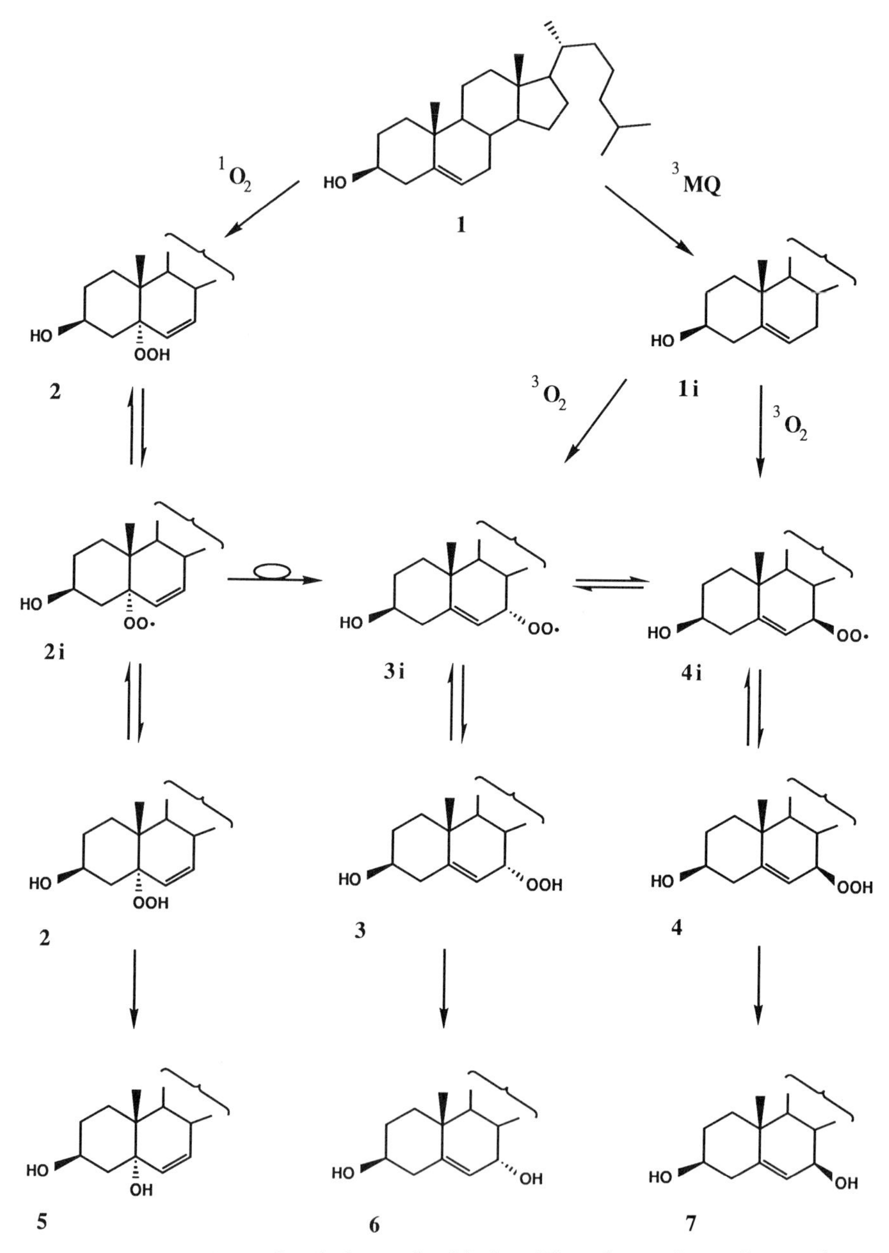

Fig. 4. Reaction scheme for cholesterol oxidation. The scheme shows the reaction products for both Type I and Type II reaction pathways.

3 and 4 are converted to the corresponding alcohols, 5, 6 and 7, with retention of configuration. These stable dihydroxy sterol products are readily separated by silica gel TLC and visualized by their characteristic blue color response to sulfuric acid upon heating at 100°C (Fig. 5). In a more complex biological system, initial hydroperoxide formation may initiate lipid peroxidation resulting in the formation of a complex mixture of oxysterols. During the autoxidation of cholesterol, the 25-hydroperoxy derivative also emerges as a major oxidation product together with a spate of hydroperoxide degradation products including hydroxy, keto and epoxy derivatives (Smith, 1981). Under our experimental conditions the initially formed hydroperoxy derivatives of cholesterol are sufficiently stable to allow for their use as probes for the mechanism of the photosensitization process. However, with time the 5α-hydroperoxide, 2, will rearrange in solution to yield the 7α-hydroperoxide, 3, and the latter will eventually epimerize to the 7β-hydroperoxide, 4. The rearrangements proceed via the intermediate allylperoxyl radicals, i.e., 2i, 3i and 4i. The 5α-OOH → 7α-OOH conversion has been identified as an intramolecular pericyclic process (2i → 3i) whereas the 7α-OOH → 7β-OOH epimerization appears to involve a dissociative mechanism (3i → 4i) in which dioxygen can intrude in the β-phase but not the α-phase (Beckwith et al., 1989). The intermediate hydroperoxyl radicals also may initiate peroxidative chain reactions resulting in the formation of many secondary hydroperoxides and epoxides, as well as more stable hydroperoxide decomposition products including hydro, keto and carboxy derivatives (Smith, 1981).

Notwithstanding the complexity of the cholesterol oxidative degradation pathways under biological conditions (Smith and Johnson, 1989; Sevanian and McLeod, 1987), the formation of the 5α- and 7-OOHs have been shown useful in distinguishing between Type I and II mechanisms in isolated cell membrane systems (Girotti, 1990).

Quenching of Singlet Oxygen

In addition to the many chemical reactions in which singlet oxygen can be consumed, several examples exist whereby $O_2(^1\Delta)$ is quenched to ground state oxygen and heat. One of the most important examples involves the protective role of β-carotene in photosynthetic organisms (Foote, 1976). Although β-carotene also effectively quenches triplet chlorophyll, for which it actually competes with oxygen in the $^3O_2 \rightarrow {}^1O_2$ process, the efficient protective role of β-carotene lies in its high reaction rate with $^1O_2$ (essentially diffusion controlled) to yield triplet β-carotene, which in turn decays to ground state β-carotene and heat. A similar balance of interactions prevails in our quenching experiment of the AlOH-TSPc sensitized photooxidation of cholesterol (1 → 2). Under our experimental conditions the concentration of β-carotene is 45 μM, which is 1 to 2 orders of magnitude less concentrated than molecular oxygen. Presuming similar reaction rates for β-carotene or oxygen with triplet AlOH-TSPc would still result in substantial formation of $^1O_2$ and thus complete inhibition of cholesterol oxidation in the presence of 45 μM β-carotene must result from quenching of $^1O_2$ by β-carotene.

EXPERIMENT 11: DETECTION OF TYPE I AND II PATHWAYS IN PHOTOSENSITIZED OXIDATION OF CHOLESTEROL

OBJECTIVES

The formation of characteristic Type I and Type II products during the photosensitized oxidation of cholesterol will be demonstrated by simple chromatographic procedures. Different photosensitizers will be evaluated including, sulfonated aluminum phthalocyanine currently under study as a photosensitizer for the photodynamic therapy of cancer, and 2-methyl-1, 4-naphthoquinone, a compound related to naturally-occurring quinones involved in biochemical electron transfer processes. The effectiveness of β-carotene as a singlet oxygen quencher will be evaluated.

LIST OF MATERIALS

Cholesterol (3β-hydroxycholest-5-ene; 5 g)
Menadione (MQ; 2-methyl-1,4-naphthoquinone; Sigma Chem. Co.; 25 g)
Sulfonated aluminum phthalocyanine (AlOH-SPc; Porphyrin Products; 10 mg)
Sodium borohydride ($NaBH_4$) from Aldrich. Alternatively triphenylphosphine (Aldrich) may be used as the reducing agent to convert hydroperoxides to the corresponding hydroxy derivatives.
Reference sterols. The 7α- and 7β-hydroxy derivatives of cholesterol 6 and 7 can be obtained from Steraloids or a mixture of the two isomers can be prepared by borohydride reduction of the 7-ketocholesterol (Steraloids). One of the three end products 5, 6 or 7 (Fig. 4) suffices as a reference since all three hydroxy sterols give similar color responses in the TLC spray test and their relative mobilities on TLC analysis are well established (see Experimental Procedure). The 5α-hydroxy derivative 5 (5α-cholest-6-ene- 3β,5-diol) can readily be prepared in larger quantities using a scale up of the photosensitization experiment with AlOH-TSPc as detailed in this laboratory exercise.
β-Carotene (Sigma Chem. Co.; 1 g)
Ethyl acetate (EtOAc), HPLC grade
Methanol (MeOH)
Hexane, HPLC grade
CH_2Cl_2, HPLC grade
Petri dish, Pyrex, 10 cm
Erlenmeyer flask, 50 ml
Test tubes, 1 ml (3)
TLC tank lined with filter paper to permit equilibration of the gas phase with the solvent
Capillary pipettes, 5 or 10 μl, to apply reaction mixtures on TLC plates
Pipette, 200 μl (or an Eppendorf pipette)
Analytical thin-layer chromatography (TLC) plates, 0.25 mm Macherey-Nagel silica gel Polygram Sil g/UV_{254} pre-coated plastic plates (5 × 20 cm)
Fume hood
Spray bottle to spray TLC plates with 5% H_2SO_4 in ethanol.
Hot plate, to heat acid treated TLC plates in order to visualize sterols.
Red light source: Slide projector with a 500 W tungsten/halogen lamp and a 600 nm cut-off filter (Corion LL-600)

UVA light source (320-400 nm): Black-ray 100 W medium pressure Hg lamp
Oxygen and nitrogen sources, regulators, and tubing
Spectrophotometer (optional, to check stock solutions)

PREPARATIONS

Cholesterol Stock Solution: Commercially available cholesterol, or samples that have been on the shelf for several years, are often contaminated with autooxidation products. Such samples can be purified by crystallization from hot methanol. To prepare the cholesterol stock solution, dissolve 387 mg of cholesterol in 100 ml of a 1:1 mixture of MeOH/EtOAc to make a 10 mM stock solution. The EtOAc is required to solubilize the cholesterol.

Menadione Stock Solution: Dissolve 34 mg of MQ in 100 ml of MeOH to make a 2 mM stock solution. If a spectrophotometer is available the concentration can be checked/adjusted by the absorbance at 330 nm ($\varepsilon = 3.1 \times 10^3$ $M^{-1} \cdot cm^{-1}$, $\lambda_{max} = 330$ nm).

Sulfonated Aluminum Phthalocyanine Stock Solution: The material used in this exercise was a tetrasulfonated sample (AlOH-TSPc) prepared via the condensation of sulfophthalic acid (Ali et al., 1989). A commercially available sample (Sigma) prepared by direct sulfonation of chloroaluminum phthalocyanine consisting of a mixture of mono- through tetrasulfonated products may also be used. To prepare the phthalocyanine stock solution, dissolve 2 mg of AlOH-TSPc in 100 ml of MeOH to make a 20 μM stock solution. Check/adjust the concentration if a spectrophotometer is available ($\varepsilon = 3.1 \times 10^3$ $M^{-1} \cdot cm^{-1}$, $\lambda_{max} = 330$ nm). AlOH-TSPc is largely monomeric at this concentration in MeOH and the same ε may be used to prepare stock solutions of mixtures of differently sulfonated AlOH-TSPc.

β-Carotene Stock Solution: Dissolve 27 mg of β-carotene in 100 ml of a 1:1 mixture of MeOH/EtOAc to make a 0.5 mM stock solution.

EXPERIMENTAL PROCEDURE

Photosensitization: Type I Process

1. Mix 5 ml of cholesterol stock solution with 5 ml of MQ stock solution in a 10 cm diameter Petri dish.

ENCLOSE THE LIGHT SOURCE AND EXPOSURE AREA IN A DARK BOX TO PROTECT THE EXPERIMENTER FROM THE UV LIGHT.

2. Expose the dish to UVA light (320-400 nm) for 1 h as follows. Place the sample about 10 cm from a Black-ray 100 W medium pressure Hg lamp. Oxygenation is ensured by the large surface of the sample which allows for sufficient O_2 uptake by diffusion.
3. After illumination of the sample, set it aside for later analysis (below).

Photosensitization: Type II Process

4. Mix 5 ml of cholesterol stock solution with 5 ml of AlOH-TSPc stock solution in a 50 ml Erlenmeyer flask (4 cm diameter).
5. Expose the flask to red light ($\lambda > 600$ nm) for 1 to 1.5 h as follows. Place the flask over a 500 W tungsten/halogen lamp of a slide projector fitted with a long-pass filter (Corion LL-600). During irradiation, slowly bubble oxygen through the solution.
6. After illumination of the sample, set it aside for later analysis (below).

7. Repeat steps 4 through 6 with a separate sample, but add 1 ml of β-carotene stock solution to quench singlet oxygen.

8. Retain the samples generated in steps in 6 and 7 for later analysis.

Hydroperoxide reduction

9. Pipette 200 µl of each reaction mixture (from steps 3, 6, and 7) into a small test tube. Add a few crystals of $NaBH_4$. The reduction will be complete within 1 min.

Sample preparation for TLC analysis

10. Before applying the reduced samples onto a TLC plate, the reaction solvent must be replaced by a less polar solvent so that the material will remain in a small spot while being applied. To this end, mark the level of the meniscus on each 200 µl sample (use tape or marker), and then blow a stream of nitrogen over the surface, until the volume has been reduced to between 1/2 and 1/3 of the original volume. Then add CH_2Cl_2 to restore the volume to its original level.

Sample Application to TLC Plate

11. With a soft lead pencil, draw a line across the width of the TLC plate, about 1.5 cm from the bottom. Mark 4 equally spaced spots along this line. Three spots will receive the reaction mixtures, and the remaining spot wil receive reference solutions of cholesterol and its 5α, 7α, and 7β-hydroxy derivatives, 5, 6, or 7 (≈ 10 nmoles; Fig. 5).

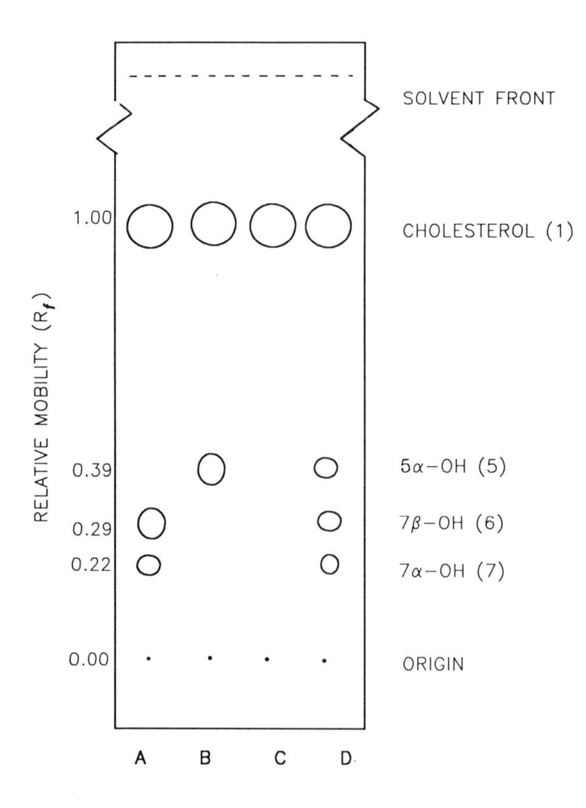

Figure 5. Tracing of a silica gel TLC developed twice in hexane: EtOAc (1:1) of products obtained from the photooxidation of cholesterol. **A**: Type I photosensitization with MQ, **B**: Type II photosensitization with AlOH-TSPc, **C**: Same conditions as **B**, in the presence of β-carotene and **D**: Reference sterols.

12. Aspirate 5 µl of each solution in a calibrated capillary pipette. Then touch the tip of the pipette to the appropriate spot on the TLC plate and allow fluid to wet the plate until a spot about 3 mm in diameter has formed.

13. Remove the pipette and allow the solvent to air-dry. Repeat this procedure until all 5 µl aliquots of each solution have been deposited on their appropriate spots.

TLC Plate Development

14. Place the TLC plate in the TLC tank containing a 1:1 mixture of EtOAc/hexane. Be sure the filter paper lining is in place as described in Preparations.

15. Allow the solvent front to migrate about 2/3 of the way up the TLC plate (6 to 7 cm). Remove the plate and allow it to dry in air.

16. Return the plate to the tank and allow the solvent to migrate to the top of the plate. Again remove the plate and allow it to air dry.

Visualization of Sterols on the TLC Plate

17. Place the plate against a cardboard backing in a fume hood and gently spray it with a 5% H_2SO_4 in ethanol mist. Little acidity is required for the subsequent color development and care should be taken that the plate is not over sprayed since this results in spot diffusion.

18. Place the TLC plate on a hot plate (100-120°C) and observe for color development, i.e. magenta in the case of cholesterol and deep blue for the 5α, 7α, and 7β-hydroxy derivatives, 5, 6, and 7.

19. Determine the relative mobility of the different sterols by measuring the distance from the center of the spot to the origin. Relative mobility (R_f) is expressed as the ratio of the migration distance of each sample to that of cholesterol.

DISCUSSION

The absolute mobility of cholesterol in this system varies between 5 and 7 cm, depending on the activity of the silica gel, which in turn is strongly influenced by the humidity of the storage environment. A single development of the TLC plate results in a poorer resolution between the isomeric 7-OHs, 6 and 7 , and the 5α-OH, 5. The expected relative mobilities are, cholesterol (1.00), 5α-OH (0.39), 7β-OH (0.29), and 7α-OH (0.22) (Fig.5).

TLC analysis of the reduced cholesterol oxidation products obtained via MQ or AlOH-TSPc photosensitization gives a clear answer as to the underlying mechanism (Fig. 4). In the case of MQ and UVA light, only the epimeric 7α- and 7β- hydroxycholesterols (7β-OH, major product), 6 and 7, are observed without a trace of the characteristic singlet oxygen product, i.e. the 5α-OH derivative, 5, while in the case of the AlOH-TSPc and red light, only the 5α-OH is observed without a trace of the epimeric 7-hydroxycholesterols. Thus, under these experimental conditions, cholesterol oxidation products serve as an unambiguous probe to assign Type I or Type II pathways. Furthermore, the addition of relatively small amounts of a known $O_2(^1\Delta)$ quencher, i.e. β-carotene, completely blocks product formation in the AlOH-TSPc photosensitized reaction, using wavelengths where β-carotene does not absorb (Fig. 3), confirming the involvement of $^1O_2$ in this typically Type II process.

Unfortunately, extrapolation of the information obtained by such chemical probes to a more complex biological system should be done with caution. Although the presence of the 5α-OOH (2) unambiguously indicates the involvement of $O_2(^1\Delta)$, its absence does not confirm a unique radical mediated Type I process, since the 5α-OOH will slowly rearrange to the 7α-OOH (3). In addition, albeit slower than the 5α-OOH → 7α-OOH (2 → 3) rearrangement, the dissociative rearrangement of the 7α-OOH → 7β-OOH (3 → 4) further complicates the picture (Fig. 4). Since the Type I pathway yields the 7β-OOH (4) as a major product, as can be noted qualitatively from our TLC analysis, the ratio between the amount of 7α- and 7β-OOH (or 7α/β-OH) could be a further measure to quantify the role of singlet oxygen in photodynamic processes in biological systems (Girotti, 1990).

Another complication is the participation of probe molecules, including cholesterol, in peroxidative chain reactions resulting in complex mixtures of hydroperoxides and stable secondary oxidation products. To date, however, no alternative method to distinguish between Type I and Type II processes in biological samples appears fool-proof, and it thus is sensible to further exploit the use of endogenous target molecules as probes in the study of the mechanism of photosensitized biological damage.

REVIEW QUESTIONS

1. Explain the differences in reactivity between ground state oxygen and singlet molecular oxygen.
2. Why does the presence of trace amounts of ionic iron enhance photosensitized oxidative damage in a biological system?
3. How do the maximum absorbance, fluorescence and phosphorescence wavelengths of a photosensitizer relate to its capacity to generate $O_2(^1\Delta)$?
4. Why is singlet oxygen difficult to detect by its luminescence in biological systems?
5. Cholesterol gives distinct Type I and II oxidation products. What complicates the use of these products as mechanistic probes in biological systems?

ANSWERS TO REVIEW QUESTIONS

1. Ground state $O_2(^3\Sigma)$ behaves like a diradical since the two highest energy electrons are in different orbitals. Singlet oxygen reveals zwitterionic reactivity due to facile polarization of its highest energy electron pair by π electrons of the substrate. $O_2(^1\Delta)$ will preferentially add to the double bond of unsaturated compounds (see also Fig. 1).
2. Propagation of photosensitized oxidative damage involves intermediate oxy-radical species. Iron ions will accelerate such reactions since they can readily donate or accept an electron ($Fe^{3+} + e^- \rightleftharpoons Fe^{2+}$) and generate oxy-radical species to initiate peroxidative chain reactions (i.e., $ROOH + Fe^{2+} \rightarrow Fe^{3+} + RO\cdot + OH^-$; see also Sies, 1986 and Girotti, 1990).
3. The capacity of a sensitizer to generate $O_2(^1\Delta)$ depends on the available energy of its triplet state (^{3}Sens), which should be higher than the energy required for the $O_2(^3\Sigma)$ ground state → $O_2(^1\Delta)$ excitation. In terms of wavelengths, the λ of the ^{3}Sens → Sens + hν transition should be shorter (more energetic) than the 1270 nm singlet → ground state emission of O_2. The fluorescence (^{1}Sens → Sens + hν) wavelength alone is not indicative of the available energy to generate singlet oxygen but needs to be corrected by the energy loss in the intersystem crossing to the triplet state (^{1}Sens → ^{3}Sens + heat).

4. The lifetime of singlet oxygen is affected by the physical environment and can be very short in a biological system. However, the difficulties encountered in detecting the 1270 nm luminescence of $O_2(^1\Delta)$ in biological systems during photosensitization experiments, likely reflects the high reactivity of $O_2(^1\Delta)$ with the abundantly available scavengers.
5. The primary product formed during the Type II photooxidation of cholesterol, i.e. the 5α-OOH, slowly rearranges to the 7α-OOH (2 → 3), which in turn will convert with time to the 7β-OOH (3 → 4). Thus, although the 7-hydroperoxides, 3 and 4, are characteristic for a Type I process, they also may be formed via a Type II process during secondary reactions of the initially formed 5α-hydroperoxide, 2.

SUPPLEMENTARY READING

Foote, C.S. (1976) Photosensitized oxidation and singlet oxygen: Consequences in biological systems. In *Free Radicals in Biology* (Edited by W.A. Pryor), Vol. II, pp. 85-133. Academic Press, New York.

Spikes, J.D. (1989) Photosensitization. In *The Science of Photobiology*, 2nd edition (Edited by K.C. Smith), pp. 79-111. Plenum Press, New York.

Turro, N.J. (1978) Singlet oxygen and chemiluminescent organic reactions. In *Modern Molecular Photochemistry* (Edited by N.J. Turro) pp. 579-614. Benjamin/Cummings, London.

REFERENCES

Amesz, J. (1973) The function of plastoquinone in photosynthetic electron transport. *Biochem. Biophys. Acta*, **301**, 35-51.

Beckwith, A.L.J., A.G. Davies, I.G.E. Davison, A. Maccoll and M.H. Mruzek (1989) The mechanisms of the rearrangements of allylic hydroperoxides: 5α-Hydroperoxy-3β-hydroxycholest-6-ene and 7α-hydroperoxy-3β-hydroxy-cholest-5-ene. *J. Chem. Soc. Perkin. Trans. II*, 815-824.

Bernauer, K. and S. Fallab (1961) Phthalocyanine in wasseriger Losung I. *Helv. Chim. Acta*, **44**, 1287-1292.

Bruce, J.M. (1967) Light-induced reactions of quinones. *Quart. Rev.*, **21**, 405-428.

Firey, P.A., T.W. Jones, G. Jori and M.A.J. Rodgers (1988) Photoexcitation of zinc phthalocyanine in mouse myeloma cells: the observation of triplet states but not of singlet oxygen. *Photochem. Photobiol.* **48**, 357-360.

Foote, C.S. (1976) Photosensitized oxidation and singlet oxygen: Consequences in biological systems. In *Free Radicals in Biology* (Edited by W.A. Pryor), Vol. II, pp. 85-133. Academic Press, New York.

Girotti, A.W. (1990) Photodynamic lipid peroxidation in biological systems. *Photochem. Photobiol.*, **51**, 497-509.

Hartley, J.A., K. Reszka and J.W. Lown (1988) Photosensitization by antitumor agents. 7. Correlation between anthracenedione-photosensitized DNA damage, NADH oxidation and oxygen consumption following visible light illumination. *Photochem. Photobiol.*, **48**, 19-25.

Langlois, R., H. Ali, N. Brasseur, R. Wagner and J.E. van Lier (1986) Biological activities of phthalocyanines. IV. Type II sensitized photooxidation of L-tryptophan and cholesterol. *Photochem. Photobiol.*, **44**, 117-125.

Lee, P.C.C. and M.A.J. Rodgers (1987) Laser flash photokinetic studies of rose bengal sensitized photodynamic interactions of nucleotides and DNA. *Photochem. Photobiol.*, **45**, 79-86.

Schenck, G.O. (1963) Photosensitization. *Ind. and Eng. Chem.*, **55**, 40-43.

Sevanian, A. and L.L. McLeod (1987) Cholesterol autoxidation in phospholipid membrane bilayers. *Lipids* **22**, 627-636.

Sies, H. (1986) Biochemistry of oxidative stress. *Angew Chem. Int. Ed. Engl.* **25**, 1058-1071.

Smith, L.L. (1981) *Cholesterol Autoxidation.* Plenum Press, New York.

Smith, L.L. (1987) Cholesterol autoxidation 1981-1986. *Chem. Phys. Lipids* **44**, 87-125.

Smith, L.L. and B.H. Johnson (1989) Biological activities of oxysterols. *Free Rad. Biol. Med.* **7**, 285-332.

Spikes, J.D. (1989) Photosensitization. In *The Science of Photobiology* (Edited by K.C. Smith), pp. 79-111. Plenum Press, New York.

Turro, N.J. (1978) Singlet oxygen and chemiluminescent organic reactions. In *Modern Molecular Photochemistry* (Edited by N.J. Turro) pp. 579-614. Benjamin/Cummings, London.

van Lier, J.E. (1990) Phthalocyanines as sensitizers for PDT of cancer. In *Photodynamic Therapy of Neoplastic Disease* (Edited by D. Kessel), Vol. I, pp. 279-291. CRC Press, Boca Raton, FL.

Wagner, J.R., J. Cadet and G.J. Fisher (1984) Photo-oxidation of thymine sensitized by 2-methyl-1,4-naphthoquinone: Analysis of products including three novel photo-dimers. *Photochem. Photobiol.* **40**, 589-597.

Wagner, J.R., H. Ali, R. Langlois, N. Brasseur and J.E. van Lier (1987) Biological activities of phthalocyanines. VI. Photooxidation of L-tryptophan by selectively sulfonated gallium phthalocyanines: singlet oxygen yields and effect of aggregation. *Photochem. Photobiol.*, **45**, 587-594.

Wagner, J.R., J.E. van Lier and L.J. Johnston (1990) Quinone sensitized electron transfer photooxidation of nucleic acids: chemistry of thymine and thymidine radical cations in aqueous solution. *Photochem. Photobiol.* **52**, 333-345.

Membrane Photomodification

Dennis Paul VALENZENO
University of Kansas Medical Center
U.S.A.

INTRODUCTION

Why is Membrane Photomodification Important?

Cell membranes are one of the most common and most important targets of cellular photomodification by many types of photosensitizers. They are critical in the inactivation or killing of a wide variety of cell types including viruses, bacteria, yeast, plant tissue, and a variety of animal cells including humans (Valenzeno, 1987). Although there are debates in the literature concerning the precise target in photomodified cells, membranes may well be the ultimate site of modification. For example, whether cells are inactivated by photomodification of lysosomes, mitochondria, or the plasma membrane, it is the membrane of each structure that is considered the most likely, critical modification site. Likewise, whether tissues such as tumors are killed by photomodification effects on the malignant cells themselves, or by effects on the vessels that supply blood to the tumor, membranes in both cases are viewed as likely targets.

The importance of membrane photomodification is also evident from an examination of the areas of scientific research in which it is important. These include photodynamic therapy of malignant tumors, purging of autologous bone marrow transplants, reperfusion injury, irreversible shock, organ transplantation, light-activated pesticides, spreading depression in the central nervous system, drug toxicity testing, and the treatment of Herpes simplex (Valenzeno and Tarr, 1990).

The key role that membranes play in cellular photomodification is perhaps not too surprising in view of the lipophilicity of many photosensitizers, including for example, many members of the porphyrins, phthalocyanines, acridines and xanthenes. For these reasons membrane photomodification has been gaining prominence and importance. In this chapter we will briefly survey the membrane preparations and techniques available for membrane photomodification studies. We will then focus on one commonly used technique that can be performed with a minimum of sophisticated equipment, namely photohemolysis, and we will see how it can be used to determine the relative effectiveness of 3 photosensitizers of the xanthene class.

DENNIS PAUL VALENZENO, Department of Physiology, University of Kansas Medical Center
39th and Rainbow Boulevard, Kansas City, Kansas, 66103, U.S.A.

How Does Membrane Photomodification Occur?

To understand photomodification of membranes you need to have at least a basic knowledge of the structure of biological membranes. In this chapter, unless otherwise stated, the term membrane will be used to refer to a biological membrane as described below. Fig. 1 shows the basic bilayer structure of a typical membrane. Phospholipid molecules are arranged in a double layer with their hydrophobic, hydrocarbon chains facing each other. These form the hydrophobic core of biological membranes. The more polar phosphate groups on the glycerol backbone of the phospholipids face the aqueous medium on either side of the membrane. The molecules of the bilayer are not covalently linked, but are held together by hydrophobic forces in a dynamic equilibrium. The whole structure is fluid. Individual phospholipid molecules can slide past one another quite easily. Cholesterol is a major lipid component of the plasma membrane of many animal cells, comprising as much as 40 mol % of the membranes of some cells, such as red blood cells.

Embedded in the phospholipid bilayer are membrane proteins. Again the proteins are held in position by hydrophobic forces. That is, the portions of the protein in contact with the membrane interior have a predominance of hydrophobic amino acid residues at their surface, while the portions in contact with the aqueous media present a preponderance of hydrophilic amino acids to those media. Some of these so-called integral or intrinsic proteins are relatively free to slide past one another, within the plane of the membrane, much like the phospholipids. Others are anchored in position by a network of intracellular proteins known as the cytoskeleton. As opposed to these integral or intrinsic proteins, there are also peripheral or extrinsic proteins which are attached only to the surface of the membrane, typically by ionic bonds. A typical biological membrane has a thickness of about 5 nm.

The photochemical and biochemical processes of membrane photomodification are thought to be similar to those that occur in solution (see, for example, the previous chapter by Van Lier). However, there are important differences related to the altered environment of the sensitizer in a membrane (Fig. 2). Sensitizers may remain in the aqueous medium adjacent to the membrane and generate excited intermediates such as singlet oxygen or superoxide which may then diffuse to the membrane to modify it. Alternately, sensitizers may be bound to the membrane surface, where their concentration and the concentration of other ions may be different from those in solution due to the surface charge present on most biological membranes (typically negative charge). Many of the most effective membrane photosensitizers are, in fact, amphipathic. Thus they are water soluble, but also adsorb to membrane surfaces. Finally, they may dissolve in the hydrocarbon interior of the membrane to varying degrees. This may affect their local concentration as well as their photophysical

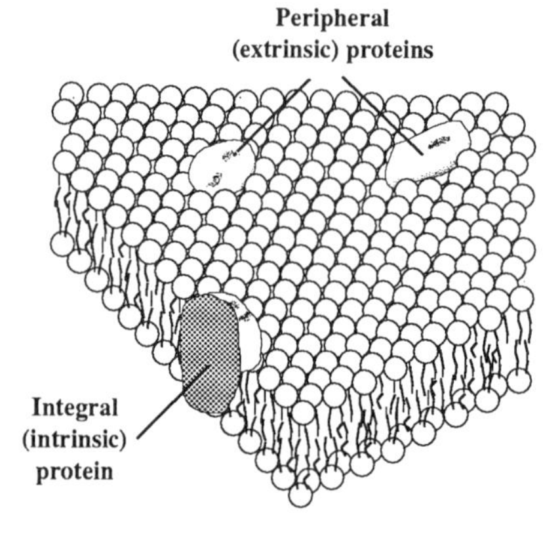

Figure 1. The figure shows a cut segment of a biological membrane. The phospholipid bilayer is indicated by the small circles. The hydrocarbon chains from the phospholipid head groups are depicted as wavy lines. The embedded proteins, both peripheral (extrinsic) and integral (intrinsic) are indicated.

and photochemical properties. Sensitizer interaction with membranes is known to affect sensitizer bleaching, electron transfer processes, hydrogen abstraction, lipid peroxidation, and excited intermediate generation (Valenzeno and Tarr, 1990).

What are the Effects of Membrane Photomodification?

There are 3 broad categories of membrane response to photomodification: morphological changes, chemical changes, and functional changes. Fig. 3 is a drawing of a photomodified cell which emphasizes changes due to membrane photomodification. The primary morphological change is an outpocketing of the surface membrane which is frequently referred to by the esthetically unpleasing, yet descriptive, term of blebbing. Membrane bound vacuoles frequently form in the cell cytoplasm also.

Chemical changes occur to both of the major classes of membrane molecules. The polyunsaturated fatty acid chains (hydrocarbon chains) of phospholipids are peroxidized. Cholesterol oxidation also occurs (see the previous chapter by Van Lier). Proteins are cross-linked and side chains of photomodifiable amino acids are modified. As in simple solution, the photomodifiable amino acids are histidine, tryptophan, tyrosine, cysteine, and methionine. These chemical changes can lead to significant alterations in membrane structure (Girotti, 1990).

There are many functional changes in photomodified membranes. A few examples may provide an idea of the scope of the functional alterations. One commonly studied functional change is alteration of the barrier function of the membrane. The most fundamental job of a cell membrane is to keep the inside in and the outside out. Photomodified membranes become leaky. They no longer exclude certain substances from the cytoplasm, such as vital dyes. Neither can they maintain the high intracellular concentrations of solutes such as potassium or, as we shall see in the laboratory experiment, hemoglobin in red blood cells. The activity of membrane bound enzymes is impaired. For example sodium-potassium ATPase is inhibited. Photomodification interferes with the ability of specific transporters in cell membranes to move solutes across the membrane. Thus amino acid and glucose transport are diminished. Finally, photomodification alters the electrophysiological properties of many membranes. Initially ion specific channels that allow current flow through the membrane are blocked. Later a non-specific leak is induced. This abolishes the normal resting membrane potential of the cell. Excitable cells, i.e. nerve and muscle cells, are thus completely inactivated.

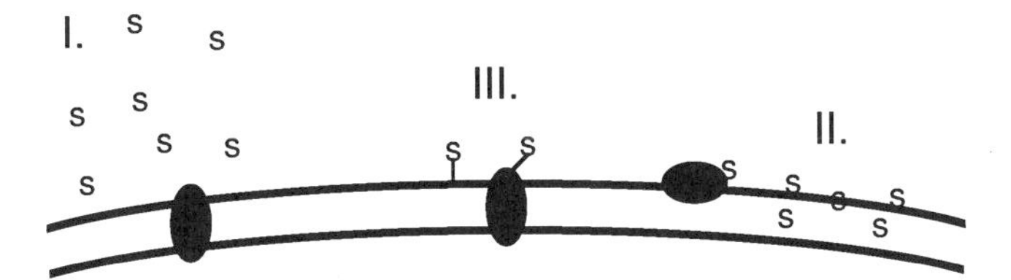

Figure 2. A small segment of a cell membrane is shown in cross section. Membrane proteins are indicated by the black ovals. Sensitizer molecules are indicated by the letter S. Three possible locations for effective membrane sensitizing molecules are indicated; I. sensitizer in the bathing medium, II. sensitizer bound to surface proteins or phospholipids, and III. sensitizer dissolved in the lipid bilayer.

How Can Membrane Photomodification be Studied?

Membrane phenomena can be studied using a variety of membrane model systems. A brief survey of these and their relative merits includes the following:

Solvents: Because the membrane interior is a hydrophobic environment, organic solvents are used as models for the membrane. Thus excited state properties are believed to correspond more closely to those in organic solvents than to those in aqueous solution (Lee and Rodgers, 1983). Immiscible solvents, such as octanol and water can be used to study partitioning behavior between the aqueous medium and the membrane interior. However, these models suffer from the lack of, or small surface area of, an interface between the two phases. This interface is an important area for many photosensitization processes in membranes.

Micelles: These are the simplest systems having an extensive interface (Fig. 4). They are composed of amphipathic detergent molecules in an aqueous medium (or vice-versa for reverse micelles). They are very simple to make and have been exploited most extensively for excited state studies. Their dimensions are small (1-2 nm diameter) compared to the thickness of a normal membrane bilayer (50 nm), so they do not scatter light and are useful in optical studies. They have a high radius of curvature which can influence accessibility of solutes to the interior of the micelle. They model only the lipid bilayer devoid of proteins and are monolayer rather than bilayer structures.

Planar Lipid Bilayers: Sometimes called black lipid membranes because of their appearance in the light microscope, these are bilayers of phospholipids formed within a small aperture separating two compartments. They are planar, so they do not suffer from radius of curvature problems. The composition of the solution on each side of

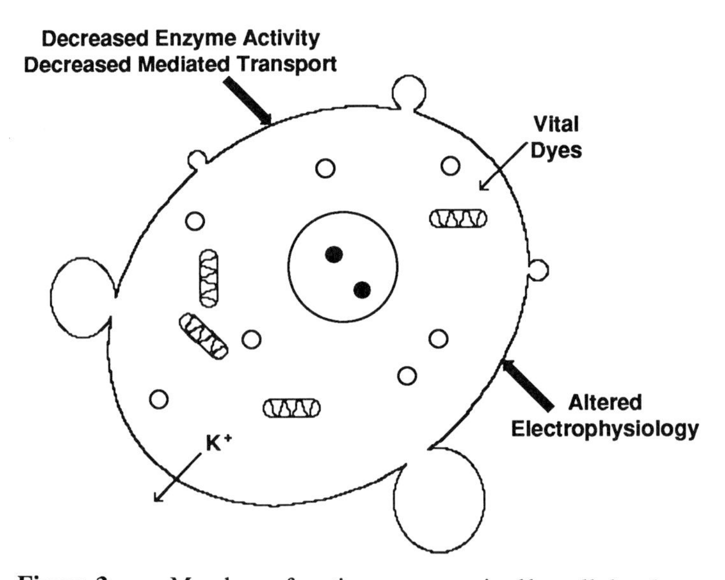

Figure 3. Membrane functions compromised by cellular photomodification are indicated in the figure. Vital dyes penetrate photomodified cell membranes, which also leak potassium. There are also alterations in enzyme activity, membrane transport and electrophysiology.

the membrane is easily controlled. Electrodes can also be used for electrophysiological studies. Although there are no radius of curvature complications, the small membrane area can be a problem.

Liposomes: Liposomes are enclosed lipid bilayers which can be prepared from different kinds of phospholipids (Fig. 4). They can be prepared with a little more difficulty than micelles. Depending on the fabrication technique, they may be either unilamellar (one bilayer) or multilamellar (onion-like layered structure). Formation technique also determines size, generally in the range of 0.05 to 1.5 μm in diameter. Liposomes can also be formed from phospholipids extracted from living cells, so that the phospholipid composition approximates the natural state. Ordinarily they do not contain proteins. However, recent techniques have made it possible to incorporate proteins into the liposomal membrane, but the techniques are considerably more complex. Liposomes, especially small ones, are still plagued by the high radius of curvature problem.

Vesicles: This term usually refers to liposome-like enclosed bilayers prepared from the membranes of a living cell. They, therefore, contain proteins, lipids, and most other normal membrane components. Like liposomes, they frequently have a high radius of curvature.

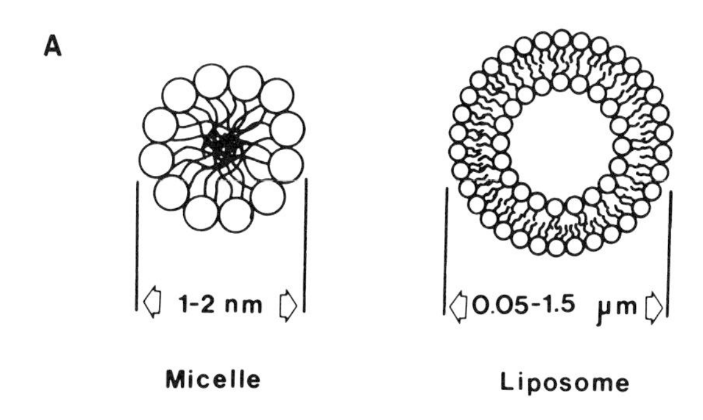

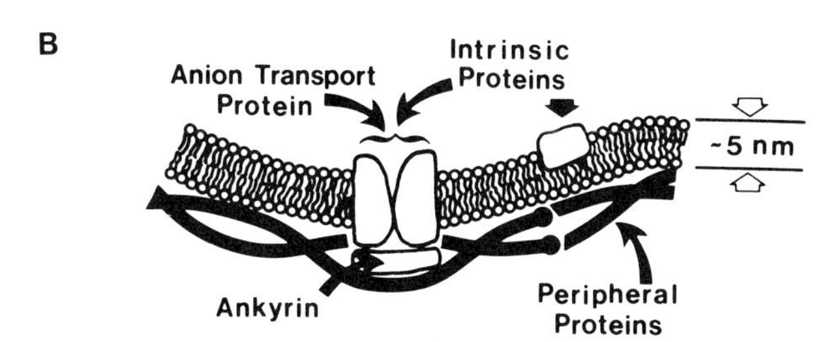

Figure 4. Three experimental membrane preparations are depicted in the figure. Two synthetic membrane models are shown in part A, a micelle and a liposome. Part B shows a highly schematic model of a segment of red blood cell membrane.

Red Cell Ghosts: The red cell ghost is a red blood cell in which the normal intracellular solution (mainly hemoglobin) has been replaced with a saline solution (Fig. 4). This is easily done by lysing the cells in a saline solution with very low salt concentration (hypotonic medium). Red cell ghosts can be prepared with most desired intracellular solutes and can be resealed after incorporation of these solutes. Though the membrane is that of a real cell, the red cell is somewhat unusual because of its high concentration of cholesterol compared to most cell membranes.

Intact Cells: Finally, membrane photomodification can be studied in intact cells which are, obviously, the most natural state. The difficulty in this case is separating primary effects on the membrane from secondary effects produced by photomodification of other cellular structures.

The techniques used to study photomodification of membranes can be categorized as were the kinds of changes induced in photomodified membranes. That is, there are techniques that measure morphological, biochemical and functional alterations.

Morphological techniques involve microscopic and electron microscopic studies of cells and their membranes. Freeze-fracture investigations are especially useful in providing information on changes in the architecture of membranes. Microspectrophotometry studies of sensitizer localization within cells might also be considered a part of this category.

Biochemical techniques include determination of cholesterol oxidation products by thin layer chromatography of cholesterol as discussed in the preceding chapter, protein cross-linking measurements by SDS (sodium dodecyl sulfate) polyacrylamide gel electrophoresis of extracted membrane proteins, and determination of lipid peroxidation by iodometric methods or by malonaldehyde generation, among others. While these tests can be used to infer reaction mechanisms (Girotti, 1990), it is difficult to determine the time course of the photomodification reactions since these methods do not easily provide a continuous measure.

Functional tests have been alluded to during the discussion of functional changes in membranes. For example, membrane barrier function can be assessed by assessing the exclusion of a vital dye, by measuring potassium release to the suspension medium, or by release of hemoglobin by red blood cells, the technique to be used in the experiments described below. These are, however, relatively late phenomena in the sequence of events following membrane photomodification. Decreases in membrane enzyme function and specific membrane transport processes are not as late, but analysis can frequently require considerable time after photomodification, much like the biochemical techniques. Electrophysiological techniques, on the other hand, are very sensitive and measure events that occur quite early in the sequence of changes due to membrane photomodification. They can be used to follow changes in ionic current flow in real time, with temporal resolution in the millisecond time domain. Their main drawback is that they require a considerable investment in equipment and significant training in the techniques of voltage and/or patch clamping.

Photohemolysis

Photohemolysis is a simple technique to monitor membrane photomodification. We will use it here to determine the relative effectiveness of three similar photosensitizers, eosin Y, erythrosin B, and Rose Bengal. This method has the advantage that blood cells are easily obtained, usually free or at minimal cost. Mature red cells have no intracellular organelles, so the effects seen are due to membrane photomodification. The deficiencies are the atypically high cholesterol content of red cell membranes, the possibility that organelle membranes with different composition may not be well modeled by red blood cell membranes, and the fact that lysis is a late event in the sequence of changes produced by membrane photomodification.

In the following experiments you will observe *colloid osmotic hemolysis*. In brief, because all living cells contain impermeable intracellular solutes (colloids), they face a constant challenge when immersed in isotonic salt solutions, namely the tendency to swell. This tendency occurs because the cells try to simultaneously maintain equal concentrations of permeable solutes inside and outside the cell, as well as equal total solute concentrations (osmolality) inside and outside. Since there are no extracellular, impermeable solutes when a cell is immersed in an isotonic salt solution, the tendency of the cell is to take up enough water (and permeable solute) to dilute the impermeable solutes inside the cell to zero, which is at infinite volume. This tendency is normally compensated by a sodium-potassium pump in the cell membrane which accomplishes a net expulsion of salt (and water) to maintain cell volume constant. [For a more complete treatment of colloid osmotic hemolysis see the discussion of cell volume regulation in any standard text of physiology.] If for any reason the leakiness of the membrane to salt is increased, the sodium-potassium pump may not be able to keep up, and the cell will swell until it bursts (colloid osmotic lysis).

As discussed above, photomodification makes membranes leaky. It may also directly impair function of the sodium-potassium pump with prolonged treatment, but this is not necessary for photohemolysis to occur. It is important to recognize that photohemolysis is *not* simply leakage of hemoglobin through holes created by membrane photomodification. Only small solutes, such as sodium and potassium ions, can leak through the membrane after photomodification. It is the colloid osmotic swelling and subsequent lysis of the cell that releases intracellular hemoglobin.

Lysis of red blood cells will be detected in the experiment in two ways. First, you will see that the red cell suspensions will become clear. The index of refraction of the interior of an intact red blood cell, with a high concentration of hemoglobin, is different from its suspension medium. Therefore, light is bent when it enters and exits the cell. That is, each cell acts as a tiny lens, bending light this way and that. The result is that nothing can be seen clearly through the suspension. When the cells lyse, the solution inside the broken membranes becomes the same as that outside the cells. Since the index of refraction is the same inside and outside the cell, light is not bent and you can see clearly through the suspension.

Alternately, the hemoglobin released to the medium can be detected using a spectrophotometer. Hemoglobin absorbs strongly at about 414 nm (Soret peak). To make an accurate measurement, any remaining cells which can scatter light must first be removed from the suspension by spinning in a centrifuge. The absorbance of the supernatant at 414 nm can then be used as a measure of hemolysis (photohemolysis in this case).

EXPERIMENT 12: RED CELL PHOTOGRAPHY

OBJECTIVE

This laboratory exercise is designed to graphically demonstrate photosensitized lysis of erythrocytes. It is a qualitative experiment which demonstrates the requirements for both sensitizer and light, and in the process produces a "photograph" in blood.

LIST OF MATERIALS

Blood (dog or other mammal), 1 ml
Fluorescent light (standard cool white fluorescent), preferably VHO (very high output)
Fluorescent fixture for bulb
Lucite sheet to support dishes over light
NaCl, 10 g
Tris hydroxymethyl aminomethane, 1 g
HCl, a fcw ml
Rose Bengal, 5mg
Pipettors, 1 ml and 5 ml (or pipettes and pipette bulbs)
Beakers, 50 or 100 ml, (4) for waste, saline, distilled water and red cells
Pasteur pipets and bulbs
Magnetic stirrer and medium/small stir bar (at least one)
Plastic (or glass) culture dishes, 60 mm or 100 mm diameter (4)
Grease pencil or waterproof marker
Aluminum foil, 1 roll
pH meter for solution preparation
Volumetric flask or graduated cylinder, 1 liter, for saline solution
Volumetric flask or graduated cylinder, 100 ml, for sensitizer stock solution
A few small (5 ml or so) test tubes for red cell preparation
Optional: Dark room or empty box from photographic film ($12.5 \times 17.5 \times 2.5$ cm minimum)

PREPARATIONS

Saline Solution: Prepare 1 liter of isotonic saline (149 mM NaCl and 5 mM Tris) as follows. Dissolve 8.71 g of NaCl and 0.61 g of Tris hydroxymethyl aminomethane in about 950 ml of distilled water. When the salts have completely dissolved, titrate the solution to pH 7.4 with HCl as required. Add distilled water to bring the volume to 1 liter.

Red Cell Preparation: A stock erythrocyte suspension is prepared from whole blood. The experiment is designed to be performed with dog blood, but blood from other species may suffice, although experimental conditions may have to be varied.

Mix 0.5 ml of whole blood in a test tube with 1.5 ml of saline. Mark the level of fluid on the side of the test tube with a water-proof marker. Spin in a centrifuge at about $1000 \times g$ for about 3 min, or until all the cells have collected on the bottom of the tube. Carefully aspirate the supernatant and resuspend to original volume with fresh saline. Repeat the spin and resuspension steps until four spins have been completed. Mix the final resuspended cells with 38 ml of saline in a 100 ml beaker. Stir this slowly (3 to 4 revolutions/s) on a magnetic stirrer until needed. Be careful that the stirrer does not get hot as this may damage the cells.

Sensitizer Stock Preparation: Prepare a 50 μM solution of Rose Bengal in saline. This requires about 5 mg of Rose Bengal in 100 ml of saline.

Illumination System: The most convenient and effective arrangement for illumination is to place the fluorescent fixture on a counter so that the light is directed upward. Support the lucite sheet 10 cm or less above the bulbs. For illumination, arrange the dishes directly above a bulb. Place dishes only over the middle 2/3 of the bulb length. This arrangement assures uniform illumination of all dishes and avoids absorption of light by sensitizer in the suspension medium between the cells and the source, as would occur with illumination from above.

A fluorescent light source is desirable since it can produce a moderately high light output over a wide area while producing little heat which could damage cells. If a fluorescent source is unavailable, the dishes may be illuminated from above with the most intense source available, taking care to avoid excessive heating. This may be checked by illuminating a suspension of cells with no sensitizer and looking for evidence of hemolysis as described below.

EXPERIMENTAL PROCEDURE

The following procedure was developed for use with 60 mm diameter glass or plastic culture dishes. Volumes in the text are appropriate for these dishes, but volumes for 100 mm diameter dishes are also given in parentheses.

1. Label the lid of one dish "Sens + Lite" and then pipette 3 ml (*12 ml*) of sensitizer stock solution into it.
2. Label the lid of a second dish "Sens– Lite" and then pipette 3 ml (*12 ml*) of sensitizer stock solution into it.
3. Label the lid of a third dish "Saline" and then pipette 3 ml (*12 ml*) of saline into it.
4. Label the lid of a fourth dish "Distilled Water" and then pipette 3 ml (*12 ml*) of distilled water into it.

AT THIS POINT BE SURE THAT THE LAB LIGHTS ARE DIMMED

5. To each of the four dishes just prepared, add 2 ml *(8 ml)* of stock erythrocyte suspension. Mix by gently moving the dish circularly in a horizontal plane.
6. The experiment will work best if the dishes are set aside on a level surface in a dark room, in a light-proof box (an old box in which photographic paper is sold works well), or wrapped in aluminum foil for 30 min before further use, but this is not absolutely necessary.
7. Wrap the dish labelled "Sens – Lite" in aluminum foil.
8. Illuminate all 4 dishes by placing them on the lucite table. Each dish should be directly above the axis of one of the fluorescent bulbs, as near to the center of the bulb axis as possible. Under the dish labelled "Sens + Lite" place a high contrast black-and-white negative (or cut a design of black paper or aluminum foil which covers about half the dish). *DO NOT MOVE THE DISHES ONCE THEY HAVE BEEN CORRECTLY POSITIONED.*
9. During illumination observe the dishes for clearing of their normally cloudy appearance. If a lucite sheet is used for a table, this is easily assessed by viewing a sharp edge or a printed or typed page below the dish through the cell suspension. (Note that if you use a printed or typed page it must be behind the light so that it does not block illumination!) Compare the clarity of the sensitizer-containing suspension with that of the completely lysed cells in distilled water and with the intact cells in the dish without sensitizer.

10. When the suspensions exposed to light have just cleared, turn off the light and immediately remove the solution from all dishes by gently aspirating from one spot near the edge of the dish, tilting just a little to collect the solution to the pipette tip. Once thoroughly aspirated, the resulting patterns will remain for many days.

DISCUSSION

This qualitative experiment demonstrates very dramatically that sensitizer alone, or light alone is much less effective than both sensitizer and light at producing lysis of red cells. Light alone can produce some hemolysis, but only after very long exposure times. Likewise, at very high sensitizer concentrations, the sensitizer can cause some lysis in the dark. You might want to test this by preparing a few dishes with sensitizer concentrations 10 and 100 times higher than those used here.

Based on the last section of the Introduction to this chapter, you should be able to explain why all of the red cells in distilled water lysed so rapidly. [Hint: At what cell volume would you expect the intracellular and extracellular osmolalities to be the same?] Likewise you should be able to explain why lysis occurred only over the clear areas of your negative or opaque pattern. If you have trouble with these explanations, try rereading the photohemolysis section in the Introduction, or consulting the Supplementary Reading on colloid osmotic hemolysis.

EXPERIMENT 13: QUANTITATIVE PHOTOHEMOLYSIS

OBJECTIVE

This laboratory exercise results in a quantitative determination of the relative effectiveness of three photosensitizers. Photohemolysis is used as an assay technique to make the determination. Controls will demonstrate that photohemolysis requires both sensitizer and light.

LIST OF MATERIALS

Blood (dog or other mammal), 1 ml
Fluorescent light (standard cool white fluorescent), preferably VHO (very high output)
Fluorescent fixture for bulb
Lucite sheet to support dishes over light
NaCl, 10 g
Tris hydroxymethyl aminomethane, 1 g
HCl, a few ml
Eosin Y, 10 mg
Erythrosin B, 5 mg
Rose Bengal, 5 mg
Pipettors, 1 ml and 5 ml (or pipettes and pipette bulbs)
Test tubes, 4-7 ml (about 50)
Test tube rack to hold at least 8 tubes (1)
Cuvette rack (or a styrofoam cuvette supply pack cut in pieces) (1)
Beakers, 50 or 100 ml, (4) for waste, saline, distilled water and red cells
Pasteur pipets and bulbs
Magnetic stirrer and medium/small stir bar (at least one)
Plastic (or glass) culture dishes, 60 mm or 100 mm diameter (8 per sensitizer used)
Grease pencil or waterproof marker (1)
Aluminum foil, (1 roll)
Dark room or empty box from photographic film (12.5 × 17.5 × 2.5 cm minimum)
Spectrophotometer: to measure absorbance at 414 nm
Plastic cuvettes (about 50)
Table top centrifuge (about 1000 to 4000 rpm)
Semi-log paper: 3 cycles (1 sheet per student or group)
pH meter for solution preparation
Volumetric flask or graduated cylinder, 1 liter, for saline solution
Volumetric flask or graduated cylinder, 100 ml, for sensitizer stock solution
A few small (5 ml or so) test tubes for red cell preparation

PREPARATIONS

Saline Solution: Prepare 1 liter of isotonic saline (149 mM NaCl and 5 mM Tris) as follows. Dissolve 8.71 g of NaCl and 0.61 g of Tris hydroxymethyl aminomethane in about 950 ml of distilled water. When the salts have completely dissolved, titrate the solution to pH 7.4 with HCl as required. Add distilled water to bring the volume to 1 liter.

Red Cell Preparation: A stock erythrocyte suspension is prepared from whole blood. The experiment is designed to be performed with dog blood, but blood from other species may suffice, although experimental conditions may have to be varied.

Mix 0.5 ml of whole blood in a test tube with 1.5 ml of saline. Mark the level of fluid on the side of the test tube with a water-proof marker. Spin in a centrifuge at about 1000 × g for about 3 min, or until all the cells have collected on the bottom of the tube. Carefully aspirate the supernatant and resuspend to original volume with fresh saline. Repeat the spin and resuspension steps until four spins have been completed. Mix the final resuspended cells with 38 ml of saline in a 100 ml beaker. Stir this slowly (3 to 4 revolutions/s) on a magnetic stirrer until needed. Be careful that the stirrer does not get hot as this may damage the cells.

Sensitizer Stock Preparation: Prepare a 100 μM solution of eosin Y (6.9 mg in 100 ml of saline), a 25 μM solution of erythrosin B (2.2 mg in 100 ml of saline), and a 1 μM solution of Rose Bengal. The Rose Bengal solution can be prepared as a 50:1 dilution of the 50 μM solution prepared for the photography demonstration (5 mg in 100 ml of saline). For each sensitizer prepare 3 serial dilutions, so as to have 100, 50, 25, and 12.5 μM solutions of eosin Y, 25, 12.5, 6.25, and 3.125 μM solutions of erythrosin B, and 1, 0.5, 0.25, and 0.125 μM solutions of Rose Bengal.

Illumination System: The most convenient and effective arrangement for illumination is to place the fluorescent fixture on a counter so that light from it is directed upward. Support the lucite sheet 10 cm or less above the bulbs. For illumination, arrange the dishes directly above a bulb. Only place dishes over the middle 2/3 of the bulb length. This arrangement assures uniform illumination of all dishes and avoids absorption of light by sensitizer solution between the cells and the source, as would occur with illumination from above.

Although a source other than a cool white fluorescent may be used, Table 1 would not apply quantitatively. However, the analysis can still be complete with the understanding that a strict quantitative determination would require an evaluation of the relative photon absorption by each sensitizer from whatever source is used.

EXPERIMENTAL PROCEDURE

The following procedure was developed for use with 60 mm diameter glass or plastic culture dishes. Volumes in the text are appropriate for these dishes, but volumes for 100 mm diameter dishes are also given in parentheses.

1. Experimental suspensions: For each sensitizer used, label the lids of each of 5 culture dishes. There should be one dish corresponding to each sensitizer concentration prepared above. Example: If you are only working with Rose Bengal, you will have 4 dishes labelled RB 1, RB 0.5, RB 0.25 and RB 0.125. The 5th dish for each sensitizer will be an unilluminated control with the highest sensitizer concentration. Label it accordingly, for example RB 1–LITE. If you are using all three sensitizers, you will label 15 dishes.
2. Control suspensions: Label the lids of 3 more culture dishes as follows: DW for distilled water, SAL+LITE for saline + light, and SAL–LITE for saline without light.
3. Pipette 4.5 ml *(18 ml)* of the following solutions into each dish.
    a. Each dish labelled with a sensitizer concentration in step 1 receives 4.5 ml *(18 ml)* of the corresponding sensitizer stock solution.
    b. Both dishes with SAL in their labels receive 4.5 ml *(18 ml)* of saline.
    c. The dish labelled DW receives 4.5 ml *(18 ml)* of distilled water.

AT THIS POINT BE SURE THAT THE LAB LIGHTS ARE DIMMED.

4. In the next step, pipette into the 2 dishes that will not be illuminated, last (labels include "–LITE"). Prepare 2 pieces of aluminum foil and wrap these two dishes, sealing the edges, immediately after they receive cells in the next step.

5. Into each of the dishes prepared above, pipette 0.5 ml *(2 ml)* of the red cell suspension from the beaker in which it is slowly stirring. Mix by gently moving the dish circularly in a horizontal plane.

6. Illuminate all dishes, including those wrapped in aluminum foil, for 20 min, or until the dish containing the highest concentration of Rose Bengal is as clear as the cell suspension in distilled water. (Do not unwrap the foil, simply expose the foil-covered dishes as controls.)

7. While the suspensions are being illuminated, label a corresponding set of test tubes, one for each dish prepared. Similarly, label a set of disposable plastic cuvettes near the top edge of the cuvette. There should be a corresponding test tube and cuvette for each dish.

8. Immediately after illumination, pour (or use Pasteur pipettes if necessary) the contents of each dish into its corresponding test tube.

9. Quickly load the tubes into a centrifuge and spin at about 1000 × g for about 3 min (or until any cells have collected on the bottom).

10. Use a Pasteur pipette to remove the supernatant into the correspondingly labelled cuvettes.

11. Measure the optical density of each supernatant at 414 nm (the Soret peak of hemoglobin).

DISCUSSION

To convert the measurements of optical density in each suspension to % hemolysis, assume that cells in distilled water are 100% hemolyzed. Also assume that the optical density at 414 nm (OD_{414}) of a suspension with no hemolysis is 0. [Note: For the highest concentrations of sensitizer this is not true. The OD_{414} of a 100 μM solution of these sensitizers is about 0.2. You could correct for this to be precise.] Now convert your OD_{414} values to % hemolysis values. On a single sheet of three cycle semi-log paper, plot % hemolysis versus sensitizer concentration for each sensitizer. Finally, determine the concentration of sensitizer required to produce 50% hemolysis. Using two very high output, 115 W cool white fluorescent bulbs and 20 min of illumination at a distance of about 10 cm, 50% hemolysis is obtained with about 75 μM eosin Y, 9 μM erythrosin B, and 0.2 μM Rose Bengal.

The concentration of sensitizer required to produce 50% hemolysis gives only an approximate indication of the relative effectiveness of the sensitizer. One might expect that sensitizers that produce lysis at lower concentrations are more effective, and this may be true under many circumstances. However, it is a considerable oversimplification, and may simply be wrong under some circumstances. To take an extreme case, imagine two sensitizers which have rather narrow absorption maxima at 500 nm and 700 nm respectively. The latter might be a rather poor photosensitizer, but if both were illuminated with red light, it could produce a considerably greater effect because it absorbs red light whereas the other does not. Thus biological effects should always be normalized to the number of photons absorbed by the sensitizer.

What is the Effectiveness per Absorbed Photon?

This is similar to a quantum yield for fluorescence or for photochemical reaction (see Chapter 4). Relative effectiveness for membrane photomodification can be expressed as shown in Eq. 1 (Pooler and Valenzeno, 1979). That is, relative effectiveness is equal to the rate of membrane photomodification per unit time, divided by the rate of photon absorption per unit time.

$$\text{Relative Effectiveness} = \frac{(d/dt)\ \text{membrane photomodification}}{(d/dt)\ \text{photon absorption}} \quad [1]$$

Since photohemolysis is a late event in the sequence of membrane photomodification, it is not possible to directly determine the rate of membrane photomodification from measurements of photohemolysis. However, we can assume that for different photosensitizers to produce the same amount of photohemolysis (say 50%) with 20 min of illumination, they must have produced the same rate of photomodification. In other words, if we had picked just the right concentrations of each of the three photosensitizers so that they each produced exactly 50% hemolysis after 20 min of illumination, the numerator in Eq. 1 would be the same for all three. In that case, the relative effectiveness would simply be proportional to the reciprocal of the rate of photon absorption by each of the three solutions that were used.

Obviously, one isn't always that lucky in choosing the sensitizer concentrations. However, by drawing a smooth curve through the data on the semi-log plots, one may determine what this concentration would be for each sensitizer. All that remains is to determine the rate of photon absorption by the three sensitizer concentrations. This is not a simple task (Pooler and Valenzeno, 1979). In general there are four steps to the determination. First, measure the energy reaching the red cells at one wavelength. Then determinethe fraction of that energy absorbed by the photosensitizer at that wavelength. Next divide by the energy of a photon to determine the rate of photon absorption at that wavelength. Finally sum the contributions at each wavelength. (A formidable task!).

For the purposes of this lab, the situation has been simplified by using the information provided in Table 1. The data in the table has been obtained by the procedure just described and is referenced to eosin Y. That is, we assume that a 1 μM solution of eosin Y absorbs 1 unit of photons per unit time. Relative to this standard, the table gives the rate of photon absorption by a 1 μM solution of each of the other sensitizers. Using the data in the table and the experimental results, one should now be able to calculate the relative effectiveness of eosin Y, erythrosin B, and Rose Bengal for membrane photomodification in red blood cells.

Related Laboratory Experiments

Many variations of this basic experiment are possible. For example, the position of the absorption maximum of hemoglobin may be altered by photosensitizer and light under some conditions. This is due to a chemical alteration of the hemoglobin molecule. The effect

Table 1

Photon absorption rates by three halogenated fluoresceins

Sensitizer	Relative rate of photon absorption by a 1 μM solution
eosin Y	1.000
erythrosin B	1.1846
Rose Bengal	1.6985

is not strong enough to obscure the results demonstrated in this experiment. If desired, the problem can be avoided by stabilizing hemoglobin after isolation using Drabkin's solution (Heatherington and Johnson, 1984). If a scanning spectrophotometer is available, it might be instructive to determine the position of the hemoglobin peak in some of the experimental samples and to demonstrate the effect of Drabkin's solution. Another interesting variant is to prepare a second set of dishes containing eosin Y. Immediately after illumination, store these dishes in a dark place (an old photographic paper box works well). Analyze them some hours later. Why are the results different? You may also wish to entertain the following questions. What are the effects of variation of the duration of illumination (Valenzeno and Pooler, 1982) or sensitizer concentration (Valenzeno, 1984) on the kinetics of photohemolysis? For these experiments, you would need to be able to analyze suspensions at various times after photomodification. What is the effect of adding about 30 mM sugar to the suspension medium (Fleischer, *et al.*, 1966)? What does this reveal about the mechanism of photohemolysis?

REVIEW QUESTIONS

1. You know that a considerable amount of light penetrates through the solutions of Rose Bengal used in these experiments. What do you expect would happen if you placed a dish with Rose Bengal solution underneath a dish containing Rose Bengal and red blood cells? Would the cells in the upper dish lyse normally? Why or why not?

2. An antioxidant is found to significantly inhibit photohemolysis. The red cell membrane is not permeable to the antioxidant which was used at a concentration of about 30 mM. The experimenter concluded that he has provided evidence that photohemolysis is an oxidative process. Recalling the origin of colloid osmotic hemolysis, what other explanation can you propose?

3. Under normal fluorescent light, which absorbs more photons, a 10 μM solution of erythrosin B, or a 7 μM solution of Rose Bengal?

4. With a high concentration of Rose Bengal, a few s of light produces no photohemolysis when analyzed immediately after light. However, when examined 24 h later considerable lysis has occurred. Chain reactions are known to occur in chemical systems. They can persist long after the primary reaction has ceased. Such reaction chains are well known in lipid systems such as membranes. Are chain reactions in lipids indicated by this experiment? What other explanation could you propose?

5. Many photosensitizers partition into membranes. That is they are present in higher concentration in membranes than they are in the adjacent aqueous medium. How would you modify the expression for relative effectiveness (Eq. 1) to account for this phenomenon? Hint: Imagine that sensitizer A and B are both present in the aqueous medium at 10 μM. However, A partitions into the membrane so that its concentration is 2 times higher than in the medium, while the partition ratio for B is 3. Which is more effective per absorbed photon?

6. Why would you have to be concerned about testing the effect on photohemolysis of an antioxidant that appeared orange-red in solution?

ANSWERS TO REVIEW QUESTIONS

1. The cells would not lyse, unless they were exposed to light for a very long time. Recall that effective photosensitization requires not just photosensitizer and light, but the

light must be of wavelengths absorbed by the photosensitizer. In this case, though a considerable amount of light penetrates the lower dish, as you can readily see, that light is not of the appropriate wavelengths to be absorbed by Rose Bengal. The wavelengths that Rose Bengal absorbs best, have already been absorbed by the lower dish. That is why the solution looks red to us. The green light, normally absorbed by Rose Bengal, has already been absorbed.

2. Remember that colloid osmotic hemolysis occurs because cells contain impermeable intracellular solutes that are not balanced by extracellular impermeable solutes. Cells try to attain equilibrium for all permeable solutes, as well as for water (i.e. osmotic equilibrium). If all extracellular solutes are permeable, as in a salt solution, the cells rely on the sodium pump in the cell membrane to expel sodium as fast as it leaks in. If the pump can't keep up, cell sodium will rise, which increases intracellular osmolality, and thus draws water into the cell to increase its volume. On the other hand, if there is a sufficient concentration of an impermeable extracellular solute (about 30 mM coincidentally), it can balance the effect of the impermeable intracellular solutes. Then, even with an increase in permeability to sodium and potassium, the cell can maintain osmotic equilibrium, and thus volume stability. The permeable solutes can attain the same concentration inside and outside the cell, and the total osmolality can be the same inside and outside, since the concentration of impermeable solutes, as well as permeable solutes, is the same both inside and outside the cell.

3. Using Table 1, a 10 μM solution of erythrosin B absorbs:

$$10\ \mu mol \cdot l^{-1} \times 1.1846\ photons \cdot (\mu mol^{-1} \cdot l)$$

$$11.846\ photons$$

relative to eosin Y as a standard. A 7μM solution of Rose Bengal absorbs:

$$7\ \mu mol \cdot l^{-1} \times 1.6985\ photons \cdot (\mu mol^{-1} \cdot l)$$

$$11.890\ photons$$

Thus, the Rose Bengal solution will absorb slightly more photons.

4. Chain reactions are not necessarily indicated by the result, although they are possible. Normal colloid osmotic hemolysis takes time to develop. By definition, the initial insult that makes the membrane permeable to small cations does not produce lesions large enough for hemoglobin to escape (i.e. hemolysis). It takes time for cations to diffuse across the cell membrane, so it takes time for the volume increase to occur, causing lysis. Thus with mild photosensitizing conditions, hemolysis may not be immediately evident, but will develop with time as the cells progressively swell and lyse. With stronger conditions (more sensitizer or light), some lysis may be apparent immediately after illumination, and with still stronger treatment, all cells may lyse during illumination.

   Chain reactions may well occur in membranes, but their existence cannot be inferred from the kind of evidence cited, since delayed photohemolysis is expected whether or not chain reactions occur.

5. If we assume that these sensitizers act from their membrane environment, then to a first approximation, you could simply determine the rate of photon absorption for the expected membrane concentration. For example, use the rate of photon absorption for twice the medium concentration of A, or 3 times the medium concentration of B. However, to be more precise, you should determine the rate of photon absorption by the sensitizer in the environment from which it acts, in this case in the membrane. That is, in the membrane environment, photon absorption may be greater or less than in

aqueous solution, even at the same sensitizer concentration. This is not a simple determination, and is usually only estimated.

6. Photosensitization uses light as an energy source to drive the resulting reactions. Anything that interferes with light penetration to the photosensitizer can reduce the reaction rate. Thus if the photosensitization reaction being studied used a sensitizer that absorbed in the orange-red region of the spectrum, the antioxidant might reduce the rate of reaction because of this optical interference (inner filter effect), and may tell you nothing about oxidative reaction mechanisms.

SUPPLEMENTARY READING

Spikes, J.D. (1989) Photosensitization. In *The Science of Photobiology*, 2nd edition (Edited by K.C. Smith), pp. 79-111. Plenum Press, New York. General photosensitization coverage.

Heatherington, A.M. and B.E. Johnson (1984) Photohemolysis. *Photodermatol.* **1**, 255-260. A good primer on photohemolysis technique.

Valenzeno, D.P. and J.P. Pooler (1987) Photodynamic action. *Bioscience* **37**, 270-275. A somewhat less technical description of photosensitization and its relevance.

Berne, R.M. and M.N. Levy, (Editors) (1990) *Principles of Physiology*. C.V. Mosby Co., St. Louis, p. 11. Further reading on colloid osmotic hemolysis and cell volume.

REFERENCES

Fleischer, A.S., B.S. Leonard, C. Harber, J.S. Cook, and R.L. Baer (1966) Mechanism of *in vitro* photohemolysis in Erythropoietic Protoporphyria (EPP). *J. Invest. Dermatol.*, **46**, 505-509.

Girotti, A.W. (1990) Lipid peroxidation in photodynamic systems. In *Membrane Lipid Oxidation*, Vol. I. (Edited by C. Vigo-Pelfrey), pp. 203-218. CRC Press, Inc., Boca Raton, FL.

Heatherington, A.M. and B.E. Johnson. (1984) Photohemolysis. *Photodermatol.* **1**, 255-260.

Lee, P.C. and M.A.J. Rodgers. (1983) Singlet molecular oxygen in micellar systems. 1. Distribution equilibria between hydrophobic and hydrophilic compartments. *J. Phys. Chem.* **87**, 4894-4898.

Pooler, J.P. and D.P. Valenzeno. (1979) Physicochemical determinants of sensitizing effectiveness of fluorescein dyes in excitable membranes. *Photochem. Photobiol.* **30**, 491-498.

Valenzeno, D.P. (1984) Photohemolytic lesions: Stoichiometry of creation by phloxine B. *Photochem. Photobiol.* **40**, 681-688.

Valenzeno, D.P. (1987) Photomodification of biological membranes with emphasis on singlet oxygen mechanisms. *Photochem. Photobiol.* **46**, 147-160.

Valenzeno, D.P. and J.P. Pooler. (1982) The concentration and fluence dependence of delayed photohemolysis. *Photochem. Photobiol.* **35**, 427-429.

Valenzeno, D.P. and M. Tarr. (1990) Membrane photomodification and its use to study reactive oxygen effects. In *Focus on Photochemistry*. (Edited by J.F. Rabek and G.W. Scott), CRC Press, Inc. Boca Raton, FL.

Spectroscopic Determination of Hematoporphyrin-Membrane Partition Parameters

Eitan GROSS
Bar Ilan University
Israel

INTRODUCTION

A derivative of hematoporphyrin IX (HP) termed hematoporphyrin derivative (HPD), is currently being used as a photosensitizer in the new technique of cancer treatment known as photodynamic therapy, PDT (Dougherty, 1987). The parent compound, HP, is a fluorescent molecule that binds selectively to cancerous cells *in vivo* (Kessel, 1986). This property of porphyrins is being used to locate the boundaries of a tumor in the body by illuminating the suspected area with weak UV light and tracing the bound porphyrin's fluorescence. Optical excitation of the bound porphyrin causes a photosensitization reaction that ultimately leads to tumor necrosis (Dougherty, 1987; Tomio et al., 1983). Selective binding of the photosensitizer to some component of tumor tissue is an important process in the use of the photosensitizer in PDT. The delivery of the photosensitizer in the blood can be facilitated by binding to serum carriers like albumin. Thus it is very important to develop methods for determination of the binding parameters of photosensitizers to biological macromolecules and biological structures. Various spectroscopic, and other, methods have been introduced to determine the binding of dye molecules to macromolecules, membranes and other biological structures (Lakowicz et al., 1977; Bashford and Smith, 1979; Visser and Lee, 1980; Brault et al., 1986).

Dye-membrane binding

Consider a system of a fluorescent dye in a liposomal suspension. At equilibrium the association of the dye with the liposomal membrane can be described by:

$$D + M \rightleftarrows B \qquad [1]$$

where D is the concentration of free aqueous dye, M is the 'free" membrane concentration, and B is the concentration of membrane-bound dye. The equilibrium constant, K, for the dye-membrane association reaction is given by:

EITAN GROSS, Department of Physics, Bar Ilan University, Ramat Gan, Israel

Photobiological Techniques, Edited by D.P. Valenzeno *et al.*
Plenum Press, New York, 1991

$$K = B / (DM) \quad [2]$$

At high membrane/dye ratio M is approximately equal to the total membrane concentration. Obviously $D = T - B$, where T is the total dye concentration. Substituting into Eq. 2 and rearranging gives:

$$B = MT / (K^{-1} + M) \quad [3]$$

In the case of a fluorescent dye the fluorescence intensity, F, at a given wavelength, is proportional to the number of photons emitted. Thus, from Beer's Law:

$$F = QI_{abs} = QI_o (1 - 10^{-\varepsilon c l}) \quad [4]$$

where Q is the fluorescence quantum yield, and I_{abs} is the number of photons absorbed by thc dye. I_o is the number of incident photons, ε is the molar extinction coefficient, and l is the path length (in cm). When the absorbance is low, $\varepsilon c l$ is small, and we can expand the right side of Eq. 4 to get:

$$F = 2.303\, QI_o \varepsilon c l = g Q \varepsilon c \quad [5]$$

All the variables in Eq. 5, except c, are wavelength dependent. g is equal to $2.303\, I_o\, l$. The range of dye concentrations, for which the fluorescence intensity is linear with respect to dye concentration will be called the linear range. For Eq. 1, the fluorescence intensity, in the linear range of the fluorophore in equilibrium with the membrane, is given by:

$$F = g\varepsilon_D Q_D D = g\varepsilon_B Q_B B = \alpha_D D + \alpha_B B \quad [6]$$

where the subscripts D and B denote free and bound dye concentrations respectively, and $\alpha = g\varepsilon Q$. Using $D = T - B$ and rearranging yields:

$$F = \alpha_D T + (\alpha_B - \alpha_D)\, B \quad [7]$$

Substituting for B from Eq. 3 into Eq. 7 gives:

$$F = \alpha_D T + MT\, (\alpha_B - \alpha_D) / (K^{-1} + M) \quad [8]$$

α_D, the proportionality factor relating the fluorescence intensity of the free dye to its concentrration, can be readily measured as the ratio of the fluorescence in the absence of liposomes, F_o, and the total fluorophore concentration T, i.e. $\alpha_D = F_o/T$. K^{-1} is equal to the dissociation constant, K_d, of the dye-membrane complex. Substituting F_o/α_D for T and K_d for K^{-1} in Eq. 8, subtracting F_o from both sides, inverting, and rearranging gives:

$$F_o / \Delta F = \{[\alpha_D/(\alpha_B - \alpha_D)]\ K_d / M\} + \alpha_D/(\alpha_B - \alpha_D) \qquad [9]$$

where ΔF is equal to $F - F_o$. The plot of $F_o/\Delta F$ vs. 1/M, should produce a straight line. The ratio of the Y-axis intercept to the slope (intercept/slope) of the straight line is equal to the equilibrium constant, K, of Eq. 1. This is the binding constant K_b. The fluorescence enhancement factor for binding (α_B) can also be calculated from the slope of the line or its intercept with the Y-axis, provided α_D is known. Remember that for non-specific dye-membrane interaction (i.e. partition) and at high membrane/dye ratio, the partition coefficient, as usually defined, is obtained from the binding constant, K_b, by multiplying it by the concentration of the membrane phase.

EXPERIMENT 14: A SPECTROSCOPIC DETERMINATION OF HP MEMBRANE PARTITION PARAMETERS

OBJECTIVE

To determine both the partition coefficient of HP to liposomal membranes, and the spectroscopic parameters of bound HP.

LIST OF MATERIALS

Hematoporphyrin IX (HPIX), 1 mg
Lecithin (from egg yolk), 10 mg
NaOH, 1N, 2 ml
HCl, 1N, 2 ml
UV-Visible double beam ratio recording absorption spectrophotometer
Fluorescence spectrophotometer, for steady state fluorescence measurements
Glass test tubes (3)
Ethyl ether, 1 ml
Phosphate buffer, 3ml at pH 7.4
Sonicator, either bath (80 KHz) or probe
Vortex shaker
Cold water bath
Ice flakes
Nitrogen cylinder with a pressure regulator valve and a one meter rubber or plactic hose to fit the regulator tip
Pasteur pipettes (5) with a bulb
1 ml and 2 ml pipettes
100 µl and 10 µl syringes (1 each)
Millimeteric chart or programmable calculator or computer for linear least squares analysis
Analytical balance
pH meter

PREPARATIONS

Liposomes: Place 100 µl of lecithin in chloroform/ethanol solution (100 mg·ml^{-1} lecithin stock) on the bottom of a glass test tube. Evaporate the solvent under a mild nitrogen stream. Add about 1 ml of ether (to extract the remaining chloroform) and evaporate *thoroughly* under a mild stream of nitrogen (this should take at least 1 h).

NOTE: It is important to get rid of any traces of organic solvent in order to avoid solubilization of the liposomes. After this step is completed you should be left with a white film on the bottom of the test tube.

Add 2 ml of phosphate buffer pH 7.4 (you can prepare the buffer by adding 39.1 ml of 0.1 M NaOH to 50 ml of 0.1 M KH_2PO_4) and vortex for 2-3 min. After this stage you should get a turbid milky suspension. Sonicate the suspension in a nitrogen atmosphere until it is clear. During sonication the suspension should be immersed in a cold water bath (add ice flakes) to avoid overheating. With a probe sonicator this should take about 4-5 min. If a bath sonicator is used, mount the test tube so that about 2/3 of the suspension is below the water level. The sonication should take 20-30 min. After this step is completed you should be left with a clear liposomal suspension at a concentration of 5 mg·ml^{-1}.

HP Solutions: Dissolve 1 mg of HPIX in 16.7 ml of distilled water to produce a 0.1 mM stock solution. In order to facilitate solubilization of HPIX, add a few drops of 1 N NaOH until a clear homogeneous solution is obtained. Readjust the pH to 7.4 by carefully adding a few drops of 1 N HCl. Use the pH meter to follow the pH as you add the HCl. Add 10 µl of the HP stock solution to 1 ml of buffer. This will be the sample for the fluorescence intensity measurements.

HP-Liposomes: Place 1 ml of the liposomal suspension in a glass test tube. Add 10 µl of the HP stock solution (0.1 mM) and vortex for a few s. After vortexing wait 10 min to allow full equilibration. This will be the sample solution for the absorption measurement.

EXPERIMENTAL PROCEDURE

Absorbance Measurements

1. Prepare a table with 5 columns labelled: Added volume of liposomal suspension (V), Fluorescence intensity (F), Membrane concentration (M), 1/M, $F_o/\Delta F$.
2. Produce a baseline for the absorbance measurement by running an absorbance scan of two matched quartz cuvettes filled with water, from 250 to 750 nm.
3. Measure the absorbance spectrum of the sample (HP pre-equilibrated with liposomes) against liposomal suspension (the remaining 1 ml of the liposomal suspension) from 250 to 750 nm, with a path length of 1 cm and with the same cuvettes which were used to produce the base line. Subtract the baseline spectrum from the absorbance spectrum of the pre-equilibrated HP-liposomal preparation. The main absorption band (around 400 nm) is called the Soret band and the four small bands on the longer wavelength side are the Q bands. The absorbance at 500 nm should be around 0.04. In the following fluorescence measurements use 500 nm as the excitation wavelength.

Fluorescence Measurements

4. Set the excitation monochromator of the fluorescence spectrophotometer to 500 nm and the emission monochromator to 630 nm. If available, place an appropriate filter at the emission channel, to minimize the amount of stray light reaching the photomultiplier. If it is possible to integrate the signal over a preset time interval (some instruments have this option), set the integration time to 10 s. Set the excitation slit width to a practical minimum value, to minimize damage to the sample.
5. Transfer the sample for the fluorescence intensity measurement (HP in a buffer solution) to an appropriate cuvette and place the cuvette in the cell holder of the fluorescence spectrophotometer.
6. Measure the fluorescence intensity and record the result in the "fluorescence intensity" column (this will be F_o).
7. Add 5 µl of the liposomal suspension to the cuvette and mix well with a Pasteur pipette *without moving the cuvette.* Allow 5 min for equilibration and again determine the fluorescence intensity *without changing any of the instrument settings.* Record the fluorescence intensity and the added volume of liposomal suspension in the table.
8. Repeat step 7 for nine successive additions of 5 µl of the liposomal suspension.

Data Analysis

9. Determine the lipid concentration at each step of the titration (do not forget to correct for dilution) and fill in the "lipid concentration" column (M). Determine and fill in the

1/M column. Correct for the fluorescence intensity (dilution factor) and fill in the column. Determine and fill in the $F_o/\Delta F$ column.

10. Plot $F_o/\Delta F$ *vs* 1/M and fit a straight line through the points (least squares or linear regression). Determine the slope and the Y-intercept of the straight line.
11. The binding constant of HP to the liposomes is equal to the ratio of the intercept to the slope (intercept /slope).

DISCUSSION

The experimental value of the HP binding constant (K_b) to lecithin liposomes is 11.54 $\pm$ 2.11 (ml·mg^{-1}). Given the binding constant we can use Eq. 2 to determine the fraction of bound HP at any membrane concentration. An alternative meaning of the binding constant is the following: at a membrane concentration equal to $1/K_b$, the concentration of bound dye is equal to the concentration of free dye (assuming the total dye concentration is negligible compared to the total membrane concentration). Note that the value of the binding constant of porphyrins to liposomes is pH dependent, presumably due to the protonation state of the carboxylic and imino nitrogens (Brault et al., 1986).

The association of a dye molecule with macromolecules such as proteins, nucleic acids, or with larger structures like membranes, is often accompanied by spectroscopic changes. In the case of IR absorption or Raman scattering, a shift in the peak or a change in the peak intensity could result from specific interactions such as binding to a specific site on a protein, or nonspecific interactions such as hydrophobic interactions inside the membrane, all of which might change the frequency of one or more vibrational modes. In the case of UV-visible absorption or fluorescence, a spectral shift and/or a change in the signal intensity can be a result of interaction of the dye's dipole moment with the reactive fields induced in the surrounding solvent, or from specific chemical interactions of the dye with one or more solvent molecules (Bakhshieve, 1962). In the case of an ESR signal, a broadening of a spectral line due to immobilization of the spin label can result from a rigid protein surrounding to which the molecule bearing the spin label is bound, or a viscous surrounding as in a membrane. The general approach described here can be applied to a wide range of processes which involve changes of spectroscopic parameters, as long as the measured property is linear with the concentration of the species being measured. Such an approach can be used to determine the binding of dyes to proteins and nucleic acids, the pK_a values of chemical groups located near a chromophore (e.g., the hydroxyl residue of tyrosine) and the binding constant of a ligand to a chromophore-containing molecule.

Two of the commonly used techniques for determining binding constants of small dye molecules to macromolecules are dialysis and centrifugation. Such techniques involve a physical separation of the free dye from the bound dye, thus disturbing the equilibrium conditions. The method described in this experiment is not perturbative.

Determination of the pK_a value of a chemical group by classical titration methods requires a lot of data points (to fit an S-shaped curve to the data). In the method described in this experiment, relatively few data points are sufficient to fit a straight line. As such, a complete titration is not required. Moreover, the present method is more accurate since the data analysis involves a linear fit, rather than a non-linear fit.

REVIEW QUESTIONS

1. Does the value of the binding constant depend on the initial concentration of the dye?
2. What are the advantages of monitoring the spectroscopic changes, following the titration, in the fluorescence spectrophotometer over the absorbance spectrophotometer? What are the disadvantages?
3. Why can the protonation state of a porphyrin affect the value of the binding constant to liposomes? Hint: The membrane matrix is hydrophobic.
4. Show that at $M = 1/K_b$ the partition coefficient of the dye equals one. Hint: The partition coefficient is defined as B/D.
5. The relation between the fluorescence intensity and the dye concentration at a certain excitation wavelength becomes linear as the dye absorbance at that wavelength goes to zero. Determine the deviation from linearity (by percent) when the absorbance is 0.04, taking into account terms up to the first nonlinear term.
6. Suppose that at the end of the titration you reached a membrane concentration at which the fluorescence of the dye is still linear with its absorbance. However, if you would have to further increase the membrane concentration it would destroy the linear relation between the fluorescence and the absorbance. Can you still use Eq. 9 to evaluate the binding constant?
7. Show that Eq. 9 for the dissociation of a proton from an acidic dye molecule can be reduced to the Henderson-Hasselbach equation, i.e.:

$$pH = pK + \log \{[\text{conjugated base}]/[\text{conjugated acid}]\}$$

   Hint: Use the relation: $K_d = K_a = DH/B$ where H is the concentration of protons.

8. The results of a spectroscopic titration of glycyl-tyrosine are described in Fig. 1. (a) Determine the pK_a of the phenolic hydroxyl residue by the present method. Hint: Convert the pH values to proton concentration values and substitute them for M in Eq. 9. (b) Should the protonation/deprotonation of the ammonium and carboxylic groups be taken into account in determining the pK value of the hydroxyl residue of the phenol?

ANSWERS TO REVIEW QUESTIONS

1. No (ignoring possible aggregation of the dye).
2. Advantages: Fluorescence intensity is more sensitive than absorption, thus the changes in fluorescence intensity due to binding of the dye are more pronounced than for absorption.

   Disadvantages: The linear range is smaller.

3. The protonation state determines the polarity of the molecule.
4. From Eq. 2: $K_b = B/(DM)$. Multiplying both sides by M and substituting for $M = 1/K_b$, yields: $B/D = 1$.
5. Expanding Eq. 4 near $\varepsilon c l = 0$ and writing terms up to the first nonlinear term yields,

$$F = Q I_o \{2-(1-2.303\varepsilon cl + (1/2)(2.303\varepsilon cl)^2)\}$$

$$F = Q I_o \{2.303\varepsilon cl - (1/2)(2.303\varepsilon cl)^2\}$$

   At absorbance 0.04 we get $2.303 \times 0.04 = 0.092$, and the deviation from linearity is $1-\{[0.092-(1/2)(0.092)^2]/0.092\} = 0.046$ or 4.6%.

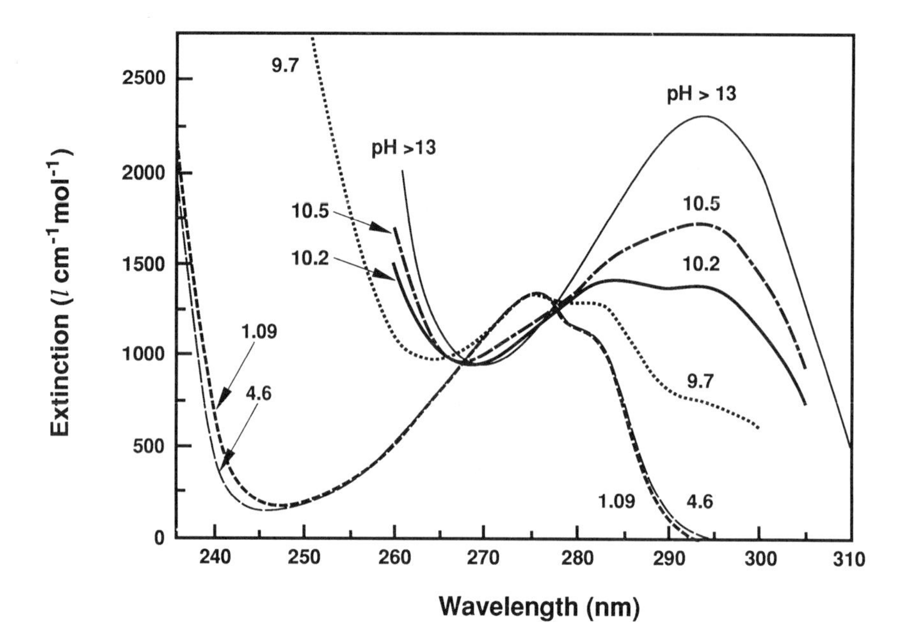

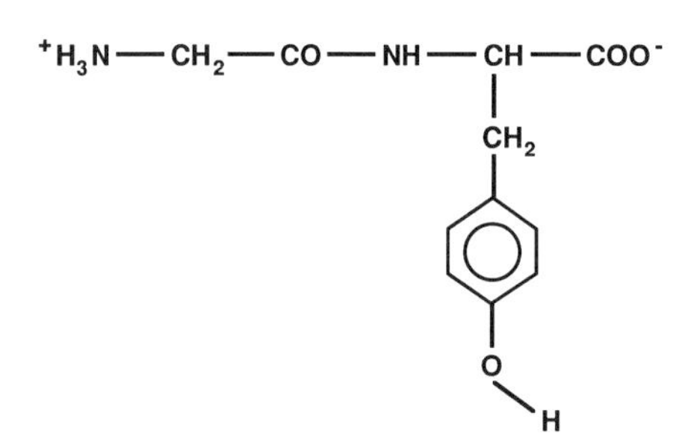

Figure 1. Spectroscopic titration of glycyl-L-tyrosine at 25°C, with an ionic strength of 0.16. Above: UV absorption spectra of the dipeptide at indicated pH values. Left: Structure of glycyl-L-tyrosine in acidic solution.

6. No. Even if the relation is linear within the membrane concentrations used in the experiment, you cannot use Eq. 9 to determine the binding constant because it assumes a linear relation up to an infinite membrane concentration. The Y-axis intercept is an extrapolation to infinite membrane concentration.

7. Substituting H for M in Eq. 9 and rearranging yields:

$$F_o/\Delta F = (\alpha_D / (\alpha_B - \alpha_D))\cdot(K_a/H) + (\alpha_D / (\alpha_B - \alpha_D)).$$

Substituting for F_o and F from Eq. 6 and rearranging yields:

$$\alpha_D\,(B+D)/(\alpha_D D + \alpha_B B - \alpha_D\,(B+D)) = (\alpha_D / (\alpha_B - \alpha_D))\cdot(K_a/H) + (\alpha_D / (\alpha_B - \alpha_D))$$

where D is the concentration of the deprotonated molecule (conjugated base) and B is the concentration of the protonated molecule (conjugated acid). After cancelling like terms and rearranging, the last equation reduces to:

$$K_a/H = D/B$$

Taking the log of both sides and rearranging yields:

$$pH = pK_a + \log (D/B)$$

where p stands for the operation –log.

8. (a) The extinction coefficient values at 290 nm, graphically evaluated from the spectra in Fig. 1, for the various pH values, are shown in Table 1. Also shown are the $1/[H^+]$ values (x-axis) and the values of the expression $\varepsilon_o/\Delta\varepsilon$ (Y-axis). The value of ε at pH=13 is taken as ε_o. The data were fitted to a straight line (by a linear regression-based computer program) whose equation is:

$$\varepsilon_o/\Delta\varepsilon = -1.0117 - 9.862 \times 10^{-11}[H^+]^{-1}$$

The graph of the fitted line and the experimental data points are shown in Fig. 2. The pK of the hydroxyl residue is given by:

$$pK = -\log K_d = -\log(\text{slope/intercept}) = -\log(-9.862 \times 10^{-11}/-1.0117) = 10.01.$$

(b) No. On a molecular scale these groups are located far from the chromophore (the phenol), thus the protonation/deprotonation of these groups does not change the absorption spectrum of the chromophore to any significant extent.

Table 1

Spectroscopic titration of glycyl-tyrosine

pH	ε (290 nm) $cm^{-1}mol^{-1}$	1/[H+] M	$\varepsilon o/\Delta\varepsilon$ at 290 nm
13	2250	1.0×10^{13}	
10.5	1714	3.2×10^{10}	–4.19
10.2	1320	1.6×10^{10}	–2.42
9.7	820	5.0×10^{9}	–1.57
4.6	71	4.0×10^{4}	–1.04

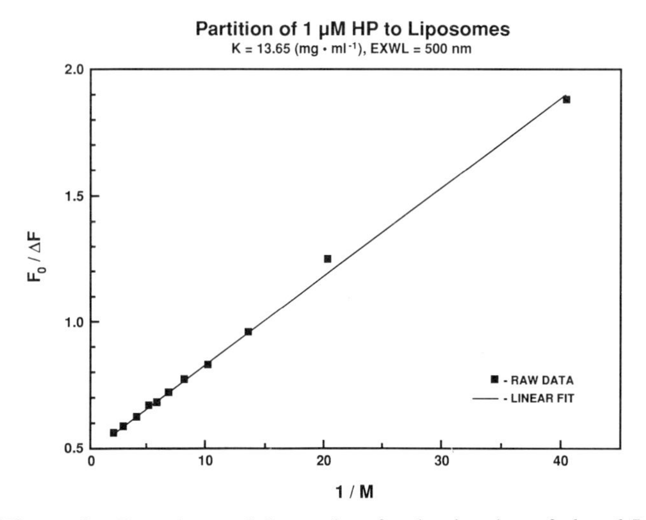

Figure 2. Experimental data points for the titration of glycyl-L-tyrosine plotted according to Eq. 9. The straight line is a least squares linear fit to the data points. pK = slope/intercept = 10.011.

SUPPLEMENTARY READING

Spikes, J.D. (1989) Photosensitization. In *The Science of Photobiology*, 2nd edition (Edited by K.C. Smith), pp. 79-111. Plenum Press, New York. General photosensitization coverage.

REFERENCES

Bakhshieve, N.G. (1962) Universal intermolecular interactions and their effect on the position of the electronic spectra of molecules in two-component solutions. *Opt. Spectrosc.* **13**, 24-29.

Bashford, C.L. and J.C. Smith (1974) The determination of oxonol-membrane binding parameters by spectroscopic methods. *Biophys. J.* **25**, 63-85.

Brault, D., C. Vever-Bizet and T. Le Doan (1986) Spectrofluorimetric study of porphyrin incorporation into membrane models – evidence for pH effects. *Biochim. Biophys. Acta* **857**, 238-250.

Dougherty, T.J. (1987) Photosensitizers: Therapy and detection of malignant tumors. *Photochem. Photobiol.* **45**, 879-889.

Kessel, D. (1986) *In vivo* fluorescence of tumors after treatment with derivatives of hematoporphyrin. *Photochem. Photobiol.* **44**, 107-108.

Lakowicz, J.R., D. Hogen and G. Omann (1977) Diffusion and partitioning of a pesticide, lindane, into phosphatidyl choline bilayers. A new fluorescence quenching method to study chlorinated hydrocarbon-membrane interactions. *Biochim. Biophys. Acta* **471**, 401-411.

Tomio, L., P.L. Zorat, L. Conti, F. Calzavara, E. Reddi and G. Jori (1983) Effect of hematoporphyrin and red light on AH-130 solid tumors in rats. *Acta Radiol. Oncol.* **22**, 49-53.

Visser, A.J.W.G. and J. Lee (1980) Lumazine protein from the bioluminescent bacterium *Photobacterium phosphoreum.* A fluorescence study of the protein-ligand equilibrium. *Biochem.* **19**, 4366-4372.

Near Infra-Red Flash Absorption: A Tool for Studying Photosynthesis

Paul MATHIS and Danièle THIBODEAU
CEN Saclay
France

INTRODUCTION

Photosynthesis

Among photobiological processes, photosynthesis occupies a unique position because it is responsible for the conversion and storage of light energy in the biosphere on a massive scale. The process can be roughly divided into primary reactions, which take place in the photosynthetic membrane, and secondary reactions which occur in a more liquid-like cellular compartment. In the membrane, light is absorbed by pigment-protein complexes which make up the so-called *antenna*, from which excitation energy is transferred to the reaction centers. These reaction centers are complex proteins where excitation energy induces charge separation (Fig. 1). The initial step of charge separation is followed by a series of electron transfer reactions, which are sometimes coupled to transmembrane proton transfer, and which have three basic functions: i) to decrease the probability of wasteful charge recombination; ii) to achieve the final steps of redox reactions, i.e. to create a useful reductant, such as NADPH, and oxidize a terminal electron donor, which is water in plants; iii) to create a gradient of H^+ concentration, which provides the free energy for ATP synthesis by the ATP-synthase. For general reviews and collective books see Clayton, 1970; Amesz, 1987; Barber, 1987; Hatch and Boardman, 1981.

Photosynthesis takes place in many classes of organisms. The process, however, can be considered as a well-defined unique process. The most obvious source of diversity resides in pigment composition and antenna structure. Photosynthetic pigments belong to three classes: chlorophylls, carotenoids and phycobiliproteins. Reaction centers all share many common properties, but their differences usually lead to a classification into four categories: purple bacteria, green bacteria, Photosystem I (PS-I) and Photosystem II (PS-II), the last two being found in organisms that use water as a terminal source of electrons (plants, algae, and cyanobacteria).

Many techniques have been used to study the diverse facets of photosynthesis. The elucidation of structural properties involves biochemical and spectroscopic methods, often

PAUL MATHIS AND DANIÈLE THIBODEAU, Service de Biophysique, Département de Biologie Cellulaire et Moléculaire, CEN Saclay, 91191 Gif-sur-Yvette, CEDEX, France

Photobiological Techniques, Edited by D.P. Valenzeno *et al.*
Plenum Press, New York, 1991

associated with other analytical tools. Recently, reaction centers from several purple bacteria have been crystallized and their structure determined at 2.3-3 Å resolution by X-ray crystallography (Deisenhofer and Michel, 1989). This achievement represented a decisive step forward in our understanding of the photosynthetic machinery. It was preceded by another important achievement, the complete amino acid sequencing of all reaction center polypeptides (this is true in purple bacteria; in PS-I and PS-II, which are much more complex, there may still be unknown polypeptides; green bacteria are still less well known). Site-directed mutagenesis now offers new perspectives in the study of structure-function relationships. Among nonoptical spectroscopic methods, special mention should be made of EPR (electron paramagnetic resonance),which is especially well adapted to study the molecular species involved in electron transfer reactions. Optical techniques are, quite naturally, of particular importance (Mathis, 1990). They have three major assets;they are highly sensitive, they are nondestructive, and they potentially offer very good time resolution. Two optical properties are of particular interest.

Electronic absorption: Each molecule has a distinctive electronic absorption spectrum. A unique spectral character is particularly evident for the photosynthetic pigments which have strong characteristic absorbance bands in the visible, and also in the near IR and UV; but most of the electron carriers (e.g., cytochromes) also share this property. Most functional studies on photosynthetic membranes have been (and are) based on these absorption properties and on their variation when a given molecule experiences some perturbation, such as oxidation or reduction.

Fluorescence: After light absorption, excitation energy is rapidly transferred to the primary electron donor in the reaction centers, where electron transfer is initiated. Fluorescence represents an energy loss, which provides information on the efficiency

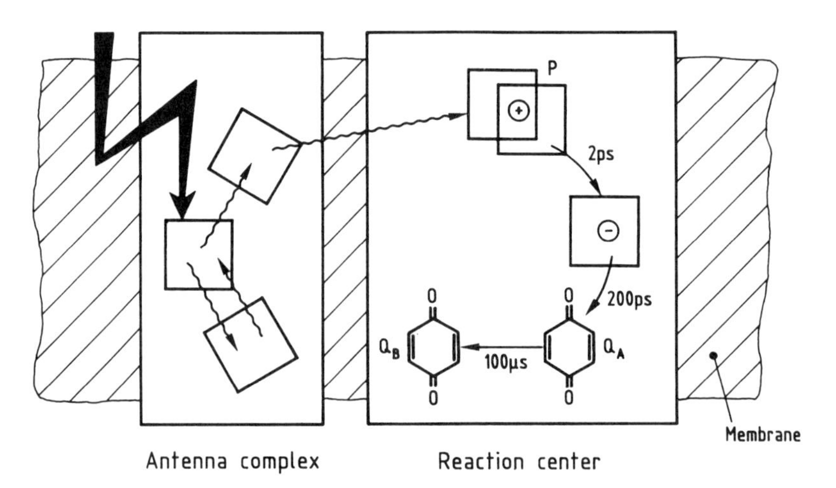

Figure 1. Schematic representation of a piece of the photosynthetic membrane with an antenna complex which absorbs light (large arrow), transfers energy among chlorophyll molecules (squares), and transfers energy to the primary electron donor (P) in the reaction center. P is a bacterio-chlorophyll dimer which starts electron transfer, first to a bacterio-pheophytin, H, then to a primary quinone, Q_A, and then to a secondary quinone, Q_B. This scheme corresponds to the actual structure in purple bacteria, with representative times for several electron transfer steps. It can also serve as a good model for PS-II, whereas PS-I and the reaction center of green bacteria probably have a different structure.

of energy transfer, on charge separation and also, by using special measurements of delayed light, on charge recombination processes.

Many other optical methods are highly informative, in particular vibrational spectroscopy (IR absorption and resonance Raman), which is in full development spurred by the technical advances in CW (continuous wave) lasers and in Fourier-transform methods. The domain of application of optical methods is very broad, in terms of biological material (from leaf to isolated molecule), of time domain (from below 10^{-12}s to over 10^{2}s), and of the kinds of properties which can be analyzed: energy transfer, electron transfer, molecular interactions, molecular orientation, etc.

Flash Absorption

In a population of reaction centers, charge separation can be induced suddenly and synchronously by a short intense flash of light (the flash is said to be saturating if all of the reaction centers in the sample under study become excited by the flash). The effect of the flash, i.e., a succession of electron transfer reactions, can be probed by measuring the change in absorbance of the sample, versus time, at selected wavelengths. Suppose that we have a cuvette with an optical path l (in cm) (Fig. 2) containing a molecule, m, which in its resting (neutral) state, does not absorb at the wavelength λ. The concentration of m is c. Suppose that flash excitation leads to the oxidation of m, and that m^+ absorbs at λ with an extinction coefficient ε (in $M^{-1} \cdot cm^{-1}$). The cuvette also contains other molecules which absorb at λ with an absorbance A, but we suppose that their absorbance is not changed by the flash. The cuvette is crossed by a beam of monochromatic monitoring light of intensity I_o. After passing through the cuvette, the beam intensity is I before the flash, and I' after the flash.

Before the flash, the cuvette absorbance is:

$$A_{cuvette} = A = \log_{10}(I_o/I) \qquad [1]$$

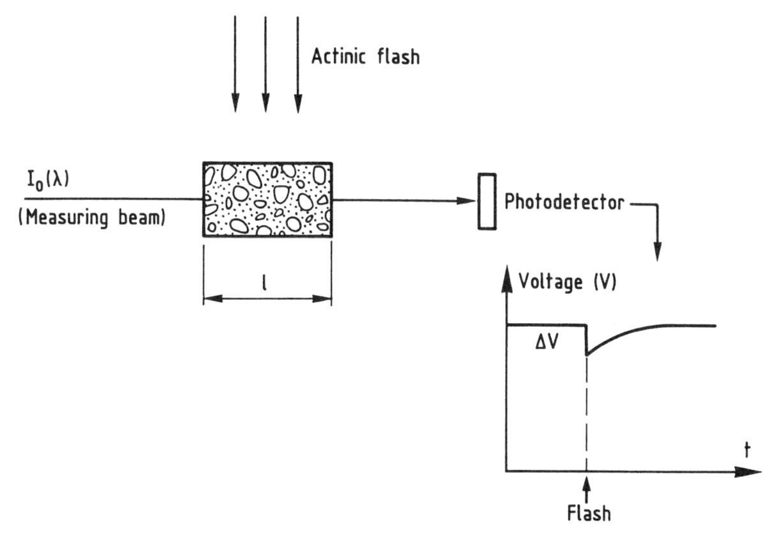

Figure 2. Principle of measurement of flash-induced absorption changes.

After the flash, the cuvette absorbance is:

$$A'_{cuvette} = A + \varepsilon cl = \log_{10}(I_o/I') \quad [2]$$

It follows that:

$$\Delta A = A'_{cuvette} - A_{cuvette} = \varepsilon cl = \log_{10}(I/I') \quad [3]$$

and that the concentration c can be measured directly from the variation of light intensity after crossing the cuvette. Neither the value of I_o nor the background absorption A matter directly. If one writes: $\Delta I = I - I'$, it can easily be shown that, if $\Delta I/I$ is small (smaller than 10%), one obtains: $\Delta A = \varepsilon cl = -(1/2.3)\Delta I/I$. The light intensity is measured by a photodetector with a linear response, and converted to a voltage, V, which varies by ΔV.

It then follows that:

$$\Delta V/V = 2.3\varepsilon cl \quad [4]$$

V has a constant value, and the time evolution of the concentration c of m^+ can be measured by recording the time variation of ΔV:

$$c = \Delta V/(2.3V\varepsilon l) \quad [5]$$

This is the general principle of absorption change measurements. A difference spectrum (post-excitation minus pre-excitation) can be recorded by repeating the measurement at various wavelengths, λ, for the monitoring beam, I_o, and then plotting $\Delta V/V$ versus wavelength. These measurements give spectral and kinetic information directly. In the case of photosynthetic reaction centers, they allow determination of the electron carriers (via the difference spectra), and the sequence and kinetics of the reactions.

Near IR Measurements: Tetrapyrrolic pigments play an important role in photoinduced electron transfer, in addition to their well known roles in light absorption and energy transfer. Chlorophyll *a*, pheophytin *a*, bacteriochlorophyll *a* and *b*, etc. can be electron donors or acceptors according to their precise positioning in reaction centers. In their neutral state, these molecules have a rather intense long-wavelength absorption band (due to the so-called Q_Y electronic transition). This band bleaches when the molecule becomes either reduced or oxidized, and a new band appears at even longer wavelength (Table 1). Similar properties are observed upon formation of excited states (e.g., the triplet state), and also upon oxidation or reduction of carotenoids. The new absorption band is characteristic and allows the identification of the product.

For kinetic measurements, using the near IR band of the product has several advantages:

(i) A wavelength can be chosen where the neutral molecules do not absorb significantly. It is thus possible to avoid any *actinic* effect of the monitoring light. This effect, which will be illustrated in the experiment below, means that the monitoring light is absorbed strongly enough by the sample, so as to signifi-

Table 1

Absorption data for the primary electron donor P in several classes of reaction centers. References: (a) Trosper et al, 1977; (b) Parson and Cogdell, 1975; (c) Mathis and Sétif, 1981.

Reaction center	Name of primary donor	Chemical constitution	max (nm) (neutral)	max (nm) (cation)
Purple bacteria:				
Rps. viridis	P-960	(BChl b)$_2$	960	1320(a)
Rh. sphaeroides	P-870	(BChl a)$_2$	870	1250(b)
PS-I	P-700	(Chl a)$_2$	700	820(c)
PS-II	P-680	(Chl a)$_?$	680	820(c)

cantly perturb the state of oxidation-reduction of the molecules in the reaction centers before an actinic flash is fired. It should be avoided as much as possible.

(ii) Modern light detectors with a very good quantum efficiency in the near infrared are available. Silicon photodiodes have excellent performance in the 700-1000 nm region, and germanium photodiodes can be used at even longer wavelengths, up to 1700 nm.

(iii) In addition to classical light sources, several types of laser diodes are very convenient sources of monitoring light in the region 740-1300 nm.

(iv) Light scattering by turbid biological samples is reduced at these long wavelengths compared to the UV-visible range.

Although the near IR spectral region is of great interest, a few disadvantages should be mentioned. The absorption bands of cationic and anionic species are rather broad. They also overlap, so that chemical identification is sometimes uncertain. Also, extinction coefficients of transitions in this spectral region are rather weak, and consequently, the absorption changes to be measured are also small. In practical terms, a $\Delta V/V$ of 10^{-4} can be measured, i.e. $\Delta A = 4.4 \times 10^{-5}$. For illustrative purposes, if $\varepsilon = 10^4\ M^{-1} \cdot cm^{-1}$ and $l = 1$ cm, measurements of $c = 4.4 \times 10^{-9}$ M are possible.

Another problem resides in fluorescence emission by the pigments. When the actinic flash excites the sample, it induces a rather intense fluorescence which goes partially toward the detector. The fluorescence can be largely blocked by a filter (e.g., a bandpass interference filter at the wavelength of interest, λ). The fluorescence spectrum, however, always includes some contribution in the near IR absorption bands under study, and the photodetector is perturbed in a manner which has to be minimized by the optical configuration.

Application to Reaction Centers: Generally speaking, flash kinetic absorbance is the method of choice to study a sequence of events that are triggered by a flash of light, as in many photobiological processes induced by rhodopsin, bacteriorhodopsin, phytochrome (after isolation), etc. In photosynthesis, the method has two important aspects in modern research (Mathis, 1990):

(i) Advanced methods are used to study very short-lived reactions (e.g., in the ps time domain) or signals of weak amplitude under difficult measurement conditions (e.g., study of the tyrosine cation-radical of PS-II, in the UV).

(ii) Classical methods, in the μs or ms time domain, are used as easy and informative tests in conjunction with functional or structural tools, for example, to study the effects of inhibitors, of specific mutation by site-directed mutagenesis, in protein engineering, etc.

Quinone Binding Sites in PS-II and Purple Reaction Centers

It has been known for quite a long time that quinones are essential constituents of reaction centers in PS-II and in purple bacteria. More recently, it was found that they participate in all kinds of reaction centers, including PS-I, green sulfur bacteria, green non-sulfur bacteria and heliobacteria. Altogether, their chemical and functional properties are rather diverse and often poorly understood. In PS-II and in purple bacteria, there are two quinones, Q_A and Q_B, per reaction center. A large body of experimental observations has established that the properties of the quinones in both types of reaction centers are rather similar.

Functional Scheme: The quinones Q_A and Q_B operate in series for electron transfer according to the scheme (Velthuys, 1981; Wraight, 1981);

Effect of 1st photon: $Q_AQ \rightarrow Q_A^-Q_B \rightarrow Q_AQ_B^-$ (stable state)

Effect of 2nd photon: $Q_AQ_B^- \rightarrow Q_A^-Q_B^- \rightarrow Q_AQ_B^{2-} \rightarrow Q_AQ_BH_2$,

with uptake of $2H^+$. The reduced quinone (hydroquinone) then leaves the reaction center, enters a pool of membrane-located quinones, and is replaced by an oxidized quinone at the Q_B site of the reaction center (McPherson *et al.*, 1990). The quinone pair has then completed one functional cycle.

It thus appears, that under normal conditions, Q_A is a one-electron carrier which remains permanently bound to the protein, whereas Q_B is a two-electron (and two H^+) carrier. Electron binding is strong in state Q_B^-, weaker in Q_B, and even weaker in Q_BH_2. Proton uptake is part of the overall proton-pumping role of reaction centers, providing free energy for the synthesis of ATP.

Chemical Structure: In many organisms, Q_A and Q_B are chemically identical. They are plastoquinone-9 in PS-II and ubiquinone-9 in most purple bacteria. These are substituted benzoquinone derivatives, with a poly-isoprenoid chain. In *Rhodopseudomonas viridis*, Q_A is a substituted naphthoquinone (menaquinone) and Q_B is a ubiquinone.

Detailed studies on extraction of natural quinones and reconstitution with other quinones have permitted great advances in understanding the relation between chemical structure and binding strength or electron transfer rates (Giangiacomo and Dutton, 1989; Gunner and Dutton, 1989).

Location of Binding Sites: In purple bacteria, it has been shown unambiguously that Q_A and Q_B are bound to the polypeptides M and L, respectively. These polypeptides make up the heart of the reaction center. The ligands to the quinones can be inferred from the analysis of X-ray diffraction by crystals of the reaction centers.

In PS-II, it can be hypothesized that quinones are located similarly, on the polypeptides D_1 (for Q_B) and D_2 (for Q_A), which constitute the core of the reaction center. The hydrophobic isoprenoid tail also contributes to keep the quinones in their hydrophobic pocket.

Quinone Inhibitors and Inhibitor Resistance: Several families of chemicals (such as ureas, triazines, phenols, cyanoacrylates) have been shown to bind at the Q_B site, in competition with the quinone. In PS-II of plants, several of these chemicals are used as agricultural weed killers on a large scale. The same kind of competitive inhibition occurs in purple bacteria, with a different relative affinity: a few potent PS-II inhibitors, such as DCMU (diuron) or phenol derivatives, have no effect on bacteria. This differential effect shows that the Q_B site, although rather similar, is not identical in both types of reaction centers.

Herbicide-resistant cultivars have appeared among crop weeds. Sequencing of the D_1 gene in these cultivars, has shown that they are mutated at a single, of several possible, amino acids belonging to the Q_B binding site. Similar mutants appear among purple bacteria when they are grown in the presence of an inhibitor. It is noteworthy that a mutant of *Rps. viridis*, which has become insensitive to triazines, has been thus rendered sensitive to DCMU, a typical PS-II inhibitor (Sinning *et al.*, 1989b). In PS-II, the chemical properties of the inhibitors and the position of amino-acid changes which confer herbicide resistance have allowed modelling of the herbicide/Q_B binding site (Trebst, 1987). This modelling process is now being refined in many laboratories.

Structural Properties and Spectroscopic Studies: Several techniques are of great interest to probe the physical state of Q_A and Q_B (see e.g., Feher *et al.*, 1989):

- Flash absorption, for kinetic properties.

- Electron paramagnetic resonance (EPR) detects the anion-radical states and shows that both Q_A^- and Q_B^- are in magnetic interaction with an Fe^{2+} atom; under reducing conditions, Q_A^- also interacts with the reduced (bacterio-)pheophytin.

- ENDOR has brought detailed information on the charge repartition in Q_A^- or Q_B^-, and on the mode of binding, essentially by hydrogen bonding between the quinone oxygens and the polypeptide backbone.

- IR absorption spectroscopy (FTIR), of more recent utilization, provides information on the mode of binding of the quinone, and on the structural changes which accompany electron transfer (Nabedryk *et al.*, 1990).

Flash-Induced Electron Transfer in Purple Bacteria

When bacterial cells are broken one obtains the so-called *chromatophores*, which are vesicular membranes containing all the reaction centers and their associated antennae (ensembles of pigment protein complexes-containing bacteriochlorophyll *b* and carotenoids). In isolated chromatophores, the reaction centers have their redox centers mainly in the state: $[\text{cytochrome}]_{ox}$...P-960...Q_A...Q_B. The four hemes of the bound cytochrome are mostly oxidized, and so are the quinones Q_A and Q_B. Flash excitation creates the state (P-960$^+$ ... Q_A^-) in less than 1 ns. Electron transfer from Q_A^- to Q_B takes place in about 20 µs, creating the state (P-960$^+$ ... Q_A ... Q_B^-). The only way for this state to decay is via a back-reaction, returning to the initial state with a $t_{1/2}$ of about 100 ms. It thus appears that flash excitation of chromatophores populates P-960$^+$ which will decay with $t_{1/2} \approx 100$ ms. P-960$^+$ has an absorbance band in the vicinity of 1300 nm (Davis *et al.*, 1979). Its formation and subsequent re-reduction can be detected by recording the flash-induced absorbance change, ΔA, around that wavelength.

Inhibitors of the quinone at the Q_B site are often called herbicides because they are used in agriculture to kill weeds: they do so by blocking PS-II function in plants, also by displacing the quinone (plastoquinone) from the PS-II Q_B site (Trebst, 1986; Velthuys, 1981; Wraight, 1981). Upon progressive addition of an herbicide (e.g. atrazine or terbutryn) to chromatophores, ubiquinone is progressively displaced from the Q_B site. Flash excitation of chromatophores with a bound herbicide molecule creates the state: (P-960$^+$ ...Q_A^-...I), where I is the inhibitor. There is no electron transfer from Q_A^- to I, and thus P-960$^+$ decays via charge recombination with Q_A^-, with a $t_{1/2} \approx 1$ ms.

EXPERIMENT 15: FLASH ABSORPTION IN PURPLE BACTERIA

OBJECTIVES

As an illustration of the use of IR flash absorption, this experiment will demonstrate the effect of an inhibitor which binds competitively with ubiquinone at the Q_B site in chromatophores of the purple bacterium *Rhodopseudomonas viridis*. With the same biological preparation, the effect of chemicals which are competitive inhibitors of the quinone at the Q_B site will also be demonstrated. This experiment allows the determination of I_{50}, the inhibitor concentration at which half of the reaction centers have the Q_B site occupied by the inhibitor.

A few control experiments will also be performed:

Variation of the excitation flash intensity - This will determine whether the flash is saturating when it is at its maximum intensity.

Addition of a reductant - The bound cytochrome will be reduced (the highest potential heme has an E_m of +360 mV) and P-960$^+$ will be re-reduced very rapidly by the cytochrome, in less than 1 μs.

Absorption spectrum of the sample - A double-beam spectrophotometer will be used to determine the antenna size, i.e. the number of antenna bacteriochlorophyll *b*'s per reaction center.

LIST OF MATERIALS

Tris-HCl, NaCl, and 0.1 N HCl.
Na^+ ascorbate
Atrazine (MW = 216; available from Chem-Service) or Terbutryn (MW = 241; available from Riedel-de-Haën): stock solutions at concentrations of 0.2 mM and 2 mM, in dimethylsulfoxide
Ethanol
K^+ ferricyanide: 20 mM stock
Tungsten lamps, 40 W (4)
Pipette or syringe, 10 μl, with 1 μl accuracy
Automatic pipettes, 1 ml and 100 μl (2)
Balance, with mg sensitivity
Flexible, thermoplastic waterproof sheet (e.g., Parafilm) to cover cuvettes for mixing
Absorption spectrophotometer covering the range 400-1100 nm

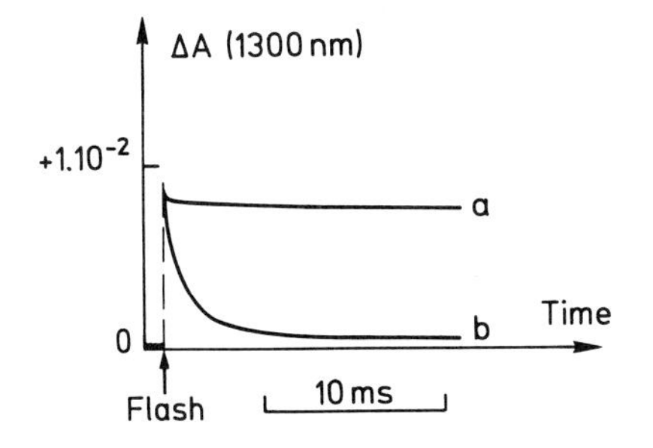

Figure 3. Kinetics of absorption change at 1300 nm induced by a flash in a cuvette containing *Rps. viridis* chromatophores (curve a) and after addition of 10 μM terbutryn (curve b). Buffer at pH 8.0; addition of 100 μM ferricyanide.

Spectrophotometer cuvette, 1 × 1 × 4.5 cm, glass or plastic (3)
Cuvette holder
Polaroid camera and film or chart recorder
Centrifuge 10 000 × g
Centrifuge 100 000 × g
French press
Pasteur pipets and bulbs
Tungsten-halogen lamp, 150 watt
DC power supply for lamp
Black box for lamp
Filters, Calflex X (2)
Filters, RG-1000-2 mm (2)
Lenses, f = 100 mm, diameter ≥40 mm (3)
Lens paper to clean optical components
Flash lamp
Plastic lightpipe
Neutral density filters, OD = 0.3 (3)
Germanium photodiode
BNC cable, 1 m long
Battery, 1.35 V
Metal box for battery
Resistor, 1 kΩ (1/4 W is sufficient)
Optical rail, 1 meter long
Adjustable mounts for lenses, filters, cuvettes and photodetector
Black paper and cardboard
Recorder
Pulse generator (see specifications below)

PREPARATIONS

Terbutryn Stock Solution: Dissolve 4.8 mg of terbutryn in 10 ml of DMSO to produce a 2 mM stock solution.

Atrazine Stock Solution: Dissolve 4.3 mg of atrazine in 10 ml of DMSO to produce a 2 mM stock solution.

Buffer Solution: Dissolve 1.21 g of Tris-HCl and 3.5 g of NaCl in 1 liter of distilled water. Adjust the pH to 8.0±0.2 by adding 0.1 N HCl.

Tris Ascorbate: Dissolve 198 mg of Na^+ ascorbate in 10 ml of buffer solution to produce a 100 mM stock solution.

K^+ Ferricyanide Solution: Dissolve 65.8 mg of K^+ ferricyanide in 10 ml of distilled water to produce a 20 mM stock solution.

Chromatophores: *Rhodopseudomonas viridis* cells can be obtained from the American Type Culture Collection (12301 Parklawn Drive, Rockville, MD, USA). To start a culture, transfer cells, under sterile conditions, to a 2 liter bottle with Hutner medium (Cohen-Bazire, et al., 1956). Incubate them for 1 day in darkness at ≈ 25°C, and then grow them for 4-5 days at ≈ 30°C under the illumination of four 40 W tungsten lamps. Collect the cells by low speed centrifugation. Resuspend them in Tris (50 mM, pH 8.0), at 0°C, and break them by passage through a French press at 1000 psi. Centrifuge the liquid at 10 000 × g for 10 min to eliminate large debris, and recentrifuge the supernatant at 100 000 × g for 1 h. Resuspend the pellet of chromatophores in Tris (50 mM, pH 8.0). Adjust the volume of buffer to give an absorbance of 100 absorbance units for a 1 cm path length, at 1015 nm (20 µl of suspension plus 1.98 ml of buffer gives an OD of 1.0). After preparation, the concentrated chromatophores may be kept for one week at 4°C. For longer durations, the chromatophores may be frozen. Upon thawing, however, they will precipitate. A brief ultrasonication will clarify the suspension. Alternatively, add 2 volumes of glycerol before freezing. The suspension then will remain clear when thawed.

Set-up for ΔA Measurement: The set-up is shown schematically in Fig. 4. It may be assembled before the experiment, or assembly can be part of the experiment. (Assembly requires about 4 h.) The *cuvette* is a standard spectrophotometric three- or four-faced, glass or plastic cuvette. It is held in a *cuvette holder* with at least three open sides. The monitoring light source is a tungsten-halogen *lamp* of about 150 watt maximum power, powered by a current-regulated DC *power supply* and inserted in a black *covering box* (wood, plastic, etc.) to block stray light. The measuring light is filtered by a two component filter, F_1: a *Calfex X* (Balzers, 40 × 40 mm) and an *RG-1000-2 mm* (Schott, diameter ≥ 40 mm) and collected by a *lens,* L_1 (f = 100 mm, diameter = 40 mm). It is then focused onto the measuring cuvette, through another *cuvette* filled with ethanol. Past the cuvettes, the monitoring light is refocused by the *lens* L_2 (same as L_1) onto the *photodetector PD*, which is protected by *filter* F_2 (same as F_1). The *actinic flash* is a Stroboslave (type 1539A, General Radio) operated at full power (or any xenon flash with an energy of 1-10 J per pulse and a duration of less than 50 μs). The actinic light is focused onto the cuvette by a *lens* L_3 (same as L_1) or preferably by a *plastic light pipe* (a piece of clear plastic about 10 cm long, shaped at both ends to fit the openings of the flash source and of the cuvette holder, and polished with a progressive evolution from one opening shape to the other). *Neutral density filters* (3, OD = 0.3 each) permit attenuation of the excitation light. According to the optics of the measuring light source, the lenses may have to be modified. The values given here should work, although they might not be optimal.

The photodetector is a 5 mm diameter germanium *photodiode* (Judson, type J16 - P2.R05 M) equipped with a BNC connector. Its output is connected by a *BNC cable* (1 m) to a 1.35-V *battery* placed in a *metal box*. The current through the battery generates a voltage V through a 1 kΩ *load resistor* (Fig. 4).

The ensemble should preferably be mounted on an *optical rail* on a laboratory bench. *Mounts* with adjustable height and orientation are necessary for the lenses, filters, cuvettes

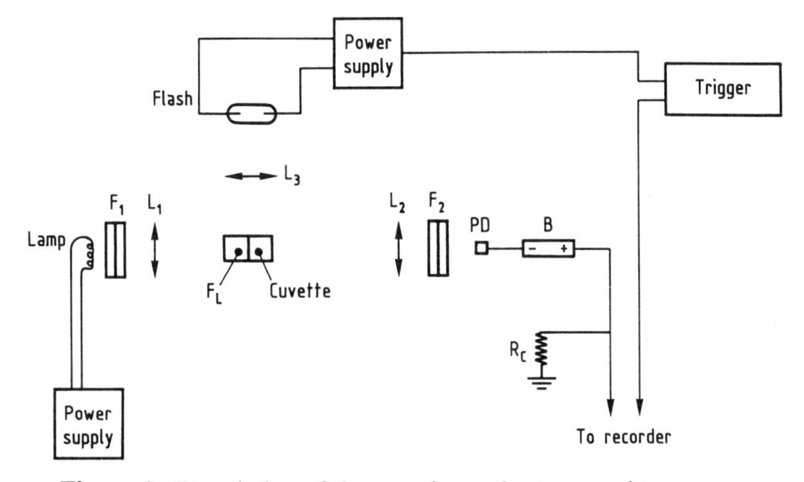

Figure 4. Description of the experimental set-up used to measure flash-induced absorption changes. F_1,F_L,F_2 = optical filters. L_1, L_2 = lenses. PD = photodetector (germanium photodiode). B = battery.

and photodetector. They should be compatible with the optical rail, and some attention should be paid to ensure that the mounts can hold these elements at the same height as the beams from the measuring and flash lamps. Pieces of *black paper* or cardboard are placed to minimize stray light from the monitoring light and from the flash. More importantly, be sure that all the measuring light that reaches the detector has passed through the measuring cuvette. For optical adjustments, leave out the filter F_1 and work with low light to protect the detector.

The experiment involves a *recorder*, for the measurement of the voltage, V, and recording of its variation, ΔV. Typically V will be 400 mV, and ΔV will be 4 mV. These measurements can be done in several ways (the required high frequency electrical bandwidth is 3 kHz or more):

(i) *A Storage or a Digital Oscilloscope*: V can easily be measured. Adequate ΔV recording requires adequate sensitivity (1 mV/div) and a DC offset. Lack of sensitivity can be solved by using a preamplifier with a gain of ≈10; a lack of offset can be solved by working with AC coupling, a solution which, however, somewhat disturbs the ΔV kinetics.

(ii) *A Transient Digitizer*: Most digitizers will suffice and most will also average the results from successive flashes (keep a time interval of about 10 s to be sure that the reaction centers have entirely recovered from the previous flash). The voltage, V, has to be measured with an ordinary oscilloscope. The problems of sensitivity and of offset can either be solved by the recorder or as described in (i), above. The digitizer can be replaced by a microcomputer equipped with an analog-to-digital conversion card capable of data acquisition at 100 μs per channel.

As an example of a recorder, the following experiment description will use a digital oscilloscope with 5 mV/div as maximum sensitivity and no offset. An AC amplifier with a gain of 30 is inserted between the diode output and the recorder (amplifier bandwidth: high frequency = 10 kHz, low frequency = 3 Hz).

If data analysis cannot be made directly with a microcomputer, final recording of the data can be made with a Polaroid *camera* or a *chart recorder*. A *pulse generator* is needed, capable of delivering 2 TTL electric pulses with a delay of 1-100 ms to trigger first the recorder and then the flash. The generator can be triggered with a manual command. This kind of generator will be found in electronics shops. If a microcomputer is used, it can easily generate appropriate pulses.

EXPERIMENTAL PROCEDURE

THE EXPERIMENT SHOULD BE DONE UNDER REDUCED ROOM LIGHT!

1. Check the alignment of the flash and of the measuring optical beam (at low lamp voltage, after removing filter F_1).
2. Adjust the detector position to optimize the DC output voltage on the oscilloscope.
3. Check the triggering circuit.

Prepare Cuvette with Chromatophores

4. Pipette 2.0 ml of buffer into the spectrophotometer cuvette to be used for flash absorption.

5. Record the base line absorption spectrum from 1100 to 400 nm with buffer in both the analytical and reference compartments of the spectrophotometer.

6. Pipette 20 μl of stock chromatophore suspension into the flash absorption cuvette and record the absorption spectrum. Note the absorbance at 1015 nm (A_{1015}); it should be around 1 (±0.2). A complete spectrum is shown in Fig. 5.

7. Add 80 μl of stock chromatophore suspension and 10 μl of the 20 mM ferricyanide stock solution. (The addition of ferricyanide is to ensure that all the cytochromes and all Q_B are oxidized.) This is the cuvette that will be used for flash absorption measurements. The extinction coefficient of bacteriochlorophyll *b* in chromatophores is about 150000 $M^{-1} \cdot cm^{-1}$ at 1015 nm (Trissl *et al.*, 1990); so the concentration in the cuvette is approximately [BChl *b*] = $A_{1015} \cdot (5/150000)$ M.

Absorbance Change Measurement

8. Place the cuvette in its holder. Adjust the lamp intensity so that the voltage, V, at the detector output is 400 mV. Note the lamp intensity. Maintain this same lamp intensity during the entire experiment.

9. Connect the diode output (with load resistor) to the amplifier input, and connect the amplifier output to the oscilloscope input. Set the oscilloscope vertical sensitivity, s, (around 100 mV/div), sweep time (around 10 ms/div) and triggering (external triggering rather than free running).

10. Push the triggering switch, to activate the recorder and the flash. Record the signal. If some oscilloscope settings are not optimal, adjust them as needed and repeat the flash. If the signal size is *h* (in vertical divisions), and G is the amplifier gain, then the absorbance change is:

$$\Delta A = 0.44\, h \cdot s / V \cdot G$$

Three Illustrations of Experimental Properties

11. *Flash artifact:* Block the measuring light before the cuvette. Record the signal: this is the flash artifact, which should not seriously interfere with the ΔA.

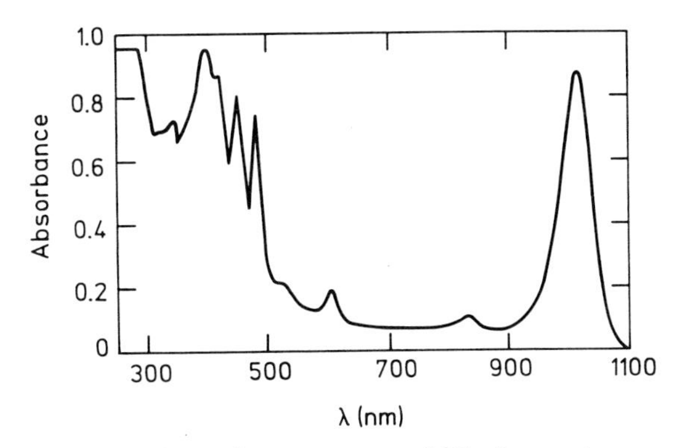

Figure 5. Absorption spectrum of *Rhodopseudomonas viridis* chromatophores.

12. *Actinic effect of the monitoring light:* Move the Calflex filter from F_1 to F_2. Record the signal. V should not be changed by filter displacement, but ΔV will be greatly reduced. This illustrates the so-called "actinic" effect of the measuring light. If there is not a good filtering of the monitoring beam before the cuvette, the photochemical reaction (here, the formation of P-960$^+$...Q_A...Q_B^-) will be significantly induced by the monitoring light. This effect can be important if the state studied is long-lived (here 100 ms). Moving the Calflex filter will allow some light around 1000-1050 nm to fall on the cuvette, and to be absorbed by the antenna pigments (peak at 1015 nm). This will induce charge separation before the flash fires.

13. *Saturation:* Replace the Calflex filter to its normal position. Remeasure the flash-induced ΔA under the following conditions: Full flash intensity (E = 100); reduce the flash intensity by one neutral density filter of OD = 0.3 (E = 50); then reduce by two filters (E = 25) and three filters (E = 12.5). Plot $\Delta A = f(E)$. The curve should tend toward saturation. Saturation indicates that all reaction centers have been hit by the flash and have reacted.

Effect of an Inhibitor

14. Repeat the measurement, with E = 100, first with no further addition, and then with progressive addition of terbutryn (or atrazine) first using the dilute stock and then the more concentrated stocks. Total concentrations of terbutryn will be (in M) : 10^{-7}, 2×10^{-7}, 4×10^{-7}, 7×10^{-7}, 1.5×10^{-6}, 5×10^{-6} (for atrazine use about 5 times higher concentrations). Record the signals for data analysis (see Discussion).

Effect of a Reductant

15. In a single cuvette, progressively add Na ascorbate, and check the evolution of the flash-induced ΔA. The signal should progressively decrease and disappear around 1 mM ascorbate. This effect is due to a reduction of the bound cytochrome. When one heme is reduced, it donates an electron to P-960$^+$ in the sub-µs time range. P-960$^+$ thus becomes undetectable by the slow-response technique used here.

DISCUSSION

Signal Size from which Antenna Size can be Extracted

Assuming an extinction coefficient of 8000 for P-960$^+$ in the wavelength band studied, one obtains [P-960$^+$] = $\Delta A/(8\times10^3)$ M. If the flash saturation curve indicates an approximate saturation, one can then assume that [P-960$^+$] is equal to the reaction center concentration. Then the average antenna size, i.e. the number of bacteriochlorophyll *b* molecules per reaction center, can be calculated as;

$$\text{Antenna size} = [\text{BChl } b] / [\text{P-960}^+]$$

A value of 30 is typical.

Decay Kinetics

Under the experimental conditions, the decay kinetics will reflect the re-reduction of P-960$^+$ by electrons coming back from Q_B^- (slow phase, $t_{1/2}$ = 100 ms) or from Q_A^- (fast phase,

$t_{1/2} = 1$ ms). In the experiment with inhibitors, the signal h can be decomposed into h_s (slow) and h_f (fast): $h = h_s + h_f$. This decomposition can be done in several ways:

(i) approximate measurement on the oscilloscope screen;

(ii) semi-log plot from a photograph or from a chart recording (plot log of h in mm versus time);

(iii) two exponential analysis by a microcomputer.

As explained in the Introduction, there is competitive inhibition between ubiquinone and the inhibitor I at the Q_B site. I_{50}, the inhibitor concentration at which half of the reaction centers have the Q_B site occupied by the inhibitor, can be determined from the data. At this point one has $h_s = h_f$. The I_{50} (in µM) is easily obtained from a graph of h_s versus inhibitor concentration. An apparent dissociation constant (neglecting ubiquinone binding) can then be extracted; $K_I = I_{50} - (1/2)[P\text{-}960^+]$. The value of K_I is useful to compare the effectiveness of various inhibitors (e.g., to compare terbutryn with atrazine). It is also useful for comparing the properties of the Q_B^- binding pocket in herbicide-resistant mutants. K_I is related to the binding affinity. Efforts have been undertaken in many laboratories to do quantitative molecular modelling in order to understand (and eventually to predict) the affinity of inhibitors according to their chemical structure, and the effect of point amino-acid substitution. The method described here permits a rather quick screening. It is an efficient way to compare reaction centers and herbicides (Sinning *et al.*, 1989a,b).

Sources of Noise

The recorded signals are not as "clean" as one would like. A fast ($t_{1/2}$= 1 ms) decaying phase is already present in the signal, even without additions. This presumably reflects damage to the Q_B site during preparation of the chromatophores. In addition, it appears that the kinetics are not completely described by two exponentials since the reaction center structure is not entirely homogeneous.

Electrical noise is a second source of noise in these measurements. The electrical noise has several causes:

(i) fluctuations of the source of the monitoring light (instability of the power supply; mechanical vibrations of the filament),

(ii) electric pick-up by the electronics (mainly 50 or 60 Hz ripple due to the main power supply),

(iii) mechanical vibrations of the optical set-up,

(iv) room light picked up by the photodetector,

(v) shot noise due to the quantum nature of light,

(vi) noise from the recording device and from the offset voltage.

Further Experiments

1. *Rps. viridis* chromatophores – The disappearance of the P-960$^+$ signal upon cytochrome reduction can be used to determine the *in situ* redox potential of the high-potential heme. ΔA can be measured with a cuvette after addition of 200 µM of each of K^+ ferri- and ferrocyanide (E_m = +420 mV). Then add ferrocyanide to decrease the redox potential, E_h, where E_h (mV) = 420 – 60 log [ferro]/[ferri]. E_m of the high-potential heme is obtained when the ΔA is half of its maximal value. The cuvette should be well protected from light to avoid uncontrolled cytochrome photooxidation.

2. Other biological materials – Many laboratories use other purple photosynthetic bacteria, such as *Rhodobacter sphaeroides*, *Rhodobacter capsulatus*, etc. The preceding experiments can be done with chromatophores from these bacteria as well. The chromatophores will differ from *Rps. viridis* in several respects:

a. Different antenna absorption: The appropriate concentration should be determined by trial and error.

b. Kinetics of charge recombination between P^+ and Q_A^- or Q_B^-: They will usually be slower than in *Rps. viridis*; wait at least 10 s between flashes.

c. Minor differences in herbicide sensitivity may be found.

d. Presence or absence of a bound cytochrome molecule: The redox titration described above may not be feasible.

3. Technical improvement – The experimental set-up described above is a minimal configuration that permits really significant measurements. For research objectives, several improvements can be suggested, to accommodate various kinds of flash absorption experiments:

a. Monitoring light: Use a laser diode at an appropriate wavelength (e.g. 820 nm to work with plant PS-I or PS-II reaction centers). Use a flash lamp to work at any wavelength with ns time resolution.

b. Filters: Use interference filters according to the wavelength of interest, or a monochromator instead of F_1.

c. Actinic flash: Use a pulsed laser flash for shorter duration and selected excitation wavelength.

d. Cuvette: Use a temperature control or a cryostat to work at low temperature, or a flow system to work with fresh material at each flash.

e. Detector: With a germanium photodiode, a pre-amplifier will allow faster time resolution (the diode area should also be optimized). To measure below 1000 nm, a silicon photodiode is preferable.

f. Recorder: A large variety of transient digitizers are available. Specifications to be considered are primarily high frequency bandwidth and time per channel, vertical dynamic range, voltage sensitivity and quality of offset, and ease of data handling.

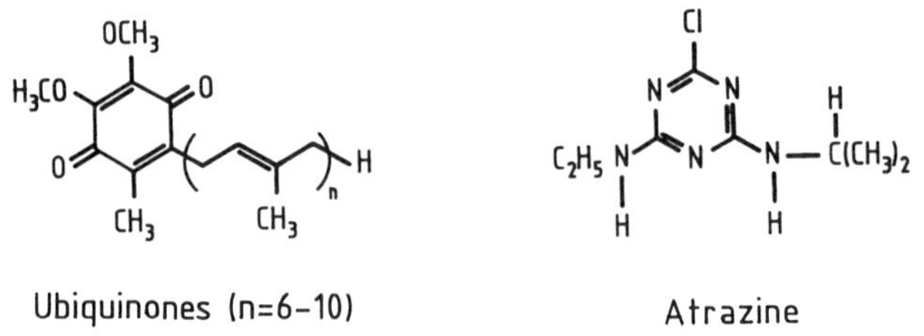

Ubiquinones (n=6–10)

Atrazine

Figure 6. Chemical structure of ubiquinone (n = 6 for UQ_6; n = 9 for UQ_9) and for atrazine (2-chloro-4-ethylamino-6-isopropylamino-s-triazine), a Q_B site inhibitor.

REVIEW QUESTIONS

1. For quite a long time, inhibitors, I, have been assumed to block electron transfer between Q_A and Q_B (this was the interpretation of the inhibition of PS-II electron transfer by herbicides in plants). Do the results obtained in this experiment support this hypothesis?

2. Based on chemical structures (Fig. 6), are ubiquinone and herbicides likely to be competitive inhibitors?

3. How can one rationalize that, in spite of the symmetry of the reaction center structure (Deisenhofer and Michel, 1989; see also Fig. 1), electron transfer to P^+ is 100-fold slower from Q_B^- than from Q_A^-?

4. What is the equilibrium constant, K, for the two states: $P^+Q_A^- \rightleftarrows P^+Q_B^-$? In *Rps. viridis*, electron transfer from Q_A^- to Q_B has a $t_{1/2}$ of about 20 µs. What is the $t_{1/2}$ for back electron transfer from Q_B^- to Q_A?

5. Demonstrate that $\Delta A \approx -\Delta I/2.3\,I$.

6. How is it possible to check whether the monitoring light has any actinic effect? What are the experimental factors that could decrease this effect?

7. In *Rps. viridis* chromatophores, the antenna bacteriochlorophyll *b* has a long wavelength absorption maximum at 1014 nm (see Fig. 5), whereas the energy trap (P-960) has a maximum at 960 nm, i.e. at higher energy. Can there be energy transfer to the trap under these conditions?

ANSWERS TO QUESTIONS

1. Most of the data can be explained by the older hypothesis. However, it does not consider a competition between ubiquinone and I. This competition can be easily demonstrated with isolated purified reaction centers where the I_{50} varies with the quinone concentration, a property which is not explainable by the older hypothesis. Recently structures have been determined with crystals of reaction centers having either ubiquinone or an inhibitor at the Q_B site (Michel *et al.*, 1986; Deisenhofer, J., O. Epp, I. Sinning, and H. Michel, unpublished). It was clearly shown that both types of molecules bind at the same site.

2. Detailed molecular modelling may answer this question in the future. However, it is currently impossible to predict the binding affinity of various chemicals at the Q_B site.

3. The electron does not return directly to P^+ through the protein, but via a series of back-steps which are possible via Q_A and the pheophytin H (this so-called right branch of the reaction center is the only active path for forward electron transfer), but not via the left branch.

4. Accepting the answer to question 3, leads to: $K = (P^+Q_A^-)/(P^+Q_B^-) = k_B/k_A = 1/100$ (where k_A and k_B are the rates of charge recombination for $(P^+Q_A^-)$ and $(P^+Q_B^-)$). Then the $t_{1/2}$ for electron transfer from Q_B^- to Q_A is 100-fold slower than the forward transfer from Q_A^- to Q_B, i.e., $t_{1/2} \approx 2$ ms.

5. $\Delta A = \log_{10} I/I' = \log_{10} I/(I-\Delta I) = -\log_{10}(1-\Delta I/I)$

   For $u \ll 1$, we have: $\log_{10}(1\text{-}u) = (1/2.3)\log(1\text{-}u) \approx -u/2.3$. It obviously follows that $\Delta A \approx -\Delta I/2.3\,I$.

6. Decrease the measuring light intensity by putting a neutral density filter (with 25% transmission, for example) between the light source and the cuvette. The amplitude of flash-induced ΔA should not change. The actinic effect can be decreased;

(i) by using a weakly absorbed wavelength,

(ii) by putting the wavelength selector (filter or monochromator) before the cuvette,

(iii) by allowing the monitoring light to fall on the cuvette a short time before firing the flash (e.g., by using a lamp booster or a shutter),

(iv) by using a large cuvette.

Note that the effect will be increased by two biological factors: the antenna size and the duration of the light-induced state.

7. In a steady-state situation, the excitation energy will reside much more in the antenna than on P-960. However, excitation of P-960 is followed by a very fast electron transfer, at most 2 ps or perhaps much less. (The rate and mechanism of the very first step are not yet elucidated.) The rapidity of electron transfer ensures an efficient drainage of energy toward charge separation. Energy transfer to a pigment of higher excitation energy is usually termed "up hill" energy transfer.

SUPPLEMENTARY READING

Fork, D.C. (1989) Photosynthesis. In *The Science of Photobiology*, 2nd edition (Edited by K.C. Smith), pp. 347-390. Plenum Press, New York.

REFERENCES

Amesz, J. (1987) *New Comprehensive Biochemistry*, Vol. 15 Photosynthesis, Elsevier, Amsterdam.

Barber, J. (1987) *Topics in Photosynthesis*, Vol. 8. The Light Reactions, Elsevier, Amsterdam.

Berger, G., M.D. Tiede and J. Breton (1984) HPLC purification of a biologically active membrane protein: The reaction center from photosynthetic bacteria. *Biochem. Biophys. Res. Commun.* **121**, 47-54.

Clayton, R.K. (1970) *Light and Living Matter*, Mc Graw-Hill, New York.

Clayton, R.K. and B.J. Clayton (1972) Relations between pigments and proteins in the photosynthetic membranes of *Rhodopseudomonas spheroides*. *Biochim. Biophys. Acta* **283**, 492-504.

Clayton, R.K. and W.R. Sistrom (1978) *The Photosynthetic Bacteria*, Plenum Press, New-York.

Cohen-Bazire, G., W.R. Sistrom and R.Y. Stanier (1956) Kinetic studies of pigment synthesis by non-sulfur purple bacteria. *J. Cell. Comp. Physiol.* **49**, 25-68.

Davis, M.S., A. Forman, L.K. Hanson, J.P. Thornber and J. Fajer (1979) Anion and cation radicals of bacteriochlorophyll and bacteriopheophytin *b*. Their role in the primary charge separation of *Rhodopseudomonas viridis*. *J. Phys. Chem.* **83**, 3325-3332.

Deisenhofer, J. and H. Michel (1989) The photosynthetic reaction center from the purple bacterium *Rhodopseudomonas viridis*. *EMBO J.* **8**, 2149-2169.

Feher, G., J.P. Allen, M.Y. Okamura and D.C. Rees (1989) Structure and function of bacterial photosynthetic reaction centers. *Nature* **339**, 111-116.

Gast P., T.J. Michalski, J.E. Hunt and J.R. Norris (1985) Determination of the amount and the type of quinones present in single crystals from reaction center proteins from the photosynthetic bacterium *Rhodopseudomonas viridis*. *FEBS Lett.* **179**, 325-328.

Giangiacomo, K.M. and P.L. Dutton (1989) In photosynthetic reaction centers, the free energy difference for electron transfer between quinones bound at the primary and secondary quinone-binding sites governs the observed secondary site specificity. *Proc. Natl. Acad. Sci. USA* **86**, 2658-2662.

Gunner, M.R. and P.L. Dutton (1989) Temperature and -ΔG^o dependence of the electron transfer for BPh$^{\cdot-}$ to Q_A in reaction center protein from *Rhodobacter sphaeroides* with different quinones as Q_A. *J. Am. Chem. Soc.* **111**, 3400-3412.

Hatch M.D. and N.K. Boardman (1981) *The Biochemistry of Plants*, Vol. 8 Photosynthesis, Academic Press, New York.

Laasch H., W. Urbach and U. Schreiber (1983) Binary flash-induced oscillations of 14C-DCMU binding to the photosystem II acceptor complex. *FEBS Lett.* **159**, 275-279.

McPherson, P.H., M.Y. Okamura and G. Feher (1990) Electron transfer from the reaction center of *Rb. sphaeroides* to the quinone pool: Doubly reduced Q_B leaves the reaction center. *Biochim. Biophys. Acta* **1016**, 289-292.
Mathis, P. (1990) Optical techniques in the study of photosynthesis. In: *Methods in Plant Biochemistry, Vol. 4: Lipids, Membranes and Light Receptors* (Edited by J.Bowyer and J.Harwood), pp. 231-258. Academic Press.
Mathis, P. and P. Sétif (1981) Near infrared absorption spectra of the chlorophyll *a* cations and triplet state *in vitro* and *in vivo*. *Isr. J. Chem.* **21**, 316-320.
Michel, H. (1982) Three-dimensional crystals of a membrane protein complex. The photosynthetic reaction center from *Rhodopseudomonas viridis*. *J. Mol. Biol.* **158**, 567-572.
Michel, H., O. Epp and J. Deisenhofer (1986) Pigment-protein interactions in the photosynthetic reaction center from *Rhodopseudomonas viridis*. *EMBO J.* **5**, 2445-2451.
Naberdryk, E., K.A. Bagley, D.L. Thibodeau, M. Bauscher, W. Mäntele and J. Breton (1990) A protein conformational change associted with the photoreduction of the primary and secondary quinones in the bacterial reaction center. *FEBS Lett.* **266**, 59-62.
Parson, W.W. and R.J. Cogdell (1975) The primary photochemical reaction of bacterial photosynthesis. *Biochim. Biophys. Acta* **416**, 105-149.
Shopes, R.J. and C.A. Wraight (1985) The acceptor quinone complex of *Rhodopseudomonas viridis* reaction centers. *Biochim. Biophys. Acta* **806**, 348-356.
Sinning, I., H. Michel, P. Mathis and A.W. Rutherford (1989a) Characterization of four herbicide-resistant mutants of *Rhodopseudomonas viridis* by genetic analysis, electron paramagnetic resonance and optical spectroscopy. *Biochem.* **28**, 5544-5553.
Sinning, I., H. Michel, P. Mathis, and A.W. Rutherford (1989b) Terbutryn resistance in a purple bacterium can induce sensitivity toward the plant herbicide DCMU. *FEBS Lett.* **256**, 192-194.
Trebst, A. (1986) The topology of the plastoquinone and herbicide binding peptides of Photosystem II in the the thylakoid membrane. *Z. Naturforsch.* **41c**, 240-245.
Trebst, A. (1987) The three-dimensional structure of the herbicide binding niche on the reaction center polypeptides of photosystem II. *Z. Naturforsch.* **42c**, 742-750.
Trissl, H.W., J. Breton, J. Deprez, A. Dobek, and W. Leibl (1990) Trapping kinetics, annihilation, and quantum yield in the photosynthetic purple bacterium *Rps. viridis* as revealed by electric measurement of the primary charge separation. *Biochim. Biophys. Acta* **1015**, 322-333.
Trosper, T.L., D.L. Benson and J.P. Thornber (1977) Isolation and spectral characteristics of the photochemical reaction center of *Rhodopseudomonas viridis*. *Biochim. Biophys. Acta* **460**, 318-330.
Velthuys, B.R. (1981) Electron-dependent competition between plastoquinone and inhibitors for binding to Photosystem II. *FEBS Lett.* **126**, 277-281.
Wraight, C.A. (1981) Oxidation-reduction physical chemistry of the acceptor quinone complex in bacterial photosynthetic reaction centers: Evidence for a new model of herbicide activity. *Isr. J. Chem.* **21**, 348-354.

Chlorophyll Fluorescence Transients in Chloroplasts and Leaves

Jean-Marc DUCRUET* and Paul MATHIS+
CEN Saclay
France

INTRODUCTION

Photosynthesis is a complex process, the study of which requires the converging use of many experimental methods (see the Introduction in Chapter 10). As a directly detectable molecular property, fluorescence is largely used in physical chemistry. In photosynthesis, however, fluorescence is mainly (but not only) utilized as an indirect method, based on the early discovery that, in plant chloroplasts and algae, the fluorescence yield of chlorophyll *a* is mainly controlled by the redox state of the Photosystem-II (PS-II) reaction center.

Fluorescence of Pigments

Absorption of a quantum of light by a molecule promotes it to an electronically excited state, which then evolves in a complex manner as outlined in Fig. 1. This evolution of the molecular state may involve many processes, such as internal conversion, intersystem crossing, energy transfer and photochemical reactions (see Chapters 2 and 4). In a complex pigment system such as the photosynthetic apparatus, several properties of fluorescence are of interest, which are summarized in the following.

Emission and excitation spectra: These properties mainly reveal energy transfer among pigment subsets. In general, in intact photosynthetic systems, energy transfer proceeds very rapidly to the reaction centers, so that "accessory" pigments (chlorophyll *b*, carotenoids, phycobiliproteins, etc.) hardly emit any fluorescence and most of the fluorescence arises from chlorophyll *a* (bacteriochlorophyll *a* in purple bacteria). The situation is different in isolated pigment-protein complexes in which the fluorescence spectra of some accessory pigments are more significant.

Fluorescence polarization: For molecules in a highly concentrated medium, such as photosynthetic pigments bound in pigment-protein complexes, fluorescence depolarization is highly efficient, largely due to rapid and extensive energy transfer. With

JEAN-MARC DUCRUET, Département de Biologie Cellulaire et Moléculaire, (INRA - Département de Phytopharmacie), CEN Saclay, 91191 Gif-sur-Yvette, CEDEX, France
PAUL MATHIS, Département de Biologie Cellulaire et Moléculaire, (URA - CNRS n°1290) CEN Saclay, 91191 Gif-sur-Yvette, CEDEX, France

Photobiological Techniques, Edited by D.P. Valenzeno *et al.*
Plenum Press, New York, 1991

oriented samples, fluorescence polarization will occur even when excited by unpolarized light; thus providing some information on the orientation of the emitting pigments.

Time-resolved emission: For a given emitting molecule, time-resolved emission studies provide information on the duration of the singlet excited state. When that state is not populated directly, the measurements can also provide the time of formation (via energy transfer) of the emitting species.

Fluorescence yield: For a given set of identical molecules, the fluorescence yield is the result of a competition between various processes which have respective rates, k: k_{IP} for internal processes (such as internal conversion and intersystem crossing), k_F for fluorescence, k_{ET} for energy transfer from the state under consideration, and k_{Ph} for photochemical processes directly induced by that state. The overall rate of excited state decay is: $k = k_{IP} + k_F + k_{ET} + k_{Ph}$ and the fluorescence yield (FY) is given by:

$$FY = k_F \times (k_{IP} + k_F + k_{ET} + k_{Ph})^{-1} \qquad [1]$$

For example, for isolated chlorophyll *a* in solution, $k_F = 6\times10^7\ s^{-1}$, $k_{IP} = 14\times10^7\ s^{-1}$ (k_{IP} is due essentially to intersystem crossing in this case) and the fluorescence yield is 30 % (approximate figures only).

Interests of Fluorescence Measurements

Fluorescence properties are sensitive to many factors and thus emission measurements should be highly informative in pigment systems as complex as the photosynthetic membranes. Experimentally, fluorescence measurements offer distinctive advantages (Mathis, 1990):

(i) Specificity. This occurs due to the fact that only pigments, and in fact usually a single class of pigments, emit in a given wavelength region. Fluorescence studies thus do not require extensive purification of the biological material.

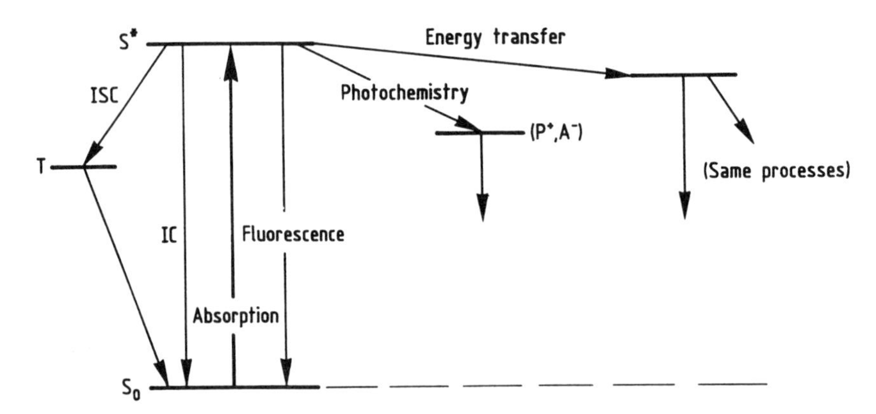

Figure 1. Energy levels of different states of a molecule. In the singlet ground state S_o, the molecule absorbs a photon that promotes it to the lowest excited singlet state, S*, from which it can evolve via various paths. T=triplet state; IC=internal conversion; ISC=intersystem crossing.

(ii) Sensitivity. In most cases a small amount of material will be sufficient to perform highly significant measurements.

(iii) Simplicity. Although very sophisticated apparatus is needed for some types of measurements (e.g., for time-resolved measurements of excited state lifetimes), many experiments are possible with rather inexpensive equipment.

On the other hand, there are a few drawbacks to this technique. These include the indirect character of most measurements, which sometimes renders the interpretations uncertain, sensitivity to a variety of factors, which accentuates the difficulty of interpretation, non-linear behavior in concentrated media, such as a leaf (or even single algal cells), etc.

Chlorophyll *a* Fluorescence and PS-II

Among the various photosynthetic structures, it is undoubtedly in the study of PS-II that fluorescence is the most useful tool (for general reviews, see Briantais et al, 1985; Govindjee et al, 1986; Lavorel and Etienne, 1972; Lichtenthaler, 1988). The photosynthetic membrane of plants includes two types of reaction centers, PS-I and PS-II. These reaction centers operate in series to transfer electrons from water to $NADP^+$ (Fig. 2). The widespread use of fluorescence in PS-II studies originates from a simple historical observation: the fluorescence yield in chloroplasts (as well as in intact cells or *a fortiori* in preparations enriched in PS-II) is largely controlled by the redox state of the primary quinonic acceptor, Q_A, of PS-II. FY is low (F_o fluorescence level) when Q_A is oxidized and FY is high (F_M fluorescence level) when Q_A is reduced to the semi-quinone form Q_A^-. The evolution from F_o to F_M is called *fluorescence induction.* A simple interpretation of this behavior is based on Eq. 1: when Q_A is reduced, the rate of photochemical reaction k_{Ph} is zero and FY is increased. This is the well-known antiparallelism of fluorescence yield and photochemical yield. In reality, the situation is not that simple, since it is now known that when Q_A is reduced, charge separation can, in principle, still take place in PS-II between the primary donor P-680 and the primary acceptor, a molecule of pheophytin *a* (see Fig. 1 in Chapter 10).

Whatever the precise reason, it is well established that FY is controlled mainly by the redox state of Q_A and, as a consequence, electron transfer from Q_A can be probed by measuring the evolution of FY in response to illumination. Three properties are studied in this manner: i) electron transfer from Q_A to Q_B and its interruption by herbicide inhibitors (see Chapter 10); ii) the size and the redox state of the pool of plastoquinones which accepts electrons from the Q_B site; iii) the functional state of the cytochrome b6/f complex and of PS-I. In fact, additional PS-II properties influence FY and may contribute to its low value. For this reason they are named "quenchers" (q), and will be discussed further, below.

H_2O...$(Mn)_4$ ...Tyr_z ...P-680 ...I ...Q_A ...Q_B ...(PQ pool)→...→to PS-I→$NADP^+$

Donor side inhibitors | DCMU-type inhibitors | rate limiting steps

Figure 2. A linear scheme of electron transfer in plant-type photosynthesis, from the water-oxidizing $(Mn)_4$ cluster to the PS-I reaction center and to $NADP^+$. The electron carriers from $(Mn)_4$ to Q_B belong to the PS-II reaction center.

Analysis of Fluorescence Transients

Many years ago it was reported that illumination of plant leaves or algae induces an initial increase of red chlorophyll fluorescence emission, followed by a decrease. This variable fluorescence intensity is a specific property of PS-II. The emission spectrum peaks at 685 nm, with a shoulder at 730 nm mainly corresponding to the PS-I constant fluorescence.

The most significant property of variable fluorescence is the fluorescence induction observable after the onset of continuous illumination onto a dark-adapted material, in which the classical OIDPMST phases may be detected (the letters correspond to successive, more or less well defined, levels of fluorescence yield). Analysis of this pattern allowed Duysens and Sweers (1963), to propose that, during the first few seconds of illumination, the fluorescence yield is controlled by the redox state of the primary electron acceptor Q_A. Approximately 300 chlorophyll molecules form an antenna which absorbs and conveys photons to one PS-II center. In the oxidized state, Q_A, the PS-II center, is able to convert a quantum of energy to perform a charge separation, so that the energy of chlorophyll-absorbed photons is rapidly trapped and the corresponding fluorescence decay is quenched. When Q_A is reduced to Q_A^-, photons absorbed by the chlorophyll antenna cannot be converted to chemical energy and their probability of decay by fluorescence emission increases. Hence, FY reflects the Q_A^-/Q_A ratio. However, there is a non-linear relationship between FY and Q_A reduction. When a PS-II center is closed (Q_A^-), excitation energy can migrate to a neighboring center and be trapped if that center is "open" (Q_A oxidized), resulting in lower than expected FY on the simple basis of the Q_A redox state (Lavorel and Joliot, 1972).

FY is the ratio of the number of photons emitted by fluorescence to the number of photons absorbed. In a given experimental set-up, one can define a relative fluorescence yield as the ratio F/I, where F is a measured fluorescence intensity and I is the excitation intensity. If I is unchanged, F can be used as a relative measure of FY. F is found to lie betwen two values:

(i) The "constant" fluorescence F_0, where FY is minimum, corresponding to entirely oxidized Q_A (open centers). This can be measured immediately at the onset of an illumination (O level), in a dark-adapted material, or under a weak light, low enough that Q_A^- is populated to a negligible extent.

(ii) The maximum fluorescence F_{max} or F_M, where FY is maximum, can be reached under a strong illumination. In the presence of an inhibitor, such as those used in Chapter 10, Q_A can be entirely reduced by illumination (100% closed centers), but the FY is slightly less than F_M because the pool of plastoquinones remains oxidized and contributes to some quenching. In PS-II studies, one often uses diuron (DCMU) as an inhibitor. It allows Q_A to be completely reduced to Q_A^- by blocking its reoxidation (the inhibitor binds at the Q_B site).

In the presence of diuron, the fluorescence increase is a simpler process corresponding to the progressive reduction of Q_A.In the absence of inhibitor, the fluorescence increase is a biphasic process, which can be explained by the reduction of two successive electron acceptor pools: the primary acceptor Q_A and the plastoquinone pool (for the purpose of interpretation of fluorescence induction, it can be estimated that the plastoquinone bound at the Q_B site is part of a homogeneous pool of plastoquinones, about 10 per PS-II center, which plays a role in transferring electrons to the PS-I center, via the cytochrome b6/f complex). The size of the electron accepting pool can be correctly measured by integrating the quenching values over the fluorescence increase (complementary area).

Factors other than the Q_A^-/Q_A ratio control the FY, and recent advances have been made in the resolution of different types of "quenching" (quenching is a well-defined property which decreases the FY). For this purpose, several types of experimental devices can be used. In a now widely used apparatus (Schreiber, 1986; Schreiber et al., 1986), fluorescence is excited by a modulated light beam (analytical light) and the FY is measured through a lock-in amplifier, in phase with the modulated analytical beam. With this instrument, additional continuous illuminations or strong flashes (actinic light) can be applied without disturbing the measurement of the fluorescence level by the modulated beam.

The quenching due to oxidized Q_A is the redox or photochemical quenching, qQ. After a strong flash, Q_A can be entirely reduced to Q_A^- and the flash-induced fluorescence increase provides an estimate of the qQ quenching. However, under these conditions, F_{max} is generally not attained and the remaining fluorescence deficit has been ascribed to a non-photochemical quenching termed qNP. This can be further resolved into:

(a) Electrochemical quenching, qE, (time domain > 1 s), associated with the influx of protons into the lumen of thylakoids during illumination (Quick and Horton, 1984).

(b) Spill-over, which is the name given to excitation energy transfer from the PS-II to the PS-I antenna. This transfer is increased by phosphorylation of the light-harvesting antenna, which is itself related to the reduction of the plastoquinone pool, via the activation of a kinase (Allen et al., 1981).

(c) Photoinhibition quenching, qI, (time domain > 1 min), which is associated with a rather severe damage to the PS-II reaction center, a process termed photoinhibition. This damage is normally reversible and can be considered as a protective mechanism against strong light.

Consider the initial increase of the fluorescence induction (OIP: see Fig. 3) which corresponds to the first seconds of the induction kinetics. This time domain is short enough to prevent the non-photochemical quenchings to build up significantly, and the kinetics essentially reflect the Q_A^-/Q_A ratio.

By careful choice of experimental conditions, the dominant qQ dependence allows simple and inexpensive devices to be used to record and analyze the fluorescence induction. Classically, fluorescence (around 685 nm) was detected by a photomultiplier with a high-voltage supply and the signal was displayed on an oscilloscope screen. The development of solid-state light detectors (silicon photodiodes) and analog to digital (A/D) converters plugged into a microcomputer have been very useful advances. The signal, stored in the microcomputer memory, can be rapidly analyzed and stored on disk.

The experimental properties discussed above make the fluorescence OIP increase, which is still useful for a global monitoring of the PS-II system in photosynthetic research, a suitable tool to study several important physiological properties (Krause and Weis, 1984), including the following.

1. DCMU-like herbicides still represent the major group of chemical weed killers in agriculture. This class of herbicides includes other substituted ureas (chlortoluron for wheat), amides (propanil for rice), uracils, triazines (atrazine for corn), triazinones (metribuzin or metamitrone for beet-root), phenolics (ioxynil) and many other chemical families. They have the common property of displacing plastoquinone at the Q_B site, thus blocking the reoxidation of Q_A, and enabling the fluorescence OIP increase to be used as a direct probe of the *in situ* herbicide concentration in whole plants or leaves. This has been used to monitor the translocation and differential detoxification of chlortoluron in susceptible or tolerant wheat cultivars (Cadahia et al., 1982). Another application is the penetration of herbicides through leaf cuticles (Voss et al., 1984). Fluorescence increase provides fast, accurate and non-de-

structive measurements which are complementary to the more time-consuming and destructive chemical analyses of herbicide in leaves. Fluorescence induction also provides a convenient assay for the activity of a given chemical or competitive inhibitor at the Q_B site, and for properties of herbicide-resistant mutants.

2. Other chemicals affecting PS-II do not act as directly on the OIP increase as DCMU-like herbicides, but their effects can provide a qualitative estimation of their presence in leaves. One example is given by viologen herbicides (diquat and paraquat), which accept electrons from PS-I to form toxic radicals. They favor a rapid reoxidation of the PQ pool (see Fig. 2) and they suppress the I→P fluorescence increase (Lavergne, 1974). This property has been used to study paraquat resistance in plants. Heavy metals, such as copper, mercury, zinc, etc. or aromatics, such as m-dinitrobenzene (Etienne and Lavergne, 1972), also modify the fluorescence properties.

3. Environmental factors such as temperature changes, photoinhibition, drought, or salinity, also change the fluorescence kinetics (see e.g. Krause and Weis, 1984; Schmidt et al., 1988; Somersalo and Krause, 1990; Yamashita and Butler, 1968). It is now well established that the ratio of the variable fluorescence to the maximum fluorescence, F_v/F_{max}, is a good indicator of the efficiency of excitation capture by open PS-II reaction centers (Genty et al., 1989). More generally, the overall fluorescence induction curve can be related to the photosynthetic capacity of a leaf (see e.g., Bradbury and Baker, 1981; Genty et al., 1989).

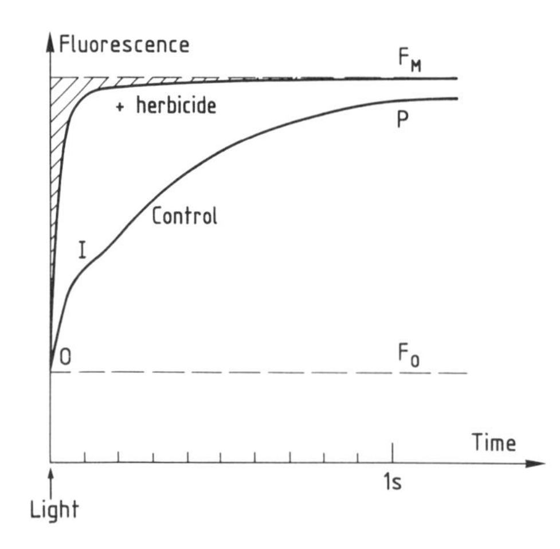

Figure 3. Typical curves recorded in fluorescence induction. At the arrow, light is turned on, and the induced fluorescence is recorded as a function of time. Measurements were done on spinach chloroplasts with no addition, or with 5 μM diuron. The cross-hatched area is the so-called "complementary area", in the presence of inhibitor. In the absence of inhibitor, there is a similar (but much larger) complementary area (not cross-hatched for clarity).

EXPERIMENT 16: MEASUREMENT OF CHLOROPHYLL FLUORESCENCE TRANSIENTS IN CHLOROPLASTS AND LEAVES

OBJECTIVES

This experiment will demonstrate light-induced variations of chlorophyll fluorescence yield in isolated chloroplasts and the influence of several perturbants: the herbicide DCMU, metal ions (Cu^{2+}), and partial inhibition of PS-II by heating. Some of the experiments will also be performed with a leaf.

Chloroplasts will be isolated from leaves from a local plant (spinach, pea, lettuce or barley are well suited) and suspended in a cuvette. The cuvette will be suddenly illuminated, using a photographic shutter (light filtered to remove wavelengths above 650 nm). Fluorescence from the sample (broadband above 680 nm), will be detected with a silicon photodiode and stored by a signal recorder. In the absence of fluorescence induction, the fluorescence signal should have a step-like shape. Fluorescence induction will be observed in the control chloroplast suspension. The experiment will then be repeated after addition of DCMU, or of $CuSO_4$, or after heating the chloroplasts to 40-45°C for 5 min. It will also be repeated using a leaf in place of the chloroplast suspension.

Fluorescence induction will be analyzed to determine the number of electrons stored on the PS-II acceptor side in the control (no addition) or after addition of DCMU. It is also possible to record the waveform of the fluorescence induction signal following addition of increasing DCMU concentrations and thus to determine the I_{50} of the inhibitor.

LIST OF MATERIALS

Spinach, pea or barley leaves, 20 g
Blender, standard kitchen variety
Cold room or refrigerator
Heated water bath for 5 ml test tubes, 40-50°C
Gauze or cheese-cloth, about 1 m^2
Centrifuge, 5 000 × g, refrigerated
Crushed ice
Potter homogenizer or small paint brush
Glycerol
Spectrophotometer
Cuvette, 1 cm path length, fluorescence, glass or plastic
Tungsten-halogen lamp, 100-200 W
Sodium dithionite
$CuSO_4$
Ethanol
HEPES (N-2-hydroxyethyl) piperazine-N'-(2-ethanesulfonic acid)
Optical filter, Calflex C (Balzers), 50 × 50 mm
Optical filter, CS-4-96 blue-green (Corning), 50 × 50 mm
Optical filter, CS-2-64 (Corning), 50 × 50 mm
Lens, 100 mm focal length, 40 mm diameter (2)
Silicon photodiode, 10 mm diameter, with amplifier (UDT, "photops" type 555D)
Batteries, 15V (2)
Switch

Light baffles (pieces of cardboard, black cloth, etc.)
Shutter, opening ≥ 15 mm (electronic triggering preferable)
Microcomputer with A/D conversion card (or digital or storage oscilloscope, DC to 3 kHz)
Graduated cylinder, 50 ml
100 ml beakers
50 ml beakers
Funnel
Test tubes, 5 ml
Bottles, 5-100 ml
Automatic pipetors (100 µl to 2 ml)
Hamilton syringes, 10 µl (2)
Diuron (DCMU, 3-3,4-dichlorophenyl-1,1-dimethylurea)

PREPARATIONS

Buffer Solutions:

Medium A — 0.3 M sucrose; 50 mM NaCl; 50 mM phosphate buffer at pH 7.5 – Dissolve 103 g of sucrose, 2.9 g of NaCl, 1.1 g of NaH_2PO_4 and 11.3 g of $Na_2HPO_4{\cdot}7H_2O$ in 1 liter of distilled water.

Medium B — 50 mM NaCl, 50 mM HEPES buffer at pH 7.5 – Dissolve 2.9 g of NaCl and 11.9 g of HEPES in 1 liter of distilled water, and adjust the pH to 7.5.

Medium C — 100 mM sucrose, 10 mM NaCl, 5 mM $MgCl_2$, 20 mM HEPES at pH 7.5 – Dissolve 34.2 g of sucrose, 0.58 g of NaCl, 1.0 g of $MgCl_2{\cdot}6H_2O$ and 4.8 g of HEPES in 1 liter of distilled water, and adjust the pH to 7.5.

Medium D — Same as A, but at pH 6.5 – Dissolve 103 g of sucrose, 2.9 g of NaCl, 4.8 g of NaH_2PO_4 and 4.2 g of $Na_2HPO_4{\cdot}7H_2O$ in 1 liter of distilled water.

DCMU Stock Solution: Dissolve 2.3 mg of DCMU in 10 ml of ethanol to produce a 1 mM solution. Take 1 ml of that solution and dilute it with 9 ml of ethanol to obtain a 0.1 mM DCMU stock solution.

$CuSO_4$ Stock Solution: Dissolve 25 mg of $CuSO_4$ in 10 ml of water to obtain a 10 mM stock solution.

Chloroplasts: In a cold room, cut 20 g of washed spinach leaves into pieces and grind for 10 s in a blender with 50 ml of cold medium A. If a cold room is not available, pre-cool the blender bowl in a refrigerator. Filter the liquid quickly, through 4 layers of gauze or cheese-cloth and centrifuge the filtrate for 5 min at 2000 × g (in a refrigerated centrifuge). Keep the pellet on ice and then homogenize it (in a Potter homogenizer or with a small paint brush) with 20 ml of medium B . In this suspension, the chloroplasts will swell and their outer envelopes will break down. Centrifuge the suspension for 10 min at 5000 × g. Resuspend the pellet in 2 ml of medium C, homogenize it and keep it on ice. To estimate the amount of chloroplasts needed, mix 2 ml of water and 2 ml of glycerol in a 1 cm path cuvette. Then add a volume, v, of the suspension, so that the optical density at 678 nm is about 0.4 absorbance units greater than at 740 nm. The measurements will then be done with 2 ml of medium C and 0.3 × v ml of chloroplast suspension (chlorophyll concentration around 4.5 µM). To study the effect of heating, use medium D.

Experimental Set-Up: An outline of the set-up is given in Fig. 4. The chloroplasts are contained in a four-window 1 × 1 × 4.5 cm cuvette (glass or plastic). The geometry shown in

Fig. 4 is chosen to be fully compatible with the absorption change experiment of Chapter 10. If only fluorescence measurements have to be performed, it will be preferable to use a 2-mm thick plexiglass cuvette placed at 45° to the incident light. When working with a leaf, place the cuvette at 45°, in the cuvette holder, attached with tape to a piece of plastic.

The same light beam has both an actinic and a monitoring role in this set-up. It is provided by a 100-200 W tungsten-halogen lamp, and is filtered by F_1 (a Calflex C (Balzers) and a CS-4-96 blue-green filter (Corning); both 50×50 mm). The light is focused by the lens L_1 (f = 100 mm, diameter = 40 mm) on the cuvette, after passing the F_L filter ($CuSO_4$ in water, 250 $g \cdot l^{-1}$). Light emitted by the cuvette is collected by the lens L_2 (same as L_1) and is focused on the photodetector, which is protected by the filter F_2 (CS-2-64, Corning); the detector is a silicon photodiode (diameter = 10 mm), equipped with an integrated amplifier (UDT, "photops" type 555 D). The amplifier is wired as indicated by the manufacturer, in the photoconductive mode (two 15 V batteries are required; a common switch should be inserted to avoid discharging the batteries between measurements). The output voltage due to fluorescence is about 10 mV. The optical part of the set-up should be covered with a black material to block room light and pieces of cardboard placed to stop the lamp light from leaking to the detector and to the cuvette.

The light from the lamp is blocked from reaching the cuvette by a shutter. The shutter can be actuated manually, but it is more convenient to start its opening by an electric pulse (as in Chapter 10) in order to synchronize the opening with the start of the recorder with an accuracy of about 1 ms or less.

The signal recorder will, preferentially, be a microcomputer equipped with an A/D conversion card (see the discussion of recorder properties in Chapter 10). A digital oscilloscope or a storage oscilloscope can be used as well. Required electrical bandwidth: DC to 3 kHz.

EXPERIMENTAL PROCEDURE

1. Check the alignment and functioning of the optics and electronics.
2. Prepare a cuvette with medium C and chloroplasts, as indicated above (Preparation, Chloroplasts).
3. Record the fluorescence and adjust the lamp voltage, recorder sensitivity and recorder time scale, if necessary.

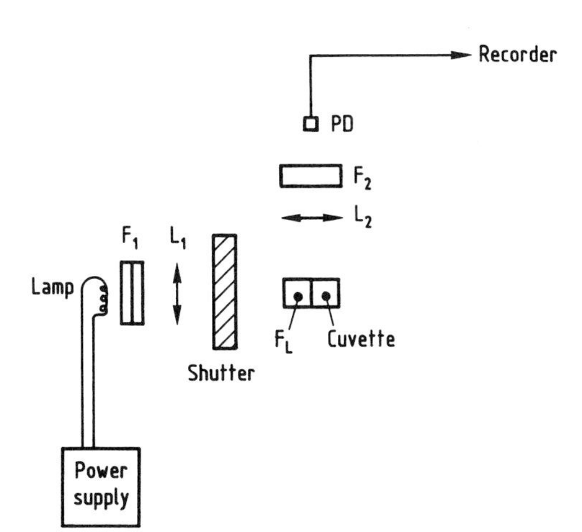

Figure 4. Schematic of the experimental set-up used for the measurement of fluorescence induction. F_1, F_2 = optical filters; L_1, L_2 = lenses; F_L = liquid filter; PD = silicon photodiode.

4. Prepare a fresh cuvette under dim light. Keep it dark for 3 min and repeat the measurement: this gives the fluorescence induction in the control. Everything should stay the same for the other measurements (lamp voltage, recorder settings, chloroplast concentration, . . .). Be careful since chloroplasts sediment. The stock suspension should be mixed before taking an aliquot.

Effect of the Inhibitor, DCMU

5. Repeat the experiment after adding 10 μM DCMU. Compare the fluorescence time course with that of the control.
6. Rinse the cuvette carefully (water, ethanol, water).
7. With a freshly washed cuvette for each concentration, repeat the experiment with different concentrations of DCMU (0.1, 0.2, 0.5 μM etc.).

Effect of $CuSO_4$

8. Repeat the experiment in a freshly washed cuvette with 10 μM DCMU and 100 μM $CuSO_4$.
9. Repeat the experiment in freshly washed cuvettes with 10 μM DCMU and increasing concentrations of $CuSO_4$ (between 20 and 200 μM).

Effect of Heating

10. Place the contents of a cuvette in a 5 ml test tube. Place the test tube in the heating bath, set at 43°C. The two parameters to be studied are the bath temperature and the heating time. Either parameter can be studied: temperature between 40 and 45°C, time between 30 s and 10 min.
11. After the heating time, dip the test tube into water at room temperature, and then transfer its contents to the cuvette for the measurement.

Control Experiments

12. Pipette 10 μl of chloroplasts into a cuvette.
13. Add 2.5 ml of ethanol. Wait a few minutes for the solubilization of chlorophyll. (The contents of the cuvette should be green and clear).
14. Record the fluorescence. It should have a step-like time profile. The rise of the step corresponds to the response time of the apparatus, which is essentially determined by the shutter opening.
15. Place a piece of white paper in an empty cuvette (at 45° to the incident beam, so as to reflect the beam toward the detector).
16. Record the "fluorescence" signal. It should be of negligible amplitude (good complementarity of filters; no light leak).

DISCUSSION

The signals which are registered display obvious properties which can be interpreted visually. A quantitative treatment is, however, more desirable. Using an oscilloscope necessitates a manual analysis of a photograph of the screen. Microcomputers offer considerable advantages in the ease and accuracy of signal analysis. The signal may first be stored in a digitizing device and then transferred to a computer through a classical data transmission line (e.g., IEEE). However, when the sampling frequency is low enough (<100 kHz or 10 μs per channel), the use of A/D conversion cards represents a much cheaper and faster solution (Ducruet et al., 1984), since the analogic (voltage) signal is digitized and stored directly in the large RAM memory.

Untreated Chloroplasts or Leaves

The fluorescence induction should behave approximately as shown in Fig. 5. It may happen, however, that the inflexion, I, is not prominent. The presence of a well-defined I level depends greatly on the excitation light intensity. It may be useful to vary the intensity (by trial and error) to reach a value that leads to a satisfactory induction curve. The 0 level is not reached immediately, because of the shutter opening time.

Effect of DCMU

This effect should be fully reproducible. The F_M/F_o ratio is sensitive to many factors such as the ionic composition of the medium (this can be checked by using a medium without Mg^{2+}: the F_M/F_o ratio will be smaller) or the integrity of PS-II and of the antenna. The latter is conditioned by the history of the leaf, which includes age (and biosynthetic course), and the environmental conditions which determine the photoinhibition and the level of antenna phosphorylation. It has also been established that the fluorescence induction in the presence of DCMU is not a simple exponential. This is due to heterogeneity in the PS-II centers (Melis and Homann, 1976; Melis and Duysens, 1979).

Effect of Cu^{2+}

This effect is studied in the presence of DCMU, to simplify interpretation of the observations. F_M decreases with progressive addition of $CuSO_4$, but F_o is not significantly modified. Control experiments would easily show that the choice of the buffer is critical in

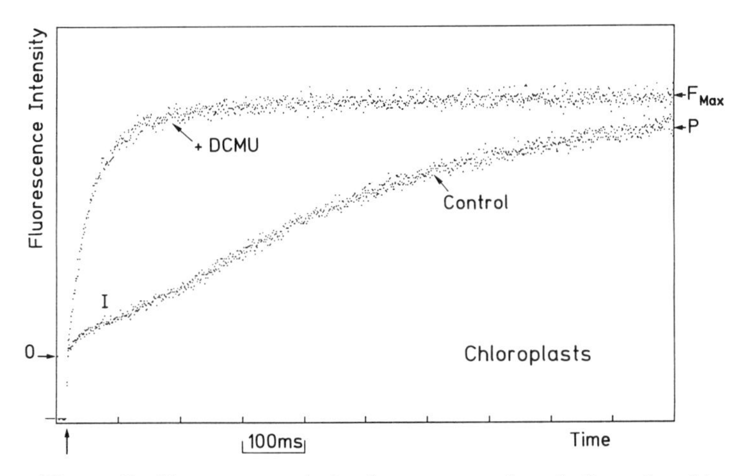

Figure 5. Fluorescence induction measured as indicated, with spinach chloroplasts without (lower curve) and with 5 μM diuron. The shutter opens at the arrow.

this experiment, because many chemicals tend to react with Cu^{2+} (Renganathan and Bose, 1990). The effect of Cu^{2+} takes place at low concentration, showing the pronounced effect of heavy metal ions. Although this has not yet been fully rationalized, it is believed that metal ions act primarily at the Q_B site (Mohanty et al., 1989).

Effect of Heating

A short and controlled heating induces a small change in the F_0 level and a more dramatic effect on the induction curve. A detailed study of the effects, however, requires careful control of the ionic composition of the buffer, the pH, and the time/temperature parameters. The effects are not fully understood (see e.g., Weis, 1982 and Ducruet and Ort, 1988). The susceptibility to heating depends on plant species and on plant history (which may include more or less heat-hardening). It has been established that herbicide-resistant mutants are more susceptible to heating (Ducruet and Ort, 1988).

Time Range of the Effects

The initial fluorescence increase has a photochemical behavior, i.e., the rate of fluorescence increase is proportional to the excitation light intensity. A few perturbations, however, may modify this simple behavior. If the light intensity is very strong (Neubauer and Schreiber, 1987; Schreiber and Neubauer, 1987), the fluorescence increase may be disturbed by two short-lived quenchers: the carotenoid triplet state (lifetime of a few µs) and the oxidized state of P-680, the lifetime of which is essentially sub-µs, but also includes slow phases up to 200 µs. It has been established that the triplet state of carotenoids and P-680^+ are efficient chlorophyll fluorescence quenchers. At low light intensity, conversely, Q_A^- may be reoxidized by mechanisms other than forward electron transfer: the back reaction between Q_A^- and the PS-II electron donors can be on the order of a few seconds.

Kinetic Distortions

In principle the measurements are very sensitive to heterogeneities in the excitation intensity reaching each part of the cuvette. These heterogeneities result in non homogeneous kinetics. These can be reduced by three means: i) keep the chlorophyll concentration as low as possible; ii) use a thin cuvette and front-surface excitation; iii) excite at a wavelength of low absorption. A compromise has to be found between these requirements and the signal-to-noise ratio.

Quantitative Analysis of the Induction Curve

O level: This value is of importance since it represents the constant fluorescence in a dark-adapted material. The O level is not always reached immediately since the opening of the optical shutter requires a few ms, and a significant induction may take place during that time. The shutter opening time can be estimated from the fluorescence measurement with chlorophyll in ethanol. An estimation of O can be obtained by computing a linear regression on the first significant points and extrapolating to the half-opening time of the shutter.

I level: This level can be defined by averaging the signal around the inflection point of control kinetics. This should also correspond to the beginning of the plateau in DCMU-saturated measurements, although a slow long-lasting increase persists under these conditions. The time position of I could be automatically computed for each measurement, but it turns out that preset limits for I averaging is more reliable.

P level: This level is theoretically close to F_{max} when illumination is high enough. It is detected when two successive values of the signal are equal. However, noise can induce a precocious detection of P, and only averaged values (10 to 100 ms, depending on noise) should be compared.

The half-time of the photochemical increase, OI, can provide information about the size and connection of the light-collecting antenna. It is nearly proportional to the reciprocal of the light intensity.

The "complementary area" is defined as the area bordered by the fluorescence induction curve, the Y ordinate and the final (maximum) fluorescence level (hatched in Fig. 3). It can be computed by integrating the signal for times between O to P levels, then substracting it from the signal-containing rectangle of height P. These measurements can also be done with photographs or paper plots. The complementary area is a useful parameter since it is linearly dependent on the size of the electron-accepting pool of PS-II.

When leaves are studied (Fig. 6), the absolute values of O, I, and P can vary considerably between samples (for example, when a vein is present!). For this reason, ratios between level values have to be used. O or P levels are generally taken as references. O is theoretically more suitable since P is not exactly equivalent to F_{max}. However, P provides better accuracy, since it is averaged over a greater time period.

Technical Alternatives and Improvements

The basic set-up described here is not fully optimized for optical performance. It is an operative set-up which utilizes a large part of the flash absorption apparatus described in the previous chapter. Many improvements can be envisioned, such as those listed below.

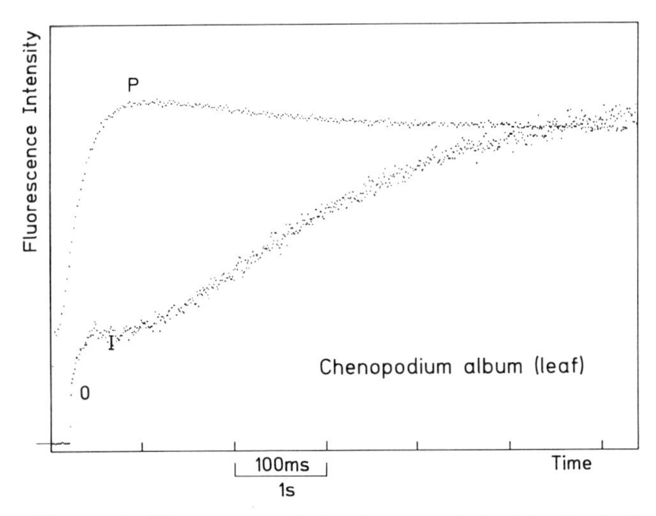

Figure 6. Fluorescence induction recorded with a leaf of the weed lambsquarter (*Chenopodium album*). The whole recording (6 s) is presented with an averaging over 20 successive data points; the first 600 ms are shown on a 10 × expanded scale (1 ms per point). Fluorescence was excited with an He-Ne laser, and measured with a red-sensitive photomultiplier. The signal was recorded and handled with a PC microcomputer.

Light source: Continuous illumination was used to both induce a fluorescence increase (actinic role) and to measure the relative fluorescence yield (analytic role). The light intensity has to be strong enough so that fluorescence rapidly reaches a P level (Fig. 5) close to F_{max}. It should be at least 100 $\mu E \cdot m^{-2} \cdot s^{-1}$ of blue or red chlorophyll-absorbed light. Since fluorescence is detected in the red (680-750 nm), illumination must be at shorter wavelengths. Three kinds of light sources can be used;

(i) A tungsten-halogen lamp (as described above) or a xenon arc lamp, filtered by a blue filter (a liquid filter with $CuSO_4$ 250 $g \cdot l^{-1}$ provides a broad illumination spectrum). The main drawback of this light source is that it requires a photographic shutter, which takes several ms for full opening. This makes an accurate estimation of the O level more difficult.

(ii) Light-emitting diodes (LEDs) with high luminosity have recently become available. It is possible to focus several LEDs (up to 50) on the sample. They provide instantaneous illumination which is noise free (when powered by batteries). However, their emission peaks near 660 nm and complementary filters are needed, shifting fluorescence detection to above 700 nm. This is a disadvantage since long-wavelength fluorescence includes a greater PS-I contribution, which does not display induction behavior.

(iii) Lasers such as helium-neon (633 nm) or argon (488 nm) provide a rather stable continuous beam. A converging lens is needed, which first focuses the beam (node) and then spreads it to a larger size (5 mm). An electromagnetic shutter placed at the beam node gives full intensity illumination in less than 1 ms. Since laser emission is monochromatic, the full fluorescence emission (670 to 750 nm) can be detected through a broadband red filter (Corning CS-2-64) or specific wavelengths can be selected by a monochromator or interference filters.

Cuvette: With a set-up like that of Fig. 4, it is possible to use a thin cuvette (1 or 2 mm) placed at 45° to both the excitation beam and the detector. With a thin cuvette, it is also convenient to use a bifurcated (Y-shaped) light pipe. One branch serves for excitation, the other branch for detection. The common branch is applied against the cuvette.

Biological material: In addition to chloroplasts or leaves, experiments can also be done with all kinds of algae or cyanobacteria.

Related Laboratory Experiments

Detoxification by corn leaves: Dip pieces of corn leaves (2 × 2 cm) in a 100 μM solution of the herbicide atrazine (in ethanol) under weak room light (there should be enough light to keep the stomata open). Let them stand for at least 2 h. Remove the pieces and rinse them with water. Then put a piece of leaf in the cuvette holder (with the bottom of the piece in water) and measure the fluorescence induction. Repeat the experiment with the same piece kept in darkness, every 30 min. The induction will be very fast at the beginning (atrazine blocks electron transfer from Q_A to Q_B), and will then slow progressively. This is because atrazine is metabolized into an inactive product. This process is operant in corn. It explains why atrazine can be used as a weed killer in corn fields.

Effect of light intensity: Study the induction at several light intensities, by placing neutral density filters in front of the cuvette. The experiment can be done with or without inhibitor.

REVIEW QUESTIONS

1. With isolated chlorophyll *a* in ethanol solution, the fluorescence yield is about 30%. In chloroplasts the fluorescence yield is about 1.5% at F_0 and 7% at F_M. Why is the FY at F_M much lower than in solution ?
2. Why is there a fluorescence induction, since primary charge separation can still take place when Q_A is reduced ?
3. Why do we suggest a fluorescence experiment with a plant material (this chapter) and an absorption experiment with a bacterial material (previous chapter) to study a very similar problem?
4. At very low temperature, e.g., 77 K in liquid nitrogen, the fluorescence induction displays a simple photochemical behavior, similar to that observed with diuron at room temperature. Explain this property.

ANSWERS TO REVIEW QUESTIONS

1. In the F_M state, many sources of quenching remain, that keep the FY rather low. The primary charge separation can still be effected: $P^* \rightarrow P^+ I^-$. Energy is transferred to the PS-I reaction center which is always a trap for energy. Fluorescence quenching takes place in antenna complexes, perhaps because of interaction between chlorophyll molecules.
2. When Q_A is reduced, the radical pair $P^+ I^-$ can be formed (P is the primary donor P-680. I is the pheophytin. See Fig. 2 of this chapter and Fig. 1 of Chapter 10). If that state had a low energy, there would be a strong fluorescence quenching. However, the energy of $P^+ I^-$ is nearly the same as that of P^*. The reaction is reversible and the excitation partly returns to the antenna where it eventually gives rise to fluorescence.
3. In bacteria, the photooxidation of P is easy to study by absorption (well-defined absorption bands, long lifetime of P^+) and difficult by fluorescence, because the fluorescence yield is very low and sensitive to contamination. It is nearly the same in plants with PS-I. In PS-II, however, fluorescence transients are very characteristic, whereas P-680 oxidation is difficult to measure, essentially because it is rereduced mainly in the sub-μs time range.
4. Below about –50°C, electron transfer from Q_A^- to Q_B is blocked, whereas electron transfer to P-680^+ still functions, although in a modified, temperature-dependent manner. With untreated chloroplasts or leaves, the fluorescence induction at 77 K is thus comparable to that observed at room temperature when electron transfer from Q_A^- to Q_B is prevented by the addition of an inhibitor.

SUPPLEMENTARY READING

Fork, D.C. (1989) Photosynthesis. In *The Science of Photobiology*, 2nd edition (Edited by K.C. Smith), pp. 347-390. Plenum Press, New York.

REFERENCES

Allen, J.F., J. Bennett, K.E. Steinback and C.J. Arntzen (1981) Chloroplast protein phosphorylation couples plastoquinone redox state to distribution of excitation energy between photosystems. *Nature* **291**, 25-29.

Bradbury, M and N.R. Baker (1981) Analysis of the slow phases of the *in vivo* chlorophyll fluorescence induction curve. Changes in the redox state of Photosystem II electron acceptors and fluorescence emission from Photosystem I and II. *Biochim. Biophys. Acta* **63**, 542-551.

Briantais, J. M., C. Vernotte, G.H. Krause and E. Weis (1985) Chlorophyll fluorescence of higher plants: chloroplasts and leaves. In *Light Emission by Plants and Bacteria* (Edited by A.J. Govindjee and D.C. Fork), Academic Press, NY.

Cadahia, E., J.M. Ducruet and P. Gaillardon (1982) Whole leaf fluorescence as a quantitative probe of detoxification of the herbicide chlortoluron in wheat. *Chemosphere* **11**, 445-450.

Ducruet, J.M., P. Gaillardon and J. Vienot (1984) Use of chlorophyll fluorescence induction kinetics to study translocation and detoxication of DCMU-type herbicides in plant leaves. *Z. Naturforsch.* **39c**, 354-358.

Ducruet, J.M. and D.R. Ort (1988) Enhanced susceptibility of photosynthesis to high leaf temperature in triazine-resistant *Solanum nigrum L.* Evidence for Photosystem II D1 protein site of action. *Plant Science*, **56**, 39-48.

Duysens, L.N.M. and H.E. Sweers (1963) Mechanism of two photochemical reactions in algae as studied by means of fluorescence. In *Studies on Microalgae and Photosynthetic bacteria*, pp. 353-372, University of Tokyo Press, Tokyo.

Etienne, A.L. and J. Lavergne (1972) Action du m-dinitrobenzene sur la phase thermique d'induction de fluorescence en photosynthèse. *Biochim. Biophys. Acta* **283**, 268-278.

Genty, B., J.M. Briantais and N.R. Baker (1989) The relationship between the quantum yield of photosynthetic electron transport and quenching of chlorophyll fluorescence. *Biochim. Biophys. Acta* **990**, 87-92.

Govindjee, A.J. and D.C. Fork (1986) *Light emission by plants and bacteria.* Academic Press, NY.

Krause, G.H. and E. Weis (1984) Chlorophyll fluorescence as a tool in plant physiology. Interpretation of fluorescence signals. *Photosynth. Res.* **5**, 139-157.

Lavergne, J. (1974) Fluorescence induction in algae and chloplasts. *Photochem. Photobiol.* **20**, 377-386.

Lavorel, J. and A.L. Etienne (1972) *In vivo* chlorophyll fluorescence. In *Primary Processes of Photosynthesis.* (Edited by J. Barber), pp. 203-268. Elsevier (North Holland Biomedical Press).

Lavorel, J. and P. Joliot (1972) A connected model of the photosynthetic unit. *Biophys. J.* **12**, 815-831.

Lichtenthaler, H.K. (1988). *Applications of Chlorophyll Fluorescence in Photosynthesis Research, Stress Physiology, Hydrobiology and Remote Sensing.* Kluwer Academic Publ., Dordrecht.

Mathis, P. (1990) Optical techniques in the study of photosynthesis. In *Methods in Plant Biochemistry:, Vol. 4: Lipids, Membranes and Light Receptors.* (Edited by J. Bowyer and J. Harwood) pp. 231-258, Academic Press.

Melis, A. and L.N.M. Duysens (1978) Biphasic energy conversion kinetics and absorbance difference spectra of Photosystem II of chloroplasts. Evidence for two different Photosystem II reaction centers. *Photochem. Photobiol.* **29**, 373-382.

Melis, A. and P.H. Homann (1976) Heterogeneity of the photochemical centers in system II of chloroplasts. *Photochem. Photobiol.* **23**, 343-350.

Mohanty, N., I. Vass and S. Demeter (1989) Copper toxicity affects Photosystem II electron transport at the secondary quinone acceptor Q_B. *Plant Physiol.* **90**, 175-179.

Neubauer, C. and U. Schreiber (1987) The polyphasic rise of chlorophyll fluorescence upon onset of strong illumination: I. Saturation characteristics and partial control by the Photosystem II acceptor side. *Z. Naturforsch.* **42c**, 1246-1254.

Quick, W.P. and P. Horton (1984) Studies on the induction of chlorophyll fluorescence quenching by redox state and transthylakoid pH gradient. *Proc. Roy. Soc. (Lond.) B* **220**, 371-382.

Renganathan, M. and S. Bose (1990) Inhibition of Photosystem II activity by Cu^{++} ion. Choice of buffer and reagent is critical. *Photosynth. Res.* **23**, 95-99.

Schmidt, W., U. Schreiber and W. Urbach (1988) SO_2 injury in intact leaves, as detected by chlorophyll fluorescence. *Z. Naturforsch.* **43c**, 269-274.

Schreiber, U. (1986) Detection of rapid induction kinetics with a new type of high-frequency modulated chlorophyll fluorometer. *Photosynth. Res.* **9**, 261-272.

Schreiber, U. and C. Neubauer (1987) The polyphasic rise of chlorophyll fluorescence upon onset of strong continuous illumination: II. Partial control by the Photosystem II donor side and possible ways of interpretation. *Z. Naturforsch.* **42c**, 1255-1264.

Schreiber, U., U. Schliwa and W. Bilger (1986) Continuous recording of photochemical and non-photochemical chlorophyll fluorescence quenching with a new type of modulation fluorometer. *Photosynth. Res.* **10**, 51-62.

Somersalo, S. and G.H. Krause (1990) Reversible photoinhibition of unhardened and cold-acclimated spinach leaves at chilling temperatures. *Planta* **180**, 181-187.

Voss, M., G. Renger, P. Gräber and C. Kötter (1983) Measurements of penetration and detoxification of PS II herbicides in whole leaves by a fluorometric method. *Z. Naturforsch.* **39c**, 359-361.
Weis, E. (1982) The influence of metal cations and pH on the heat sensitivity of photosynthetic oxygen evolution and chlorophyll fluorescence in spinach chloroplasts. *Planta* **154**, 41-47.
Yamashita, T. and W.L. Butler (1968) Inhibition of chloroplasts by UV-irradiation and heat-treatment. *Plant Physiol.* **43**, 2037-2040.

Response to Ultraviolet Radiation in a Simple Eukaryote (Yeast): Genetic Control and Biological Consequences

Dietrich AVERBECK and Ethel MOUSTACCHI
Institut Curie, Section de Biologie, CNRS
France

INTRODUCTION

UV radiation can damage many cellular components including DNA. In order to provide qualitative and quantitative evidence for the photobiological effects of short wave UV (254 nm) radiation and the involvement of DNA repair systems, a few simple tests have been designed to evaluate the photosensitizing effects in a unicellular eukaryotic cell system, the yeast *Saccharomyces cerevisiae*.

UV-Induced DNA Lesions and their Repair

UV irradiation at 254 nm is known to induce primarily pyrimidine dimers, as well as pyrimidine (6-4) pyrimidone adducts and some hydration products of pyrimidines in cellular DNA (for review see Friedberg, 1985). The induction of these lesions is dose dependent. Pyrimidine dimers are considered to be the major type of lesions. The proportion of 6-4 photoproducts, for example, has been estimated to be only 1/10 of that of the dimers (Brash and Haseltine, 1982). If left unrepaired, these lesions constitute a block for DNA replication and result in the inhibition of colony forming ability (clonogenic survival). About 27 000 UV-induced pyrimidine dimers per cell constitute a mean lethal hit. That is, if all cells have 27 000 pyrimidine dimers, only 37% (1/e) of the cells will survive. Pyrimidine dimers can be removed (excised) from nuclear DNA, but not from mitochondrial DNA, as demonstrated in wild-type yeast (Moustacchi, 1987).

Three kinds of repair of UV-induced DNA damage will be of importance for the experiments to be performed here; 1) nucleotide excision repair, 2) post-replication repair and mutagenic repair, and 3) recombinational and double strand break repair. The first repair process, excision repair, is a removal of photoproducts and subsequent replacement with undamaged nucleotides. It involves the incision (cutting) of DNA containing a pyrimidine dimer, for instance, excision (removal) of the dimer, filling of the excision gap by the action

DIETRICH AVERBECK AND ETHEL MOUSTACCHI, Institut Curie, Section de Biologie
URA 1292 du CNRS, 26 rue d'Ulm, F-75231 Paris cedex 05, France

Plenum Press, New York, 1991

of a DNA polymerase, and resealing of the cut strands by a DNA ligase. This repair mechanism reconstitutes the integrity of DNA and is consequently error-free. Post-replication repair concerns gaps occurring in daughter strands in front of unrepaired photolesions during DNA replication. The gaps are refilled by an exchange with the correct template from the undamaged parental strand. The new gap formed in the parental strand is then refilled using the intact daughter strand of DNA as a correct template. The persisting pyrimidine dimer in the parental DNA strand is then removed by the nucleotide excision repair system. An alternative pathway, taking place during replication, involves the bypass of the lesion by a DNA polymerase leading to replication errors (mutations). Recombinational and double strand break repair involves exchanges between homologous damaged DNA strands taking, undamaged parts of the DNA strands as a template. In addition to these repair processes another, but light-dependent, mechanism exists called photoreactivation. Following 254 nm UV exposure, an enzyme, the photolyase, binds specifically to pyrimidine dimers. After activation of the complex by visible light, the dimers are monomerized *in situ*.

Yeast as a Model Eukaryotic System

The baker's yeast, *Saccharomyces cerevisiae*, is a unicellular eukaryotic organism that has been extensively studied with respect to the DNA repair system involved in the response to UV radiation and other genotoxic agents. Its main advantages can be enumerated as follows:

1. It is easy to handle and inexpensive.
2. It shows a mitotic (vegetative) cell cycle in stable n, 2n and polyploid conditions. In complete medium, the doubling time for wild type cells is around 90 min. Conditional cell cycle mutants (temperature sensitive) are available.
3. There are 2 mating types, *a* and α, allowing conjugation (mating) and a meiotic cycle (sporulation) (Hershowitz, 1988). The spores (tetrads), which result from sporulation, can be easily analyzed and tested for the distribution of genetic markers. Conditional sporulation-defective mutants (temperature sensitive) are available.
4. The nucleus shows a typical nucleosome-type chromosomal organization. In the haploid (n) condition, there are 17 chromosomes ranging in size from 245 to 2200 kbp (kilobase pairs = 1000 nucleotide base pairs) with an average DNA content per chromosome of 800 kbp. Repetitive DNA sequences are present in RNA genes and histone genes.
5. Several nuclear genetic events can be studied including forward and reverse mutation, intra- and intergenic recombination (gene conversion and cross-over) and non-disjunction (aneuploidy or monosomy).
6. The mitchondria are well characterized. One finds approximately 50 mitochrondrial DNA molecules per cell. Each contains circular 25 µm duplex DNA of 75 kbp with a base composition of 82% adenine and thymine. Cytoplasmic genetic events can be studied. Spontaneous or induced mitochondrial deletions give rise to respiratory deficient mutants, so-called cytoplasmic *petite colonie* or *rho*$^{-}$ mutants. Point mutations lead to specific mitochondrial deficiencies (*mit* mutants).
7. Circular (2 µm) and linear plasmids are available that provide powerful tools for genetic engineering and molecular studies including the transformation of yeast with cloned genes and the use of shuttle vectors.

When analyzing cellular responses to UV-induced DNA damage in yeast, remember that, in comparison to most photophysical and photochemical responses, biological responses take a relatively long time. In contrast to photon absorption (10^{-13} s) and the induction of DNA damage (10^{-6} s), DNA fixation during replication, and the expression of biological (including genetic) effects may take 1 min to 4 hours, and 1 day to a week, respectively.

Repair of UV-Induced DNA Lesions in Yeast

The yeast *Saccharomyces cerevisiae* has been a useful system to study the different mechanisms of DNA repair that are relevant for eukaryotic cells.

DNA dark repair mechanisms have been elucidated by the isolation and characterization of mutants in yeast that are hypersensitive to UV irradiation and other genotoxic agents (for reviews see Moustacchi, 1987; Friedberg, 1988). These repair studies involved several steps. After the isolation and characterization of the UV-hypersensitive mutants from mutagenized cell cultures, their cross-sensitivity to different types of genotoxic agents was assessed. Phenotype analysis (appearance) simplified the identification of mutants with specific defects. Subsequent genetic analysis of the mutants confirmed the genetic control of hypersensitivity and the dominance or recessiveness of gene action. By studying the biological response of double mutants carrying different genes, the interaction of repair processes could be assessed. Epistatic (masking, by one gene, of the expression of another gene at a different location), synergistic or additive responses are observed. In this way, three main epistasis groups have been identified in yeast: *RAD3, RAD6* and *RAD52*, named after prominent genetic loci (mutants) in each group.

Also extensive knowledge of light-dependent repair mechanisms in yeast has been achieved by the isolation and characterization of specific mutants. Photoreactivation-less mutants (*phr*⁻) have been isolated in yeast by Resnick (1969) and MacQuillan et al. (1981). Photoreactivation has been shown to be mediated by two closely liked genes in yeast. These genes mediate the direct reversal of UV-induced pyrimidine dimers. One of these, the *PHR1* gene, has been cloned (Yasui and Chevallier, 1983; Schild et al., 1984) and shown to encode a 66 kDa photolyase containing both flavin and folate chromophores (Sancar and Sancar, 1988). This yeast photolyase enzyme is similar to at least one other photoreactivating enzyme, that of *Escherichia coli*, with which it shows a 36.5% amino acid homology (Sancar, 1985).

Repair Deficient Mutants in Yeast

When repair deficient mutants are compared to normal wild type cells, specific repair defects can be identified. The role of genetic control in correcting UV-induced DNA lesions can be determined by analyzing the biochemical and molecular steps that are impaired in these mutants. This approach has shown that mutants of the *RAD3* type are defective in nucleotide excision repair (Table 1). Mutants of the *RAD6* type are defective in post-replicational (mutagenic) repair (Table 2). Finally, mutants of the *RAD52* type are defective in the repair of double strand breaks and recombinational repair (Table 3).

Among the genes involved in nucleotide excision repair (*RAD3* group), *RAD1, RAD2, RAD3, RAD4* and *RAD10* have been cloned. They are involved in the incision of pyrimidine dimers and of other bulky DNA lesions. The *RAD3* gene appears to be essential for cell viability. Using *RAD-lacZ* fusions, it was shown that the *RAD2* gene is inducible by UV treatment and other genotoxic agents. Other genes, such as *RAD7, RAD14, RAD16, RAD23,RAD24, MMS19, CDC8* and *CDC9* are apparently not required for the incision of DNA damage, but are required for further steps in the excision repair pathway.

Table 1

Excision repair-deficient mutants (type *RAD3*)

Mutant	Gene	Protein	Comments
rad1	3.3 kbp	126 kDa	–
rad2	3.093 kbp	118 kDa	inducible gene
rad3	2.334 kbp	90 kDa	essential gene
rad4	2.262 kbp	87 kDa	–
rad10	630 kbp	24 kDa	–

Mutants *rad1, rad2, rad3, rad4,* and *rad10* are dificient in the incision of damaged DNA. Other mutants of the group *rad7, rad14, rad16, rad23,* and *nms19* affect excision repair (see Friedberg, 1988).

Table 2

Mutants defective in post-replication repair and the mutagenic pathway (type *RAD6*)

Mutant	Gene	Protein	Comments
rad6	516 kbp	19.7 kDa	ubiquination
rad18	2.471 kbp	55.512 kDa	zinc finger protein

Mutants *rad9, rad15, rev1, rev2, rev3, pso1, pso2, and snm1* belong to this epistasis group (see Friedberg, 1988).

Table 3

Mutants deficient in recombination and double strand break repair (type *RAD52*)

Mutant	Gene	Protein	Comments
rad50	1312 kbp	157 kDa	pure binding domains
rad52	1512 kbp	65.1 kDa	–
rad54	–	–	inducible

Mutants *rad51*, *rad55*, and *rad57* belong to this group (see Friedberg, 1988).

Several mutants have been isolated that are defective in the post-replication repair pathway. The most prominent include *RAD6, RAD18, UMR* and *REV* mutants. Some amongst them are defective in error-prone (mutagenic) DNA repair: *REV, UMR* genes and *RAD6*. The latter gene is involved in the ubiquitination of histones and may be important for nucleosome assembly.

Mutants of the *RAD52* epistasis group have been shown to be defective in recombinational and double strand break repair. The genes appear to be especially important for the repair of X-ray induced damage, and the mutants are defective in recombination (for review see Haynes and Kunz, 1981).

In addition, several types of repair mutants have been isolated that are specifically sensitive to the action of mono- and bifunctional agents, such as psoralen photosensitizers (*pso*$^{-}$ mutants) and alkylating agents (*snm* mutants).

The specific sensitivity of repair-deficient mutants makes them valuable tools for the detection of specific types of induced DNA lesions. In other words, the sensitivity of different mutants to a given genotoxic agent (for example, UV radiation or photosensitizing agents like psoralen plus UVA radiation) can reveal the relative importance of different repair pathways. Interactions between the different repair pathways can also be assessed in this way. For instance, it has recently been reported that the integrity of the excision-repair pathway appears to be necessary for completion of specific recombinational events (Aguilera and Klein, 1989).

A similar approach may be used to test the response of yeast cells to other photosensitizing (and genoxotic) agents, for example psoralens or other mono- and bifunctional furocoumarins such as angelicin and 8-methoxypsoralen (8-MOP) plus UVA and photodynamic substances like hematoporphyrin derivative and Rose Bengal plus visible light exposure. By comparing the responses of different repair-deficient mutants to that of repair-proficient wild type cells, the contribution of different pathways to the repair of lesions induced in DNA can be assessed (DNA repair test). This general approach is also useful when testing photosensitizing agents whose mode of action is unknown. The reponse of repair deficient mutants defective in the repair of specific types of lesions may provide indirect evidence for the type of DNA damage induced by these agents. This approach has been successfully used for determining the mode of action of antitumor drugs such as BD-40 and neocarzinostatin (Moustacchi et al., 1983; Moustacchi and Favaudon, 1982).

Yeast as an Experimental Tool in UV Photobiology

The responses of yeast cells to UV-induced DNA damage can be measured either qualitatively or quantitatively. The qualitative approach generally involves exposure of a relatively large number of yeast cells (10^7 cells/ml) that are plated (i.e., evenly distributed on the surface of a solid complete growth medium in a sterile Petri dish) to UV radiation. An unirradiated dish, or a part of the experimental dish, is kept shielded from UV exposure as a control and reference for normal growth. In this way, UV-induced growth inhibition can be estimated. The quantitative approach (Table 4, see flow diagram) usually involves the preparation of a cell suspension containing a defined number of cells in a known phase of growth (stationary or logarithmic phase). The cell suspension (or a well defined aliquot of it), plated on a suitable medium, is exposed to UV radiation or the photosensitizing treatment. In the former case, treated cells as well as untreated cells (as controls) are plated on specific media allowing normal growth of all surviving cells and the detection of forward or reverse (back) mutations, mitotic, intragenic or intergenic recombination, or non-disjunction (Zimmermann et al., 1975, 1984).

The biological consequences of UV radiation always depend on the endpoint monitored. For example, a UV irradiation protocol sufficient to completely inhibit colony forming ability, might not affect membrane transport at all. Thus, the genetic responses that can be measured in yeast, are an important consideration. For quantitative mechanistic studies, the exact relationship between UV fluence and the biological effect is critical. This relationship usually takes the form of a typical fluence-response curve. The fluence here is the total amount of UV radiation (expressed as energy/area or photons/area). The genetic responses can be any of the following:

1. Colony-forming ability or clonogenic cell survival (genetic death).
2. Induction of cytoplasmic mutations (*rho*$^-$) indicating mitochrondrial DNA damage associated with repiratory deficiency.
3. Induction of forward mutations which can be detected either by visible changes in color of colonies due to a loss of biosynthetic function or by the achievement of resistance to a chemical analogue. The first case is illustrated by mutations from *ADE*$^+$ to *ade1* or *ade2* which lead to a block in the purine pathway and the accumulation of red pigments (polyribosylaminoimidazol derivatives) (Roman, 1956). An example of the second case can be found in the mutation from canavanine sensitivity (*can*^{S}) to canavanine resistance (*can*^{R}) (Lemontt, 1977). Indeed, normal wild type cells are sensitive to canavanine, an analogue of arginine which suppresses growth of yeast, due to the normal function of the arginine permease gene. Arginine permease allows arginine and its analogues to cross cell membranes and enter cells. After UV-induced mutation, mutant cells arise in the cell culture that are resistant to canavanine due to the loss of normal arginine permease gene function.
4. The induction of reverse mutations is tested by the selection of mutational events that abolish a particular auxotrophy (for example, mutation from histidine-dependent (his$^-$) to histidine-independent (*HIS*$^+$) growth).
5. The induction of mitotic gene conversion and crossing-over, or intragenic and intergenic recombination can be revealed by testing for the exchange of suitable genetic markers (alleles) inside a gene or between the centromere and a gene, respectively [e.g., exchange between *trp5-12* and *trp5-27* or between the *ade*^{2} gene and the centromere in the diploid strain D7 (Zimmermann, 1975; Zimmermann et al., 1975)].
6. Induction of mitotic non-disjunction of chromosomes, i.e., chromosome loss or gain, aneuploidy, monosomy (Zimmermann et al., 1984; Dixon and Mortimer, 1986). Strains have been especially designed for studying this important end-point (e.g., strain D6 of Parry and Zimmermann, 1976).

In addition to these biological endpoints, we will also consider the survival of UV-damaged plasmids in transformed yeast strains as relevant for the evaluation of genotoxic effects of UV irradiation and the repair of UV-induced DNA damage.

This chapter provides general information for yeast photobiological experiments. Qualitative and quantitative tests for UV-induced photocytotoxicity and genotoxicity are described. In addition, a DNA repair test including host cell repair of exogenous DNA (plasmid) and modification of the photosensitizing reponses by DNA repair genes is reported.

EXPERIMENT 17: PHOTOCYTOTOXICITY TEST IN YEAST

OBJECTIVE

The results of this experiment will establish a dose-effect relationship for UV photocytotoxicity in yeast by a *qualitative* approach.

LIST OF MATERIALS

Sterile tube or plastic dish containing 5 ml of liquid growth medium or solidified YEPD medium, respectively
NaCl, 10 g
Sterile distilled water or sterile phosphate buffer, pH 6.5
Centrifuge tubes, sterile, 10 ml plastic (Falcon) or glass (Corex) (4)
Bench centrifuge (5000 rpm max)
Pipettes, sterile, plastic or glass, capacity 5 ml or 10 ml (5)
254 nm UV source, UV germicidal lamp low pressure mercury HNS 12 (Osram) or General Electric G8T5
Latarjet dosimeter or radiometer UVX with probe UV-25 (Ultraviolet Products Inc., San Gabriel, CA 91778, USA)
White fluorescent light tubes Philips TLD 36/33
Gloves and eye goggles (glasses) for UV protection
Sterile loops, 10 ml, plastic, nichrome or platinum
Bent glass rods and/or sterile cotton wool pads
Ethanol, pure (Merck)
The haploid strain: N 123 *a his1* (Moustacchi, 1969)
The diploid strain: D7 a/α, *ade2-40/ade2-119, trp5-12/trp5-27, ilvl-92/ilvl-92* (kindly provided by Professor F. Zimmermann, Technische Hochschule, Darmstadt, FRG) (Zimmermann, 1975)
Other suitable wild type (repair-proficient) or repair-deficient strains of *Saccharomyces cerevisiae* can be obtained from the Yeast Genetic Stock Center.

PREPARATIONS

Saline Solution: Dissolve 9 g of NaCl in 1 liter of distilled water to make a 0.9% saline solution.

Yeast Suspensions: One day before the start of the experiment prepare yeast cells in the following manner. With the help of a sterile loop, transfer 2.5×10^7 yeast cells in stationary phase of growth from stock slants, from single colonies or from a cell lawn on solid growth medium (YEPD) into 5 ml of sterile liquid growth medium (YEPD) in sterile glass tubes. Incubate them for 24 h at 30°C. This will produce a suspension in stationary phase (2×10^8 cells/ml).

If yeast strains were received from the Yeast Genetic Stock Center, streak the small filter pads carrying the yeast cells across solid YEPD medium or directly into 5 ml of liquid growth medium with a sterile pincette (flame or ethanol sterilized). Allow them to grow to stationary phase as just described.

EXPERIMENTAL PROCEDURE

CAUTION: YOU WILL USE A 254 nm UV SOURCE IN THESE EXPERIMENTS. YOU MUST WEAR GOGGLES THAT BLOCK THIS RADIATION TO PROTECT YOUR EYES, AND GLOVES TO PROTECT YOUR SKIN. AVOID EXPOSURE AS MUCH AS POSSIBLE.

Day 1

1. Harvest yeast cells from liquid growth medium by centrifugation (5 min, 2300 × g) in sterile glass or plastic centrifuge tubes.
2. Wash twice with 5 ml sterile distilled water, saline or phosphate buffer (pH 6.5) by centrifuging as in step 1.
3. Resuspend the washed pellet in 5 ml of sterile distilled water to obtain approximately 2×10^8 cells/ml.
4. Check the cell density with a hemocytometer (Burker) and a light microscope (400 ×). You will need to dilute the cells about 100-fold for counting.
5. Using the results of your cell count in step 4, prepare a 4×10^7 cells/ml suspension in distilled water.
6. Spread 0.5 ml of the dilute suspension homogenously on the surface of solid, half-dried YEPD agar plates (95 mm diameter), using a bent sterile glass rod (washed in ethanol and dried or flamed), or sterile cotton pads.
7. Use the dosimeter to set the far UV source (mercury low-pressure lamp) to obtain a dose rate of 2 $J \cdot m^{-2}$.
8. Fold a piece of aluminum foil to produce a quadruple-layered light shield 12 cm long. Cover the middle of the plates (lids off) with the foil, and expose the dishes to increasing doses of UV radiation according to the following table (for strains N123 and D7 [wild type]).

Time (s)	Energy ($J \cdot m^{-2}$)
25	50
50	100
100	200
150	300

9. To test photoreactivability, expose both UV-treated and control plates to white fluorescent light for 20 min at a distance of about 6 to 7 cm.
10. Incubate all plates for 24 h in the dark at 30°C.

Day 2

11. Visual examination of cell growth on the plates. The difference in the density of the cell lawn in the UV-exposed and the UV-protected part gives a qualitative indication of the UV-dependent growth inhibition. Place the plates in increasing order of growth inhibition in the UV-exposed part of the plates. No growth inhibition is observed in the control plate. A slight growth inhibition is observed after small UV doses, whereas no or very little growth is seen after high UV doses. The relative UV sensitivity of different yeast strains may be established in this way (see Fig. 1) by comparing the degrees of growth inhibition obtained by identical UV doses. The effects of photoreactivation can be assessed as well by the increased survival obtained in photoreactivated as compared to non photoreactivated samples.

A dose-dependent increase in UV-induced growth inhibition is seen for the haploid yeast strain N123 (Fig. 1, upper lane) and the diploid yeast strain D7 (Fig. 1, lower lane). A comparison of growth inhibition clearly indicates that the diploid cells are more resistant to UV than the haploid cells. This is consistent with the higher repair capacity of diploid as compared to haploid cells.

DISCUSSION

Photocytotoxic effects in yeast can be recognized by the inhibition of cell growth, i.e., inhibition of colony-forming ability (clonogenic cell survival). Note that in wild type untreated yeast cells, the plating efficiency is 100%, i.e., the number of colonies derived from single cells corresponds to the number of cells counted (hemocytometer or electronic Coulter counting). This is not the case for certain repair-defective mutants which demonstrate simultaneous alterations in cell cycle parameters leading to lethality in a fraction of the cell population. After UV irradiation, the inhibition is a consequence of the induction of specific lesions such as pyrimidine dimers and pyrimidine (6-4) pyrimidone adducts in cellular DNA. The induction of these lesions is dose-dependent. The establishment of the relationship between UV dose and biological effect is necessary for the assessment of the biological effectiveness of UV radiation and for mechanistic studies.

As described above, UV-induced pyrimidine dimers can be monomerized by the action of photoreactivation enzyme (photolyase). Such photoreactivation can take place in photoreactivation-competent cells (*PHR*$^+$) when they are exposed to visible light that activates the photoreactivation enzyme, after UV-exposure. This process leads to an increased survival after UV-irradiation. For this reason, experiments aiming at the analysis of UV effects alone should be performed under yellow light. Treated suspensions and cells on plates should be shielded from visible light exposure by suitable covers (for example, aluminum foil).

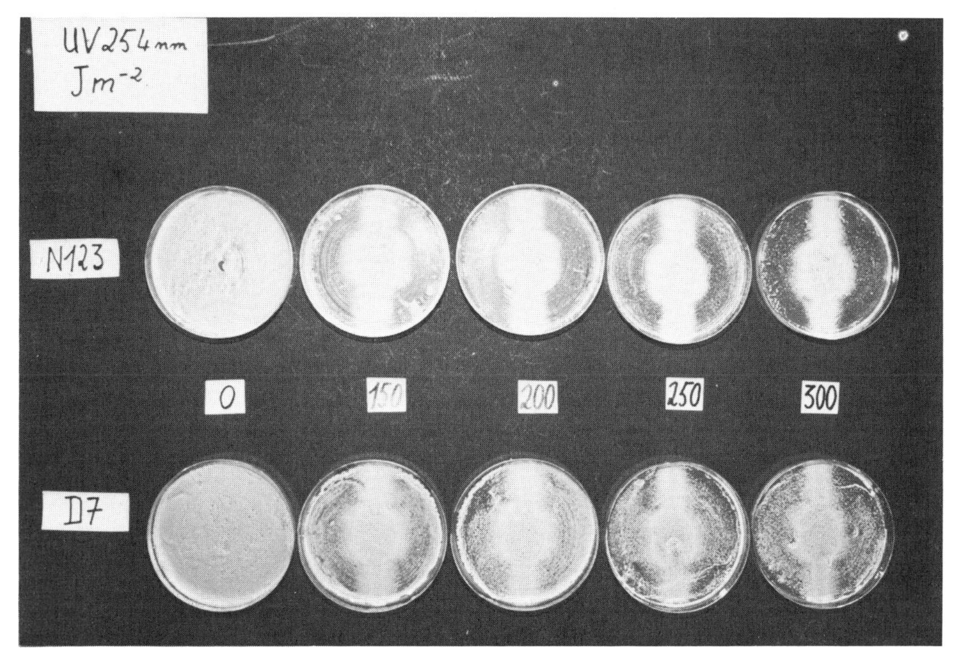

Figure 1. Photocytotoxicity in yeast. The figure shows the typical result of exposure of 2 strains of yeast, N123 and D7, to different fluences of 254 nm UV radiation, 0-300 J·m^{-2}.

Related Laboratory Experiments

Quantitative dose-response relationship: In contrast to the qualitative test, in this test a defined number of cells from a UV-irradiated cell suspension is placed on the surface of YEPD agar plates. Ideally, suitably diluted cell samples are plated on each plate, giving rise to the growth of 50 to 200 colonies. In this way, an exact relationship between surviving cells able to form colonies and UV dose (full survival curves) can be established. The procedure is similar to that just described with the following variations. Use saline instead of distilled water for all suspension media. After harvesting the cells in step 1, wash them once and resuspend them in saline. Then sonicate them to disrupt cell clumps and wash them once more before counting the cells. Cell density should be diluted to 2×10^7 cells/ml and UV fluence rate should be 1 or 0.1 $J \cdot m^{-2} \cdot s^{-1}$ for wild type strains or repair-deficient mutants, respectively. Exposure to UV radiation for different periods of time, can be done either on YEPD agar plates, as above, or in stirred liquid suspension. Photoreactivation can be tested by exposing duplicate samples to white fluorescent light (Philips TLD 36/33 400-700 nm; peak emission at 440 nm and 550 nm) for 20 min at a distance of 6 to 7 cm. Otherwise, shield the cells from visible light. A quantitative dose-response curve is obtained by counting the number of colonies for each fluence.

EXPERIMENT 18: SPOT TEST FOR DNA REPAIR

OBJECTIVE

This experiment is designed to rapidly and simply demonstrate UV-induced lethality in wild type and repair-deficient yeast strains. Strains that can repair DNA lesions are much more resistant to UV-induced lethality. The test consists of measuring the response of wild type and correspoinding repair-deficient mutant strains of *Saccharomyces cerevisiae* to photosensitizing treatments. Three main DNA repair pathways are distinguished: nucleotide excision repair (*rad3* mutants), post-replication repair (mutagenic repair; *rad6* mutants) and double strand break (recombinational repair; *rad52* mutants).

LIST OF MATERIALS

Liquid YEPD medium, 30 ml
Sterile distilled water or sterile phosphate buffer, pH 6.5
NaCl, 10 g
Sterile tubes containing 1 ml of sterile water, saline or phosphate buffer (6)
Eppendorf tubes (12)
Hemocytometer (Burker)
Microscope (10 × 40)
Plate of YEPD solid growth medium (1)
UV source: germicidal low pressure mercury lamp (OSRAM)
Latarjet dosimeter or radiometer UVX (Ultraviolet Products Inc., San Gabriel, CA 91778, USA)
Pipettor and sterile tips, 0.1 ml
Sterile loops, 10 ml, plastic, nichrome or platinum
Incubator, 30°C
Gloves and eye goggles (glasses) for UV protection
Yeast strains:
- N123, a *his1* (wild type) RAD
- N123, a *his1 rad1-3* mutant
- *rad2D*
- JG *rad18*
- *rad52*
- rad2 rad6 rad52

PREPARATIONS

Saline Solution: Dissolve 9 g of NaCl in 1 liter of distilled water to make a 0.9% saline solution.

Yeast Suspensions: One day before the start of the experiment prepare cells from each of the 6 strains of yeast in the following manner. With the help of a sterile loop, transfer 2.5×10^7 yeast cells in stationary phase of growth from stock slants, from single colonies or from a cell lawn on solid growth medium (YEPD) into 5 ml of sterile liquid growth medium (YEPD) in sterile glass tubes. Incubate them for 24 h at 30°C. This will produce a suspension in stationary phase (2×10^8 cells/ml).

If yeast strains were received from the Yeast Genetic Stock Center, place the small filter pads carrying the yeast cells directly into 5 ml of liquid growth medium with a sterile pincette (flame or ethanol sterilized). Allow them to grow to stationary phase as just described.

EXPERIMENTAL PROCEDURE

CAUTION: YOU WILL USE A 254 nm UV SOURCE IN THESE EXPERIMENTS. YOU MUST WEAR GOGGLES THAT BLOCK THIS RADIATION TO PROTECT YOUR EYES, AND GLOVES TO PROTECT YOUR SKIN. AVOID EXPOSURE AS MUCH AS POSSIBLE.

Day 1

1. Turn on the UV lamp and determine the fluence rate at the position where cells will be irradiated, about 6 to 7 cm from the lamp.
2. Prepare 6 Eppendorf tubes with 1 ml of cell suspension (2×10^8 cells/ml), one for each strain of yeast.
3. Transfer 10 µl of each well agitated cell suspension into 1 ml of sterile distilled water in 6 plastic tubes. This will dilute the suspensions 100-fold.
4. Transfer a small aliquot of each diluted suspension to a hemocytometer and count the number of cells (Burker hemocytometer: depth = 0.1 mm, width = length = 0.2 mm, volume = 0.004 mm^3) in each. Determine the concentration of cells [cells per cm^3=number of cells/number of squares $\times$ 250 $\times$ 1000 $\times$ dilution factor (100)].
5. With a sterile pipette or suitable plastic loop, drop 10 to 30 µl of each cell suspension onto a solid YEPD growth medium agar plate at geometrically defined places. With a permanent marker, mark the position of each strain on the plastic lid. Also, put a small dot on the side of both the lid and bottom of the dish so that the lid can be realigned properly with the bottom of the dish.
6. When all strains have been spotted onto the plate, expose the plate (without the lid) to 100 $J \cdot m^{-2}$ of UV radiation. To avoid photoreactivation, keep the plate shielded from visible light after irradiation.
7. After irradiation, replace the lid and incubate the plate for 24 h in the dark at 30°C.

Day 2

8. Determine cell growth at the different spots corresponding to the wild type and the repair-deficient strains by evaluating visually the density of growth in each spot.

DISCUSSION

The *rad3* type mutants (*rad1-3*) and *rad6* type mutants (*rad18*) show no growth in this test because of their extreme sensitivity to UV, due to their repair deficiencies. The spots corresponding to the wild type cells and *rad52* type mutants are little, or not, affected by the UV treatment. Their growth is unaffected due to the fact that the *rad52* cells are deficient in pathways that do not concern UV-induced lesions; they are hypersensitive to ionizing radiation and to DNA cross-linking agents (including psoralen plus UVA photosensitization).

Related Laboratory Experiments

Quantitative dose-response relationship: The DNA repair capability of different repair-deficient strains can be assessed quantitatively by following the general procedures outlined in Experiment 17, Related Laboratory Experiments, but using the repair-deficient strains indicated in this experiment (18). Determination of the dose-response curves allows a quantitative comparison of UV effects on repair-competent and repair-deficient strains.

Host cell repair of exogenous plasmid (DNA): Repair competent yeast cells can repair exogenous DNA, that is DNA introduced from another source. Comparisons of the repair of exogenous and endogenous DNA may give some hints on the importance of chromatin structure for the DNA repair systems involved. The assay involves the *in vitro* treatment of plasmid DNA with 254 nm UV light and its subsequent introduction by transformation into competent recipient cells defective for a given genetic marker. The latter can be complemented by the corresponding marker on plasmid DNA. The method can be sketched as follows (for complete details see White and Sedgwick, 1985, 1987; Averbeck et al., 1990; Keszenman-Pereyra, 1990). Exogenous plasmid DNA can be isolated from *Escherichia coli* according to Birnboim and Doly (1979) and Maniatis et al. (1982). Plasmid DNA should be irradiated before incorporation into the yeast (2-10 $kJ \cdot m^{-2}$). An exponentially growing yeast culture should be suspended in Li acetate buffer, before addition of carrier DNA (sonicated and denatured calf thymus DNA) and the exogenous plasmid DNA. Polyethyleneglycol 4000 is added and the suspension is heat shocked in a water bath at 42°C. The cells are then washed, plated and analyzed 5 days later. Transformants are recognized by their ability to grow on a selective minimum medium. Plot plasmid survival against UV dose. To some extent, transformation efficiency can be equated with plasmid survival and repair proficiency within the transformed cells (White and Sedgwick, 1985, 1987).

EXPERIMENT 19: PHOTOSENSITIZED UV LETHALITY

OBJECTIVE

Most of the methods described above provide general procedures which are also applicable to chemical photosensitizing agents. In this experiment we will obtain evidence for the activity of a photosensitizing agent (the furocoumarin 8-methoxypsoralen; for reviews see Ben-Hur and Song, 1984; Averbeck, 1989) using a rapid filter assay developed by Daniels (1967). This test is often employed as a preliminary screening to indicate the presence of a photosensitizing agent (see also Chapter 13, Experiment 20). Usually, the experiment is performed with a wild type strain. However, the sensitivity of the test can be increased by using rad mutant strains.

LIST OF MATERIALS

NaCl, 10 g
8-methoxypsoralen, 6 mg
Eppendorf tubes with 1 ml of sterile saline
Hemocytometer (Burker)
Microscope (10 × 40)
Plates of YEPD solid growth medium (4)
Glass pipettes, sterile, 0.1 ml and 1 ml, or Pipetman (200 µl) with sterile yellow tips
Whatman glass fiber filter disks (GF/A)
Sterile loops, 10 ml, plastic, nichrome or platinum
Glass rod, sterile, for plating, or sterile cotton pads
UV source: HPW 125 Philips (black lamp), peak emission at 365 nm
UVA radiometer, UVX plus UV-36 probe (Ultraviolet Products Inc., San Gabriel, California, USA)
Incubator, 30°C
Ruler, clear plastic
Yeast strains:

N123, a *his1* (wild type) RAD
N123, a *his1 rad1-3* mutant
rad2D
JG *rad18*
rad52
rad2 rad6 rad52

PREPARATIONS

Saline Solution: Dissolve 9 g of NaCl in 1 liter of distilled water to make a 0.9% saline solution.

8-MethoxypsoralenStock Solution: Dissolve 5 mg of 8-methoxypsoralen (8-MOP; Sigma Chemical Co.) in 5.78 ml of ethanol (Merck) to produce an 8-MOP stock solution at a concentration of 4 mM.

Yeast Suspensions: One day before the start of the experiment prepare yeast cells in the following manner. With the help of a sterile loop, transfer 2.5×10^7 yeast cells in stationary phase of growth from single colonies on solid growth medium (YEPD) into 5 ml of sterile liquid growth medium (YEPD) in sterile glass tubes. Incubate them for 24 h at 30°C. This will produce a suspension in stationary phase (2×10^8 cells/ml).

EXPERIMENTAL PROCEDURE

Day 1

1. Harvest yeast cells from liquid medium by centrifugation (5 min, 2 300 × g) in glass or plastic centrifuge tubes.
2. Wash twice with 5 ml sterile distilled water, saline or phosphate buffer (pH 6.5) by centrifuging as in step 1.
3. Resuspend the washed pellet in 5 ml of sterile distilled water to obtain approximately 2×10^8 cells/ml.
4. Check the cell density with a hemocytometer (Burker) using a light microscope (400 ×). You will need to dilute the cells about 100-fold for counting.
5. Using the results of your cell count in step 4, prepare a 4×10^7 cells/ml suspension in distilled water.
6. Spread 0.5 ml of the dilute suspension homogenously on the surface of solid, half-dried YEPD agar plates (95 mm diameter), using a bent sterile glass rod (washed in ethanol and dried or flamed), or sterile cotton pads. Prepare a total of 4 plates.
7. Using a sterile pipette, transfer 8-MOP stock solution to 4 fiber filter disks, 200 µl per disk. Place each disk in the middle of one of the agar plates from the previous step.
8. Incubate the plates for 1 h in the dark at room temperature to allow diffusion of the 8-MOP.
9. Use the UVA radiometer to determine the distance from the UVA source at which a fluence rate of 15 $J{\cdot}m^{-2}{\cdot}s^{-1}$ occurs. This will be the distance at which agar plates will be irradiated. Expose 2 plates (without lids) to UVA radiation for 20 min.
10. Incubate the 2 irradiated plates and the 2 unirradiated control plates at 30°C for 24 to 48 h.

Day 2 or 3

11. Using the clear plastic ruler, measure the diameter of the zone of growth inhibition around the filter paper disk in each dish.

DISCUSSION

The relative diameter of the zone of inhibition in the irradiated dishes compared to the unirradiated controls is an indicator of the phototoxicity of the photosensitizer. If different photosensitizers are compared, the size of the zone of growth inhibition gives an indication of the comparative activity of the sensitizers.

Figure 2 shows the results obtained with a haploid wild type strain N123 after treatment with ethanol (200µl) or 200 µl of a 4 mM solution of 8-MOP in ethanol and different fluences of UVA (0, 20, and 40 $kJ{\cdot}m^{-2}$). No effect is seen with ethanol alone or with 8-MOP alone. However, with increasing doses of UVA, the zone of growth inhibition increases significantly providing evidence for a dose-dependent cell killing by lesions induced by 8-MOP plus UVA.

Related Laboratory Experiments

Pre-incubation of sensitizer with yeast cells: Sensitivity can be increased by pre-incubating the sensitizer with the yeast cells before plating. In this way a qualitative dose-response curve can be generated. Monofunctional furocoumarins such as angelicin or 3-

carbethoxypsoralen can be used. The cells, at 4×10^7 cells/ml, should be incubated with sensitizer (50 µM) for 15 to 30 min in the dark. After incubation, plate the cells as described above and expose different dishes to UVA fluences in the range of 5 to 20 kJ·m^{-2}. Remember control plates without UVA and without sensitizer. It is also helpful to shield the center of each dish with a small screen of 5 thickness aluminum foil. Incubate the dishes in the dark for 24 h and then note the differences in growth inhibition between the shielded and unshielded portions of each dish. The difference in growth inhibition is a measure of the photocytotoxic effect of the treatment and can be used to construct a qualitative dose-response curve.

Quantitative dose-response relationship: A quantitative dose-response relationship can be obtained by combining the techniques given in this experiment and those given in Experiment 17, Related Laboratory Experiments (Averbeck, 1985; Averbeck et al., 1990). In principle, these experiments can be conducted in two different ways: 1) change the UVA fluence and keep the furocoumarin concentration constant, 2) change the furocoumarin concentration and keep the UVA fluence constant. Experimentally it might be easier to adjust UVA fluences by extending exposure times, rather than preparing solutions of furocoumarins at different concentrations and using a constant UVA fluence. In any case, note that for a given biological effect (e.g., survival), the relationship between UVA fluence and furocoumarin concentration is not a 1:1 relationship. In other words, increases in UVA fluence always appear to be more effective than increases in furocoumarin concentration (Averbeck and Averbeck, 1978; Averbeck et al., 1990). If different furocoumarins are compared, use equimolar concentrations or, better still, concentrations that give the same absorbance (in a standard cuvette) at the wavelength of peak emission of the UVA source.

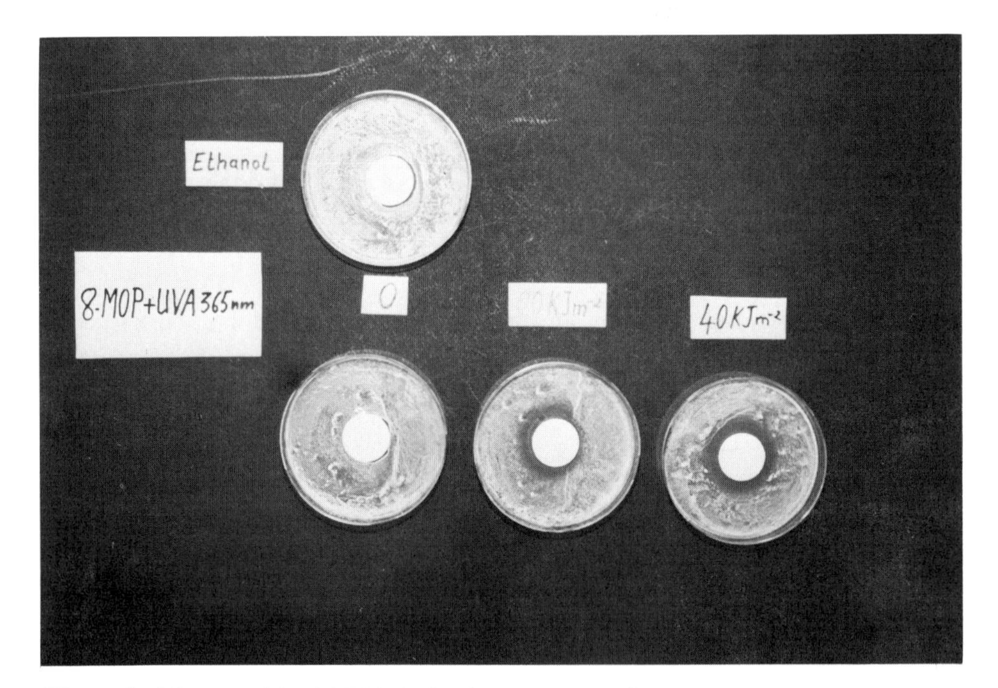

Figure 2. Photosensitized UV lethality in yeast. The figure shows the typical result of exposure of the haploid strain N123 to 0, 20 and 40 J·m^{-2} of UVA radiation after exposure to 4 mM 8-MOP. The upper plate is a control with no 8-MOP.

Wavelength specificity: The wavelength specificity of 8-MOP plus UVA photosensitization of yeast can be demonstrated by exposing yeast cells to either 365 nm or 405 nm light, and noting the difference in growth inhibition. To do this, pre-incubate yeast cells (2×10^7 cells/ml) with 8-MOP (0.01 mM) for 30 min in the dark. Plate 0.5 ml on YEPD dishes and shield the middle with aluminum foil, as described above. Expose different dishes to 5 $kJ \cdot m^{-2}$ of 365 nm UVA [e.g., 2500 W xenon arc lamp plus water filter (LH 162) and a Kratos 252 monochromator (Kratos Analytical Instruments, Ramsey, NY 07446, USA)] or to the same fluence of 405 nm UVA [e.g., 2500 W xenon arc lamp plus water filter (LH 162), a Kratos 252 monochromator, and a Schott glass filter WG 385]. Fluence rate can be determined with a YSI-Kettering Model 65 A Radiometer (Yellow Springs Instruments, Ohio, USA). At the same light dose, 8-MOP plus 365 nm UVA inhibits cell growth whereas 8-MOP plus 405 nm light does not. This is in part due to the low light absorption of 8-MOP as well as to the induction of monoadducts alone at 405 nm in contrast to the induction of interstrand cross-links by 8-MOP plus 365 nm UVA.

Chemical filters: In the variant just described, it is instructive to add 13 μl of a 0.5% (wt/vol in ethanol) solution of the "solar filters" Parsol MCX or Parsol 1789 before irradiation (Averbeck et al., 1990). Cells are protected, that is there is less growth inhibition, when the chemical filters are present.

DNA repair tests: For a spot test assay, follow the procedures of Experiment 18 with the following exception. Pre-incubate the yeast suspension with 8-MOP or other furocoumarins before pipetting onto the dishes. Use a fluence of 20 $kJ \cdot m^{-2}$. After 24 h of incubation at 30°C, treated wild type cells show normal growth, whereas repair-deficient strains show inhibited growth. A qualitative dose-response relationship can be obtained by plating wild type and repair-deficient cells on separate plates and exposing them to different fluences of UVA (0, 5, 10 and 20 $kJ \cdot m^{-2}$). Before plating the cells (2×10^7 cells/plate), they should be pre-incubated with 8-MOP (50 μM) for 15 to 30 min. In both wild type and repair-deficient strains, a UVA dose-dependent decrease in cell growth will be observed. At the same UVA dose, growth inhibition will be more pronounced in the repair-deficient strains than in the wild type. The triple mutant *rad2rad6rad52*, deficient in all three types of repair pathways is most sensitive. This clearly indicates that all three repair pathways are involved in the repair of 8-MOP plus UVA induced lesions. A quantitative dose-response relationship can also be obtained by combining the techniques given here with those given in Experiment 17, Related Laboratory Experiments.

SUPPLEMENTARY READING

Smith, K.C. (1989) UV radiation effects: DNA repair and mutagenesis. In *The Science of Photobiology*, 2nd edition (Edited by K.C. Smith), pp. 111-133. Plenum Press, New York.

REFERENCES

Aguilera, A. and H.L. Klein (1989) Yeast intrachromosomal recombination: Long gene conversion tracts are preferentially associated with reciprocal exchange and require the *RAD1* and *RAD3* gene products. *Genetics* **123**, 683-694.

Averbeck, D. (1985) Relationship between lesions photoinduced by mono- and bifunctional furocoumarins in DNA and genotoxic effects in diploid yeast. *Mutation Res.* **151**, 217-233.

Averbeck, D. and S. Averbeck (1978) Dose-rate effects of 8-methoxypsoralen plus 365 nm irradiation on cell killing in *accharomyces cerevisiae. Mutation Res.* **148**, 47-57.

Averbeck, D. (1989) Yearly review: Recent advances in psoralen phototoxicity mechanisms. *Photochem. Photobiol.* **50**, 859-882.

Averbeck, D., S. Averbeck, L. Dubertret, A. Young, and P. Morlière (1990) Genotoxicity of bergapten and bergamot oil in *Saccharomyces cerevisiae*. *J. Photochem. Photobiol. B*, **7**, 209-230.

Ben-Hur, E. and P.S. Song (1984) The photochemistry and photobiology of furocoumarins. *Adv. Radiat. Biol.* **11**, 131-171.

Birnboim, H.C. and J. Doly (1979) A rapid alkaline extraction procedure for screening recombinant plasmid DNA. *Nucleic Acids Res.* **7**, 1513-1523.

Brash, D.E. and W.A. Haseltine (1982) UV-induced mutation hotspots occur at DNA damage hotspots. *Nature* **298**, 189-192.

Daniels, F. (1965) A simple microbiological method for demonstrating phototoxic compounds. *J. Invest. Dermatol.* **44**, 259-263.

Dixon, M.L. and R.K. Mortimer (1986) A yeast screening system for simultaneously monitoring multiple genetic endpoints. *Mutation Res.* **161**, 49-64.

Friedberg, E.C. (1985) *DNA Repair*, pp. 1-614, W.H. Freeman Co., New York.

Friedberg, E.C. (1988) Deoxyribonucleic acid repair in the yeast *Saccharomyces cerevisiae*. *Microbiol. Rev.* **52**, 70-102.

Haynes, R.M. and F. Eckardt (1979) Analysis of dose-response patterns in mutation research. *Can. J. Genet. Cytol.* **21**, 277-302.

Haynes, R.M. and B.A. Kunz (1981) DNA repair and mutagenesis in yeast. In: *The Molecular Biology of the Yeast Saccharomyces* (Edited by J.N. Strathern, E.W. Jones and J.R. Broach), pp. 371-414. Cold spring Laboratory Publications, NY.

Haynes, R.H., F. Eckardt and B.A. Kunz (1984) The DNA damage repair hypothesis in radiation biology: Comparisons with classical theory. *Brit. J. Cancer* **49, suppl. VI**, 81-90.

Herskowitz, I. (1988) Life cycle of the budding yeast *Saccharomyces cerevisiae*. *Microbiol. Rev.* **52**, 536-553.

Keszenman-Pereyra, J.R. (1990) Repair of UV-damaged incoming plasmid DNA in *Saccharomyces cerevisiae*. *Photochem. Photobiol.* **51**, 331-342.

Lemontt, J.F. (1977) Pathways of ultraviolet mutability in *Saccharomyces cerevisiae*. IV: The relation between canavanine toxicity and ultraviolet mutability to canavanine resistance. *Mutation Res.* **43**, 339-355.

MacQuillan, A.M., A. Herman, J.S. Coberty and G. Green (1981) A second photoreactivation defined mutation in *Saccharomyces cerevisiae*. *Photochem. Photobiol.* **34**, 673-677.

Maniatis, T., E.F. Fritsch and J. Sambrook (1982) *Molecular Cloning*, Cold Spring Harbor Laboratory, Cold Spring, Harbor, New York.

Moustacchi, E. (1969) Cytoplasmic and nuclear events induced by UV light in strains of *Saccharomyces cerevisiae* with different UV sensitivities. *Mutation Res.* **7**, 171-185.

Moustacchi, E. and V. Favaudon (1982) Cytotoxic and mutagenic effects of neocarzinostatin in wild-type and repair-deficient yeasts. *Mutation Res.* **104**, 87-94.

Moustacchi, E., V. Favaudon and E. Bisagni (1983) Likelihood of the new antitumoral drug 10-(α-diethylaminopropylamino)-6-methyl-5H-pyrido (3',4':4,5) pyrolo (2,3-g) isoquinoline (BD-40), a pyridopyroloisoquinoline derivative, to induce DNA strand breaks *in vivo* and its non-mutagenicity in yeast. *Cancer Res.* **43**, 3700-3706.

Moustacchi, E. (1987) DNA repair in yeast. Genetic control and biological consequences. *Adv. Radiat. Biol.* (Edited by J.T. Lett), **13**, pp. 1-30. Academic Press Inc.

Ogur, M., R. St. John and S. Nagai (1957) Tetrazolium overlay technique for population studies of respiratory deficiency in yeast. *Science* **125**, 928-929.

Parry, J.M. and F.K. Zimmermann (1976) The detection of monosomic colonies produced by mitotic chromosome non-disjunction in the yeast *Saccharomyces cerevisiae*. *Mutation Res.* **36**, 49-66.

Resnick, M.A. (1969) A photoreactivationless mutant of *Saccharomyces cerevisiae*. *Photochem. Photobiol.* **9**, 519-531.

Roman, M. (1956) A system selective for mutations affecting the synthesis of adenine in yeast. *C.R. Travaux Lab. Carlsberg* **26**, 299-314.

Sancar, A. and G.B. Sancar (1988) DNA repair enzymes. *Ann. Rev. Biochem.* **57**, 29-67.

Sancar, G.B. (1985) Sequence of the *Saccharomyces cerevisiae* PHR1 gene and homology of the PHR1 photolyase to *Escherichia coli* photolyase. *Nucleic Acids Res.* **13**, 8231-8245.

Schild, D., J. Johnston, C. Chargand and R.K. Mortimer (1984) Cloning and mapping of *Saccharomyces cerevisiae* photoreactivation gene PHR1. *Mol. Cell. Biol.* **4**, 1864-1870.

Tobias, C.A., E.A. Blakely, Q.H. Ngo and T.C.H. Yang (1980) The repair-misrepair model of cell survival. In: *Radiation Biology in Cancer Research* (Edited by R.E. Meyn and H.R. Withers), pp. 195-230. Raven Press, NY.

Tubiana, M., J. Dutreix and A. Wambesie (1986) *Radiobiologie*. (Edited by J. Herman), pp. 1-295. Paris, France.

White, C.I. and S.G. Sedgwick (1985) The use of plasmid DNA to probe DNA repair functions in the yeast *Saccharomyces cerevisiae*. *Mol. Gen. Genet.* **201**, 99-106.

White, C.I. and S.G. Sedgwick (1987) Induced cellular resistance to ultraviolet light in *Saccharomyces cerevisiae* not accompanied by increased repair of plasmid DNA. *Curr. Genet.* **11**, 321-326.

Yasui, A. and M.R. Chevallier (1983) Cloning of photoreactivation repair gene and excision repair gene of the yeast *Saccharomyces cerevisiae*. *Cur. Genet.* **7**, 191-194.

Zimmermann, K.K. (1975) Procedures used in the induction of mitotic recombination and mutation in the yeast *Saccharomyces cerevisiae*. *Mutation Res.* **31**, 71-86.

Zimmermann, K.K., R. Kern and H. Rasenberger (1975) A yeast strain for the simultaneous detection of induced mitotic crossing-over, mitotic gene conversion and reverse mutation. *Mutation Res.* **28**, 381-388.

Zimmermann, K.K., R.C. Von Borstel, E.S. Von Halle, J.M. Parry, D. Siebert, G. Zetterberg, R. Barale and N. Loprieno (1984) Testing of chemicals for genetic activity with *Saccharomyces cerevisiae*: A report of the U.S. Environmental Protection Agency Gene Tox. Program. *Mutation Res.* **133**, 199-244.

APPENDIX 1: Technical Information for Yeast Photobiology Experiments

Yeast Strains

Most yeast strains of *Saccharomyces cerevisiae* can be ordered from the Yeast Genetic Stock Center, Department of Biophysics and Medical Physics, University of California, Berkeley, CA 94270 USA. A catalogue including strain designation, genetic markers and references is available on request from the Yeast Genetic Stock Center (Sixth Edition 10/1/1987).

The yeast cultures sent out are prepared by spotting a heavy suspension of cells in evaporated milk onto a small folded section of sterile filter paper which is then wrapped in sterile aluminum foil. To revive the culture, the filter paper has to be streaked across on enriched agar medium, such as YEPD, see below. Growth should be visible two or three days after incubation at 30°C. If applicable, individual clones (where present) should be tested for genetic markers.

Preservation of Yeast Strains

For short periods (5 months), strains can be preserved on agar slants stored at 4°C. Use the following medium to prepare slants (adenine is added to inhibit the reversion of the *ade1* and *ade2* mutants which are often used as markers).

Dissolve the following in distilled water (%'s are by weight). Keep the solution heated in a bath of boiling water.

Bacto yeast extract	1%
Bacto-peptone	2%
Glucose	2%
Adenine sulfate	0.003%
Bacto-agar	2%

Dispense 1.5 ml aliquots into one dram (3.7 ml) vials. Loosely screw on the caps and autoclave the vials in a rack. After autoclaving, incline the rack so that the agar is just below the neck of the vial. Tighten the caps one day later.

For long periods (>1 year), strains can be preserved in "glycerol-YEPD" (see below) at –20°C. To preserve cells in this way, suspend about 10^8 cells, either from a liquid suspension in YEPD or from a YEPD plate streak (use a sterile toothpick) in 1 ml of "glycerol-

YEPD" in a 1 dram vial. Store the vial at –20°C. To start a new culture, inoculate 5 ml of YEPD with the 1 ml suspension in "glycerol-YEPD" and incubate at 28°C with aeration by rotary shaking.

Media

Prepare media for Petri dishes in 1 liter Roux culture bottles following one of the recipies given below. Each bottle should contain 500 ml of medium. This is sufficient for approximately 18 plates. Autoclave the media for 25 min at 120°C (250°F and 15 lbs pressure). Before use, allow the plates to dry at room temperature for 2 days after pouring. Plates can be stored in sealed plastic bags for over 2 months. For long periods of storage, it is best to pour plates under a sterile hood. For liquid media, omit the agar from the recipies below.

YEPD: A complete medium for routine growth. Dissolve the following in distilled water:

Bacto yeast extract (Difco)	0.5%
Bacto-peptone (Difco)	2%
Glucose	2%
Bacto-agar (Difco)	2%

Glycerol-YEPD: YEPD containing 50% (v/v) of glycerol.

YEPG: A complete medium containing a non-fermentable carbon source (glycerol) that does not support the growth of mitochondrial (*rho*$^-$) or nuclear (*pet*) respiratory deficient mutants. Dissolve the following in distilled water:

Bacto yeast extract	0.5%
Bacto-peptone	2%
Glycerol	3% (v/v)
Bacto-agar	2%

MM: A synthetic medium that supports growth of only prototrophic strains. It contains salts, vitamins, trace elements, and a nitrogen source (BYNB without amino acid). Dissolve the following in distilled water:

Bacto yeast nitrogen base without amino acids (Difco)	0.67%
Glucose	2%
Bacto-agar	3%

YPDG: A complex medium which can be used to differentiate between *rho*$^+$ and *rho*$^-$ colonies. Dissolve the following in distilled water:

Bacto yeast extract	1%
Bacto-peptone	2%
Glycerol	3% (v/v)
Glucose	0.1%

TTC: A medium used to distinguish between respiratory competent (*rho*$^+$), respiratory deficient (*rho*$^-$) and sectored colonies (Ogur et al., 1957).

Dissolve 15 gm of Bacto-agar in 900 ml of 0.067 M phosphate buffer at pH 7. Autoclave this medium at 110°C for 20 min. Dissolve 1% of 2,3,5 triphenyl-tetrazolium chloride in distilled water. Mix 100 ml of the latter solution with 900 ml of the melted Bacto-agar buffer medium at 55-60°C. Gently add 20 ml of this hot mixture to each plate containing yeast colonies. Leave the plates for 1 h at 30°C before checking for mutants (see below).

Methods Used to Distinguish between Respiratory Competent (*rho*$^+$) and Respiratory Deficient (*rho*$^-$) Cells

Several methods have been successfully employed for distinguishing *rho*$^+$ and *rho*$^-$ colonies. One simple method is to plate 100-200 cells on a YPDG dish (3% glycerol and 0.1% glucose as carbon sources) which is then incubated at 30°C for 3 to 4 days. At this time, the *rho*$^+$ colonies will be several times larger than the *rho*$^-$ colonies, owing to the fact that the *rho*$^-$ mutants cannot utilize glycerol for growth and that glucose is present in growth-limiting amounts.

When the colony size is modified by agents such as ultraviolet light, high temperature, genotoxic chemicals, etc., or if the detection of sectored colonies is desirable, other methods must be used. These include the following:

(1) Overlaying with 2,3,5-triphenyltetrazolium chloride. This method is very convenient. Employing TTC medium, respiratory-competent cell colonies turn red, whereas respiratory-deficient mutants stay white (Ogur et al., 1957).

(2) Replica-plating or streaking from YPDG or similar media containing non-fermentable carbon sources.

(3) The inclusion of various dyes in the growth medium.

(4) Employing *ade1* or *ade2* strains, which produce a red pigment when the cells are respiratory competent and which turn white when they are respiratory deficient.

References – Appendix 1

M. Ogur, R. St. John and S. Nagai (1957) Tetrazolium overlay technique for population studies of respiratory deficiency in yeast. *Science* **125**, 928-929.

APPENDIX 2: Analysis of Survival Data

The graphical representation of survival data in the form of survival curves is based on mathematical models defining parameters that are convenient for detailed descriptions and comparisons (for a review see Haynes and Eckardt, 1979).

The simplest case described by target theory assumes that a single event causes the effect, for example the action of a single photon on a critical cellular target. It also assumes that the initial cell population is homogenous with regard to sensitivity to cytotoxic and stochastic effects that are responsible for the induction of lethal effects. The number of lethal events is proportional to the fluence, and the surviving fraction N/N_0 is an exponential function of the fluence, where N is the number of surviving colonies and N_0 the number of viable cells in the untreated cell population.

In the surviving fraction of cells, there are by definition, no lethal events. The surviving fraction, S_F (defined as N/N_o), can be expressed, according to Poisson statistics: as;

$$S_F = N/N_o = e^{-\alpha D} \quad \text{or} \quad e^{-D/D_o} \qquad [A1]$$

where e is the base of the natural logarithms (=2.7). α defines the slope $1/D_o$, D_o is the fluence that reduces survival to 37% (1/e) which corresponds to a mean of one lethal hit per cell.

Eq. A1 can be rewritten as:

$$\ln S_F = -\alpha D = D/D_o \qquad [A2]$$

Equation A2 indicates that a semi-logarithmic plot of S_F versus fluence should be linear with a slope of α (= $1/D_o$). For example, exponential survival curves are obtained with haploid yeast cells after exposure to ionizing radiation.

However, survival curves obtained with eukaryotic cells after UV irradiation are not of the single hit, exponential type, but of the multi-hit type. That is, the survival curves are not linear on a semi-logarithmic plot, but rather demonstrate more complex kinetics. Several models have been developed to describe these cases (Tubiana et al., 1986). Following target theory, cell death may arise from inactivation of several (n) targets. If this is the case, the surviving fraction can be expressed as:

$$S_F = 1 - (1 - e^{-D/D_o})^n \qquad [A3]$$

and a shouldered survival curve will be obtained. In this model n targets are susceptible to lethal hits, but the final slope of the survival curve is still $1/D_o$.

Another possible interpretation of shouldered survival curves is based on the assumption of two independent events leading to lethality, either by a single lethal event or by the addition of two independent sublethal events. In such a linear quadratic model, the surviving fraction can be expressed as:

$$S_F = e^{-(\alpha D + \beta D_2)} \qquad [A4]$$

Here α is the probability of induction of lethal lesions, and β is the probability of induction of sublethal lesions. The slope of the survival curve changes from $-\alpha$ to $-\beta$ with increasing fluence. The ratio α/β is the fluence at which cell death is equally due to irreparable lethal lesions and to the accumulation of sublethal lesions.

Similar models based on the repair of lesions have been proposed by Haynes et al. (1984) and Tobias et al. (1980). Although most of these models underline the stochastic nature of lethal events, keep in mind that this may hold for ionizing radiation, but is not quite accurate for UV radiation which is known to be absorbed by nucleic acids and to induce specific lesions (i.e., pyrimidine dimers) dependent on nucleotide sequence.

Isolation and Biological Activity of Plant Derived Phototoxins

John Thor ARNASON, Robin J. MARLES and Richard R. AUCOIN
Ottawa Carleton Institute of Biology, University of Ottawa Canada

INTRODUCTION

While it has been known for some time that certain species of plants contain photosensitizers, it is only recently that the *raison d'être* of light activated toxins has been demonstrated. All of these photosensitizers fall into a diverse group of secondary compounds with no known function in primary metabolism such as photosynthesis, nitrogen assimilation etc., but have been shown to act as defenses against phytophagous insects and fungal pests. Evolution of the biosynthetic pathways leading to these compounds has probably occurred through a co-evolutionary process where pressure by herbivores or other pests has led to the selection of chance mutations which have produced novel, increasingly complex and toxic secondary substances. For example, work by Berenbaum (1983) has shown that while the widely distributed phytochemical umbelliferone is relatively non-toxic to armyworms, its derivative, xanthotoxin of the *Apiaceae*, formed biosynthetically by prenylation and cyclization, is phototoxic to this insect when ingested. Expression of toxicity was clearly shown to be dependent on irradiation by photosensitizing wavelengths. Other studies have shown that ingested plant photooxidants such as alpha terthienyl or phenylheptatriyne (Champagne *et al.*, 1986; Downum *et al.*, 1984) can cause black necrotic lesions and pupal deformities to develop in phytophagous larvae at acute doses, or at lower doses lengthen their development time and reduce final weights.

The few insects adapted to feeding on phototoxic plants react to phototoxins by avoiding light through behaviors such as leaf rolling or night-time feeding (Fields *et al.*, 1990; Sandberg and Berenbaum, 1989). Forced exposure to light has been shown to be toxic to several of these species. Other adapted insects may employ biochemical defenses against photooxidants, such as large amounts of the singlet oxygen quencher beta-carotene found in *Chrysolina* beetles or high levels of activity of superoxide dismutase in *Anaitis plagiata* (Aucoin *et al.*, 1990). Phototoxic plants appear to have several counteradaptation strategies

JOHN THOR ARNASON, ROBIN J. MARLES, and RICHARD R. AUCOIN, Ottawa Carleton Institute of Biology University of Ottawa, Ottawa, Ontario, K1N 6N5, Canada

Photobiological Techniques, Edited by D.P. Valenzeno *et al.*
Plenum Press, New York, 1991

as well, such as clear glands in the genus *Hypericum* that can transmit phototosensitizing wavelengths to an insect hiding under a leaf or in a leaf roll (Fields *et al.*, 1990) or synthesis of phototoxins in *Tagetes* following attack by fungi (Kourany *et al.*, 1988)

A very diverse group of secondary metabolites have been shown to have photosensitizing properties (Table 1) (Towers, 1984; Knox and Dodge, 1985). Evidently phototoxic activity has evolved independently in many families of plants in at least 7 different biosynthetic groups of secondary compounds.

Among the most active plant derived phototoxins are the polyacetylenes and their sulfur derivatives. These secondary compounds are found in fungi and several groups of higher plants but find their greatest diversity in the largest plant family, the *Asteraceae*. Over 500 distinct chemical structures containing conjugated double and triple bond systems have been identified by Bohlmann *et al.* (1973). Biosynthetically, these structures are derived from oleic acid by desaturation and chain shortening. C13, C14, C17 and C18 compounds are all common in higher plants. Frequently adjacent triple bond systems can be cyclized to form a thiophene ring. While the role of these substances as natural protectants is important to our understanding of environmental photobiology, their application is also envisaged as insecticides, especially as mosquito larvicides (Arnason *et al.*, 1989) or antiviral compounds (Hudson, 1990).

Table 1

Phototoxins from higher plants

Biosynthetic Class	Representative Families	Mode of Action
Acetylenes and Thiophenes	*Asteraceae*	Photooxidants
Extended Quinones	*Guttiferae*	Photooxidants
Furanocoumarins	*Apiaceae* *Rutaceae* *Fabaceae* *Moraceae* *Beta-carbolines*	Photogenotoxins
Furoquinolines	*Rutaceae*	Photogenotoxins
Furanochromones	*Fabaceae*	Photogenotoxins
Isoquinolines	*Berberidaceae* *Papaveraceae* *Anonaceae*	Photooxidants
Beta carbolines	*Rubaceae* *Simaroubaceae*	Photooxidants Photogenotoxins

EXPERIMENT 20: SURVEY OF PLANTS FOR PHOTOTOXICITY

OBJECTIVE

This laboratory experiment explores a simple yeast screening method (Daniels, 1965) that can be used for detecting phototoxicity of plants using small tissue samples or extracts, and is especially sensitive to the polyacetylenes (Camm *et al.*, 1975). Plant extracts and 8-methoxypsoralen will be assayed for phototoxic potency in the yeast lawn bioassay system.

LIST OF MATERIALS

Common brewers yeast *Saccharomyces cerevisiae* obtained from grocery store or microbiological culture of same, if available.
Petri plates, plastic
Sabaraud nutrient agar (or other yeast growth medium)
Aluminum foil
Cotton-tip swabs in test tube with autoclave closure
Micropipette or Hamilton syringe
Glass rod
Filter paper disks, diameter = 5 mm, Whatman #1
Forceps
8-methoxypsoralen
UV-A lamps (e.g., Westinghouse BLB-40W; bank of 4)
UV filter (Kodak CP 2B) or other cut-off filter at 400 nm

PREPARATIONS

Plant collection: Field collect and identify locally available species of plants, especially members of the sunflower family and other phototoxic families (one plant per student). Examples are given in Table 2. Plants must be fresh and in good condition to demonstrate activity. Prepare phototoxic tests using the yeast lawn test for each plant organ (leaf, flower, seed, stem, root) in duplicate and prepare a table of results organized by family, species, and plant part. (Or buy a variety of plants listed in Table 2.)

Table 2

Common Phototoxic Plants

Species	Common Name	Leaves	Flower	Stem	Root
Bidens pilosa	Beggar's Tick	+	+	+	+
Tagetes erecta	Marigold		+		+
Rudbeckia hirta	Blackeyed Susan		+	+	+
Chrysanthemum leucanthemum	Daisy		+		
Centaurea nigra	Knapweed	+	+	+	+
Dahlia variabilis	Dahlia	+			+
Flaveria trinerva	—	+	+	+	+
Porophylum edulis	Papaloquilote	+	+	+	+

Reference: Camm et al, 1975

Yeast Cultures: Two days before starting the experimental procedures, prepare an active yeast culture by sprinkling brewers yeast onto sabaraud agar plates or subculturing microbial culture.

8-Methoxypsoralen Stock Solution: Dissolve 10 mg of 8-methoxypsoralen in 10 ml of ethanol. Keep tightly capped when not in use.

EXPERIMENTAL PROCEDURE

1. Prepare in duplicate:
   a) Sabaraud plate with yeast lawn: Inoculate a swab by touching an active yeast culture. Use the sterile swab to spread a thin layer of active yeast on a clean plate as evenly as possible. The agar surface must be dry for the test to work.
   b) Disks of filter paper treated with 10 μl of 8 methoxypsoralen stock solution or fractions from the phytochemical extraction above. Be sure to allow any solvent to evaporate.
   c) Small pieces of plant material from *Tagetes, Bidens,* etc. Cut 2-3 mm pieces of root, leaf and stem and crush each lightly with a clean glass rod or a dental pick in a clean Petri dish to express juices.
2. Prepare 2 duplicate plates by placing disks and tissue on the Petri plates. Press the samples down lightly but firmly with forceps.
3. Wrap a dark control plate in aluminum foil. Place the duplicate for UV treatment under the banks of blacklights.
4. One or two days later assess phototoxicity by holding each plate up to a window. Measure the zone of inhibition around the disks (distance from disk to the yeast) in mm using a ruler. Phototoxicity is manifested by inhibition on UV-treated but not on unilluminated plates.

DISCUSSION

Many species of *Asteraceae* are highly phototoxic in this test, but the species listed in Table 2 should give reliable positive results. For other phototoxic plants in other families, good results have been obtained if concentrated alcoholic extracts of the plants are tested on filter paper disks. Another strategy to enhance the activity of photogenotoxins in the test is to use yeast or bacteria that are DNA repair deficient.

EXPERIMENT 21: EXTRACTION AND BIOASSAY OF PLANT PHOTOTOXINS

OBJECTIVE

A bioassay guided phytochcmical isolation of phototoxins is attempted and the precise level of phototoxicity of the isolated material to an invertebrate model (brine shrimp) (Alkofahi *et al.*, 1989) is determined by probit analysis.

LIST OF MATERIALS

Phytochemical Extraction

Mortar and pestle, or blender
Petroleum ether (30-60°C)
Ethanol (95%)
Separatory funnel, 500 ml
Sodium sulfate (anhydrous)
Flash evaporator
2 × 30 cm glass column with teflon stopcock
Silica gel G for column chromatography
Alumina for column chromatography
UV-visible spectrophotometer
Beakers
Erlenmeyer flasks
Small vials

Shrimp Bioassay

Sea salt (synthetic, from aquarium supply store, or natural if a supply of standard composition can be obtained)
HCl, 1N, 25 ml
Brine shrimp eggs from aquarium supply store (will keep with refrigeration for >1 year)
Distilled water
Solvent for sample (e.g., acetone, methanol, or ethanol)
Beakers, 1000 ml, and cover glass for beakers (2)
Funnel, large, glass and filter paper, #1, large (e.g., 40 cm)
Air supply (e.g., small aquarium pump)
Plastic air hose with thumb-screw clamp to regulate air pressure
Lamp with incandcsccnt 60 W bulb
Beakers or wide mouth vials, clear glass, 10 ml (20) and rack
Test tubes, small, 2 ml or less (10) and rack
Pipettes, 5 ml, 1 ml (automatic if available; tip opening should be wide and smooth enough not to injure shrimp during pipetting)
Hamilton syringes, 100 μl and 10 μl
Plastic wrap
Vortexer and/or sonicator to help dissolve samples
pH meter
Magnifying glass or dissecting microscope with substage lighting
Magnetic stirrer and stirring bar to dissolve sea salt
Warm water bath to help dissolve samples
Buchner funnel and Whatman No. 1 filter paper, receiving flask

PREPARATIONS

Brine Shrimp: Dissolve 76 g of sea salt in 2 liters of distilled water. Stir using a magnetic stirrer to make "brine". Adjust the pH of the brine to approximately 7.6 using dilute HCl. Filter the brine through Whatman No. 1 filter paper using a Buchner funnel and aspiration. Pour 1000 ml into one beaker for hatching the eggs. Reserve the other 1000 ml. Add 100 mg of brine shrimp eggs to the brine, insert the air hose to aerate the water gently, cover the beaker, and turn on the lamp about 50 cm above the beaker. Hatching should start within 24-48 h, depending on room temperature. Allow the shrimp to mature for another 24 h after hatching.

Phototoxin Standards: Use the extinction coefficients of α-terthienyl (α-T) and phenylheptatriyne (PHT; 22 500 $M^{-1}\cdot cm^{-1}$ at 350 nm, mol. wt.= 248 and 17 320 $M^{-1}\cdot cm^{-1}$ at 310 nm, mol. wt.= 164, respectively) to prepare 0.2 $mg\cdot ml^{-1}$ stock solutions (preferably diluted in ethanol). This isolated material can be used for accurate determination of phototoxicity in the brine shrimp bioassay or a pure phototoxin such as 8-methoxypsoralen may be used.

EXPERIMENTAL PROCEDURE

Extraction of α-T from *Tagetes Erecta* (Marigold) or PHT from *Bidens Pilosa* (Beggar's tick)

1. Grind in a mortar and pestle 100 g of cleaned roots (*Tagetes*), or blend 50 g of leaves (*Bidens*) in a blender in 50-100 ml of ethanol. Filter the extract through Whatman No. 1 paper using a Buchner funnel and aspiration.
2. Add an equal volume of H_2O and extract 3 times in a separatory funnel with 100 ml of petroleum ether.
3. Combine petroleum ether fractions. Pour them into a clean dry beaker to remove residual water. Add dry sodium sulfate to dry solvent. (Save the aqueous fraction separately).
4. Reduce the volume of the extract to 5-10 ml on a flash evaporator, or if unavailable, use a stream of air. Do not take to dryness.
5. Prepare a slurry of 100 g of alumina (for α-T) or silica gel G (for PHT) in petroleum ether. Place a small amount of cotton wool in the bottom of a column. Add the slurry and then more solvent. Pack the column by running solvent through it.
6. Bring the solvent line down to the top of the silica gel or alumina. Carefully add extract to the top, to produce an even layer of extract on the head of the column and allow it to run into the alumina or silica gel – then stop. Add 5-10 ml more solvent and run it into the alumina or silica gel again.
7. Top up the column with solvent and maintain it filled. For *Tagetes*, follow the 2 blue fluorescing bands with a UV lamp as they separate and descend the column (α-terthienyl and bithienyl). Use 1% acetone in petroleum ether to increase the elution rate. Collect both bands in separate flasks. For *Bidens*, PHT is usually eluted in pure petroleum ether as the first fraction following the carotenoid band (yellow-orange).
8. Concentrate the eluentto near dryness and place it in a small vial. Note: phenylhepatatriyne (but not alpha-terthienyl) is unstable if taken to complete dryness.
9. After suitable dilution (for example 1/1000) of a sample, determine a UV spectrum from 300-400 nm. Compare this with authentic spectra (Figs. 1-3). Note: For a laboratory exercise, it is normally convenient to stop the procedure at this point and continue the experiment in a subsequent session.

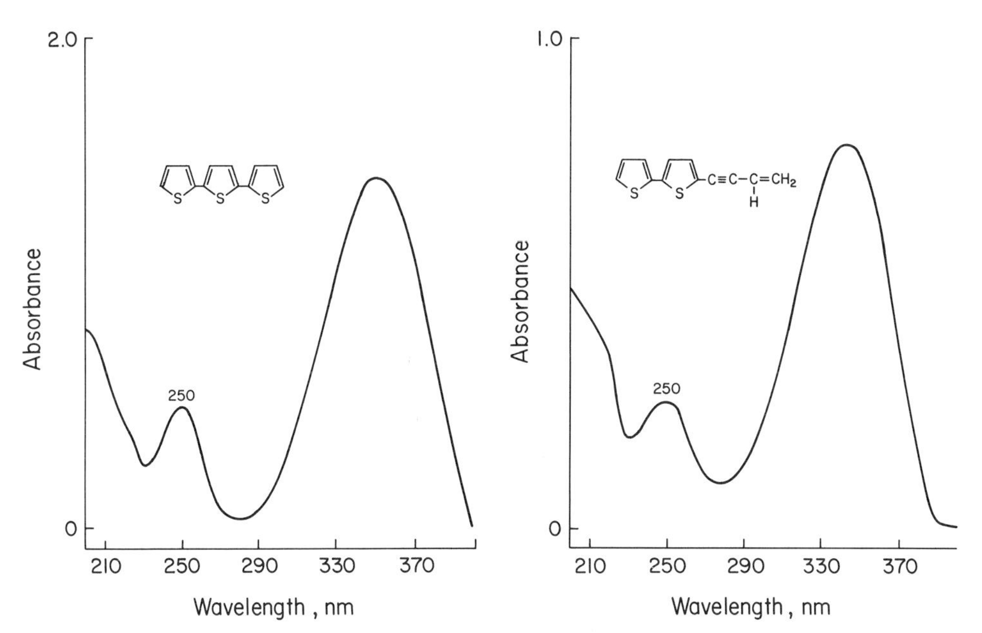

Figure 1. Absorption spectrum of α-terthienyl. The figure above shows the absorption spectrum of α-T in the UV. The structure of α-T is shown in the inset.

Figure 2. Absorption spectrum of bithienyl. The figure above shows the absorption spectrum of bithienyl in the UV. The structure of bithienyl is shown in the inset.

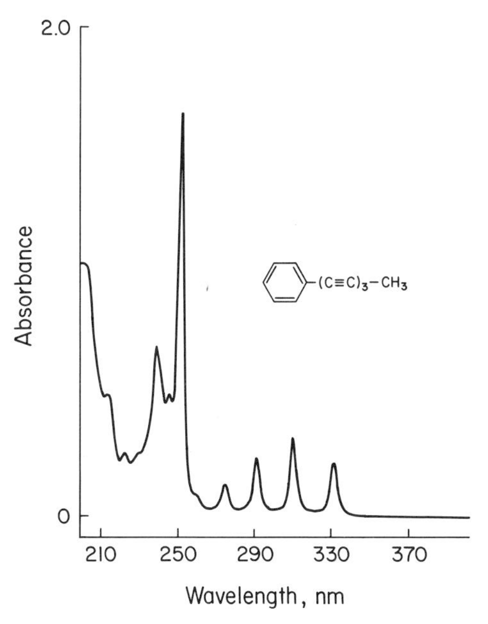

Figure 3. Absorption spectrum of phenylheptatriyne. The figure above shows the absorption spectrum of PHT in the UV. The structure of PHT is shown in the inset.

10. The aqueous fraction from the separatory funnel and all fractions collected from the columns should be tested for phototoxicity in the yeast bioassay to confirm the isolation of the active principles.

Brine Shrimp Bioassay

Day 1: (When brine shrimp are ready)

1. Label 10 ml beakers with sample and concentration information, then add 4 ml of brine from the reserved liter to each vial.
2. Using the glass syringes and small test tubes, prepare serial logarithmic dilutions (1/10 dilution) of samples in their solvent (e.g., 10 μl of stock + 90 μl solvent, vortex). Later, when an approximate lethal dose for 50% mortality (LD_{50}) is known, a geometric dilution series (half dilutions) can be used to get more accurate results. Rinse out the syringe several times with solvent between samples.
3. Add 25 μl of the sample dilutions to beakers containing brine, using a syringe, and vortex. The final concentration for the stock will be 0.2 $mg{\cdot}ml^{-1} \times 0.025$ ml/5 ml = 10 $\mu g{\cdot}ml^{-1}$ and serial dilutions: 1, 0.1, 0.01, 0.001 and 0.0001 $\mu g{\cdot}ml^{-1}$. The final concentration of solvent will be 0.025 ml/5 ml = 0.5%. It may be necessary to alter this dilution scheme to observe toxicity under different conditions.
4. Remove the air hose from the hatching beaker and allow at least 15 min for the unhatched eggs to settle to the bottom and empty egg shells to float to the very top of the brine. The brine shrimp larvae (nauplii) can then be seen swimming near the surface and wherever the light is brightest, since they are attracted to light.
5. Take 500 ml of the reserved brine in the second beaker and aerate it vigorously to stir it (magnetic stirring tends to injure the nauplii). Using a pipet or dropper with a wide smooth opening (e.g., the end of a disposable 1 ml Pipetman tip cut off with scissors), transfer some of the swimming nauplii from the hatching beaker to this second beaker of brine. Do this several times and then count the concentration of nauplii in the second beaker by taking some random 1 ml aliquots in test tubes and seeing if there is an average of 10 $nauplii{\cdot}ml^{-1}$, which is the optimum concentration. Add more nauplii or brine as needed to achieve this concentration in the second beaker.
6. You can now add 1 ml of brine with nauplii from the second beaker to each of the sample beakers. This will give an average of 10 nauplii per treatment, without having to count them out for every beaker. This saves time if there are several samples at several concentrations and several replicates.
7. Each beaker now has 5 ml (actually 5.025 ml) total volume, with sample or solvent for the control, and approximately 10 nauplii. The rack of beakers can be lightly covered with a piece of plastic wrap to prevent significant evaporation. For the study of phototoxins, two sets of samples are prepared, and one is irradiated for 4 h at 300-400 nm (near UV) 5 $W{\cdot}m^{-2}$, then held for 20 h while the other set is protected from UV by a sheet of UV filter or aluminium foil. Different concentration ranges are required for activity in irradiated and non-irradiated treatments (100-10 000 × higher in nonirradiated groups).

Day 2

8. The number of dead nauplii (motionless at bottom of beaker) is counted with the aid of a dissecting microscope or magnifying glass. Substage lighting is best because then there are no reflections off the surface of the water or vial. Moribund nauplii (only slight uncoordinated twitching with no propulsion) are counted as dead.

9. It is usually too difficult to count live nauplii because they are swimming too fast, so after the number of dead has been counted, 1 ml of methanol is added to each vial. After 1 h all nauplii should be dead, and the total number can be determined.

10. Calculate the average percent mortality for each treatment. If the solvent control shows more than 5% mortality, results (up to 20% mortality) can still be salvaged by correcting each sample percent mortality by use of Abbot's Formula:

$$\frac{\text{Sample \% Mortality} - \text{Control \% Mortality}}{(100 - \text{Control \% Mortality})} \times 100\%$$

11. To determine the LD_{50}, use the probit transformation of the data (log{%mortality/%survival}) plotted on the ordinate versus log concentration (abscissa), and perform a linear regression. The intercept on the abscissa is the LD_{50}. For determination of 95% confidence limits, if desired, the probit procedure can be found in most statistical software programs.

12. Determine the ratio of the toxicity in the irradiated group to the non-irradiated group, which is the phototoxicity ratio.

DISCUSSION

The bioassay guided isolation described here can be used for other phototoxins such as harman alkaloids (Arnason *et al.*, 1983). While other parts of *Tagetes* (marigold) contain phototoxins, the reason for undertaking the isolation from roots is to avoid the problems of separating the photosynthetic pigments. For an explanation of the possible role of phototoxic substances in roots see Gommers and Bakker (1988). While yeast and brine shrimp are not natural pests of plants, they provide a convenient microbial and invertebrate assay which can be used to indicate phototoxicity without attempting much more complex bioassays with phytophagous insects and phytopathogenic fungi.

REVIEW QUESTIONS

1. What is the photophysical reason that the majority of plant derived phototoxins are activated in the near UV range rather than other areas of the electromagnetic spectrum?
2. What are the possible advantages to a plant of chemical defense using a phototoxin over a non-phototoxin?
3. Did the phototoxic tests of the fractions from the isolation confirm the identification of the phototoxin? Is it possible that there are others?
4. What factors are likely to influence the LD_{50} of a photooxidant?

ANSWERS TO REVIEW QUESTIONS

1. This is one area of the spectrum where photosynthetic pigments do not absorb strongly.
2. Phototoxicity allows plants to use high energy excited state processes to defend themselves from pests and they do so with a minimum investment of metabolic energy, using instead the light available in their environment as an energy source.

3. Yes. Only the fractions containing phenylheptatriyne from *Bidens pilosa* or thiophenes from *Tagetes erecta* should be phototoxic.
4. The toxicity of a photooxidant is influenced by the quantum yield of singlet oxygen generation, the total number of photons absorbed from the available light source and other factors such as partition coefficient and biological target.

SUPPLEMENTARY READING

Heitz, J. and K. Downum, Editors (1987) *Light-Activated Pesticides*, Amer. Chem. Soc. Symp. Ser. Vol. **339**, 1987.

Robberecht, R. (1989) Environmental photobiology. In *The Science of Photobiology*, 2nd edition (Edited by K.C. Smith), pp. 135-154. Plenum Press, New York.

REFERENCES

Alkofahi, A., J.K. Rupprecht, J.E. Anderson, J.L. McLaughlin, K.L. Mikolajczak and B.A. Scott (1989) Search for new pesticides from higher plants. *Amer. Chem. Soc. Symp. Ser.* **387**, 25-43.

Arnason, J.T., P. Morand, J. Salvador, I.Reyes and G.H.N.Towers (1983) Phototoxic substances from *Flaveria trinervis* and *Simira salvadorensis*. *Phytochem.* **22**, 594.

Arnason, J.T., B.J.R. Philogène, P. Morand, K. Imrie, B Hasspieler and A.E.R. Downe (1989) Naturally occurring and synthetic thiophenes as insecticides. *Amer. Chem. Soc. Symp. Ser.* **387**, 164-172.

Aucoin, R., J.T. Arnason (1990) The protective effect of antioxidants to a phototoxin sensitive insect herbivore *J. Chem. Ecol.*, **16**, 2913-2924.

Berenbaum, M. (1983) Coumarins and caterpillars: a case for co-evolution. *Evolution* **37**, 163-179.

Bohlmann, F.,T. Burkhardt, and C. Zdero (1973) *Naturally Occurring Acetylenes*. Academic Press.

Camm E.L., G.H.N. Towers and J.C. Mitchell (1975) UV mediated antibiotic activity of some Compositae species. *Phytochem.* **14**, 2007-2011.

Champagne, D.E., J.T. Arnason, B.J.R. Philogène, P. Morand and J. Lam. (1986) Light mediated allelochemical effects of naturally occurring polyacetylenes and thiophenes from the *Asteraceae* on herbivorous insects. *J. Chem. Ecol.* **12**, 835-858.

Downum, K.R., G.A. Rosenthal and G.H.N. Towers (1984) Phototoxicity of the allelochemical, alpha-terthienyl to larvae of *Manduca sexta Pest. Biochem. Physiol.* **22**, 14-109.

Fields, P., J.T. Arnason and B.J.R. Philogène (1989) The behavioral and physical adaptation of 3 insects that feed on the phototoxic plant *Hypericum perforatum. Can. J. Zool.* **68**, 339-346.

Fields, P.G., J.T. Arnason and R.G. Fulcher (1990) The spectral properties of *Hypericum perforatum* leaves: The implications for its photoactivated defenses. *Can. J. Bot.* **68**, 1166-1170.

Gommers, F. and J. Bakker (1988) Mode of action of alpha-terthienyl and related compounds may explain the supression effects of Tagetes species on populations of free living endoparasitic nematodes. *Bioactive Molecules* **7**, 61-70.

Hudson, J.B. (1989) Antiviral properties of photosensitizers used in dermatology: Recent advances and alternative interpretations. *Photodermatol.* **6**, 155-165.

Kourany, E., J.T. Arnason and E. Schneider (1988) Accumulation of phototoxic thiophenes in *Tagetes erecta* elicited by infection with *Fusarium oxysporum. Physiol. Molec. Plant Pathol.* **33**, 287-298.

Knox, J.P. and A.D. Dodge (1985) Singlet oxygen and plants. *Phytochem.* **24**, 889-896.

Sandberg, S. and Berenbaum, M. (1989) Leaf-tying by tortricid larvae as an adaptation for feeding on phototoxic *Hypericum perforatum. J. Chem. Ecol.* **15**, 875-885.

Towers, G.H.N. (1984) Interaction of light with phytochemicals. *Can. J. Bot.* **62**, 29.

Visual Pigments: Absorbance Spectra and Photoproducts

James K. BOWMAKER
University of London
U.K.

INTRODUCTION

For an animal to see, to detect the light reflected from objects in its environment, the light must first be absorbed. In vertebrates this is achieved through eyes that focus, through a single lens, an image of the environment onto the retina, a light sensitive neural layer lining the back of the globe of the eye. Within the retina, photoreceptor cells contain high concentrations of molecules that selectively absorb light and are therefore colored. These are the visual pigments that bleach on the absorption of light, becoming pale yellow, and in the process, trigger an enzyme cascade that leads to excitation of the neural network of the retina and ultimately to a visual sensation.

Photoreceptors

Photoreceptors in vertebrates are divided into two major categories, rods and cones. The retinas of most vertebrates contain both types. In nocturnal species rods predominate while in diurnal species cones make up the majority of the photoreceptors. This simple observation led to the conclusion that rods are used in dim light for scotopic vision, whereas cones are used in bright light for photopic vision. Since we can perceive colors only in bright light, it follows that cones and not rods must underlie our ability for color discrimination.

Rods and cones are morphologically rather similar in that they are elongated cells consisting of an inner and outer segment (Fig. 1). The inner segment, closest to the lens, contains the normal cellular constituents, including the nucleus and mitochondria, and connnects chemically at a synapse with retinal interneurons, principally with bipolar and horizontal cells. Extending from the inner segment is the outer segment, furthest from the lens, that developmentally is a highly modified cilium. The two segments are connected via a narrow ciliary stalk that contains the typical nine pairs of ciliary filaments.

The outer segment is made up of a stack of membrane discs. It is in their formation that rods and cones differ most markedly (Young, 1967). The discs are formed by invaginations of the plasma membrane. In rods they become pinched off and are free within

---

JAMES K. BOWMAKER, Department of Visual Science, Institute of Ophthalmology, University of London, Judd Street, WC1H 9QS, London, U.K.

the cytoplasm. This is a dynamic process and as more discs are formed they migrate along the length of the outer segment to the distal end where the tips of the outer segments are removed by phagocytosis. This is carried out by processes of the cells of the pigment epithelium, a cellular layer containing melanin that forms the black non-reflecting inner lining of the eye. The discs have a regular spacing of about 30 nm so that an average-sized rod about 30 μm long will contain about 1000 discs. In cones, the discs do not become isolated within the cell, but remain as a series of invaginations throughout the length of the outer segment (Fig. 1). The visual pigments are incorporated into the disc membranes and make up about 80% of the membrane proteins. The enormous number of discs enables the cell to achieve a high concentration of visual pigment, clearly necessary if the cell is to be highly photosensitive.

Most vertebrates possess only a single class of rod, normally maximally sensitive in the green region of the spectrum around 500 nm. But if the species has color vision, then at least two spectrally distinct classes of photoreceptor (normally cones) are found. In this way the nervous system is able to compare the rate of quantum catch from one class of receptor with the rate of quantum catch from the other. In many species color vision involves more than the basic dichromatic system, with trichromacy as in humans, with cones maximally sensitive in the red, green and blue spectral regions, and tetrachromacy in many freshwater fish and birds. In species where four spectrally distinct classes of cone subserve color vision, the spectral range of the species is often extended beyond that of trichromats into the near UV with a cone class maximally sensitive around 360 nm (for a review, see Bowmaker, 1991).

Visual Pigments

The structure of the visual pigment molecules contained within the photoreceptors is now known in great detail. All those in vertebrates, and indeed many of those in invertebrates, are similar, consisting of a large protein moiety, opsin, that has embedded within it, a prosthetic group, the aldehyde (retinal) of Vitamin A (retinol). Since there are two forms of

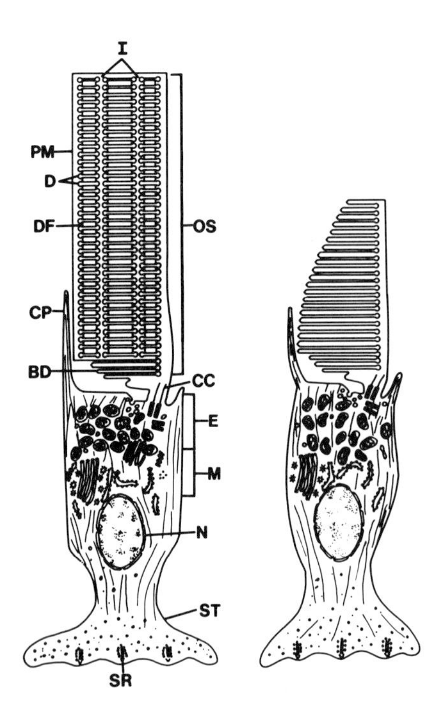

Figure 1. Schematic diagram of the structure of a mammalian rod (left) and cone (right) photoreceptor (from Bok, 1985). OS, outer segment; D, discs with membrane-bound visual pigment; PM, plasma membrane; BD, basal discs; I, incisures (rods only); DF, disc filaments; CC, connecting cilium; E, ellipsoid region containing mitochondria; M, myoid region; CP, calycal process; N, nucleus; ST, synaptic terminal; SR, synaptic ribbons.

Vitamin A, A_1 and A_2, there are two great families of visual pigments, the rhodopsins based on retinal from Vitamin A_1 and the porphyropsins based on 3-dehydroretinal from Vitamin A_2 (Fig. 2). Although these terms were originally given to the readily extractable rod pigments, rhodopsin with λ_{max} (wavelength of maximum absorbance) around 500 nm and porphyropsins with λ_{max} around 530 nm, they are often used to describe any visual pigment either from a rod or cone and irrespective of its λ_{max}. Thus the four visual pigments in humans, the rod pigment with λ_{max} at about 496 nm and the cone pigments with λ_{max} at about 563, 530 and 420 nm (Dartnall et al., 1983) are all rhodopsins.

Opsins are trans-membrane proteins packed into the discs within the outer segments of the visual receptor cells. Typically, mammalian rod opsins are composed of a polypeptide chain of about 348 amino acids that is thought to traverse the membrane seven times (Hargrave et al., 1984). The trans-membrane segments are thought to be α-helices and to comprise about 50% of the molecular mass, with a further 25%, including the carboxyl terminal region, exposed on the external aqueous surface of the membrane and the remaining 25%, including the amino-terminal end, exposed on the internal aqueous surface (Fig. 3). Retinal is bound by a Schiff's base linkage to a lysine (at position 296 in the polypeptide chain of the rod opsin) lying midway in the seventh helix.

Opsins absorb maximally in the UV below 300 nm, whereas retinal and 3-dehydroretinal have λ_{max} at about 380 and 400 nm respectively (Knowles and Dartnall, 1977). It is the formation of the chromophoric group, when retinal is incorporated into opsin, that leads to a broad absorbance band in the visible region of the spectrum. The exact spectral location of the λ_{max} of a visual pigment is established by the genetically determined amino acid sequence of the specific opsin and its interactions with retinal embedded within the trans-membrane helices (Nathans et al., 1986). For any given opsin there can be two pigments with different λ_{max} because of the two forms of retinal, one a rhodopsin and the other a porphyropsin, forming what is often referred to as a "pigment pair". Overall, within both vertebrates and invertebrates, rhodopsins are found with λ_{max} ranging from about 350 nm in the near UV to about 570 nm in the red, whereas porphyropsins range from about 360 nm to above 620 nm.

In mammals and birds porphyropsins are not present since the animals do not manufacture Vitamin A_2, but in other vertebrate groups both of the pigment pairs may be synthesized. The additional double bond in the carbon ring of 3-dehydroretinal (Fig. 2) is reflected in the longer λ_{max} of the porphyropsin in a pigment pair, though the effect is wavelength dependent, the longer the λ_{max} of the rhodopsin, the greater the long-wave displacement of the porphyropsin. This is illustrated by the visual pigments of the tadpole

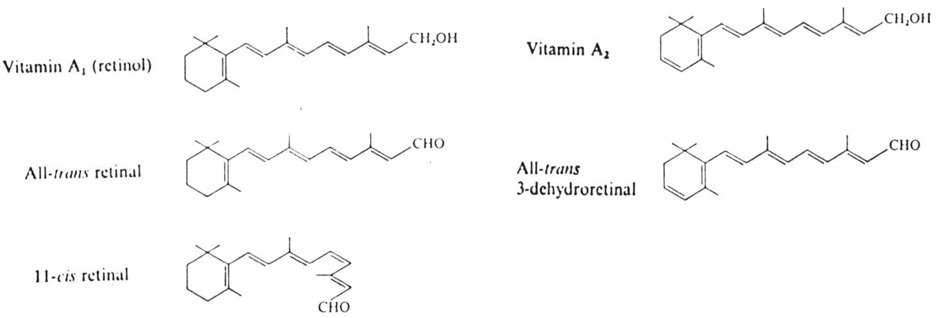

Figure 2. Structural formulae of Vitamin A_1 and A_2, all-*trans* retinal, 11-*cis* retinal and all-*trans* 3-dehydroretinal.

and adult frog, *Rana pipiens*, in which the rods contain a rhodopsin with λ_{max} = 502 nm in the adult, but a porphyropsin with λ_{max} = 527 nm in the tadpole, but both based on the same opsin. On the other hand the long-wave (red-sensitive) cones contain a rhodopsin $P575_1$ and a porphyropsin $P620_2$ in the adult and tadpole respectively (Liebman and Entine, 1968).

The absorbance of a visual pigment in the visible region of the spectrum consists of a broad a-band that covers a wide range of the spectrum. A typical rhodopsin with maximum absorbance at 500 nm has a bandwidth at 50% absorbance of about 102 nm. Visual pigments with λ_{max} at longer wavelengths have even broader spectra, whereas pigments with λ_{max} at shorter wavelengths have narrower spectra. The porphyropsins, as a class of pigments, have slightly broader spectra than their paired rhodopsins: $P500_2$ has a bandwidth at 50% absorbance of about 117 nm (Knowles and Dartnall, 1977).

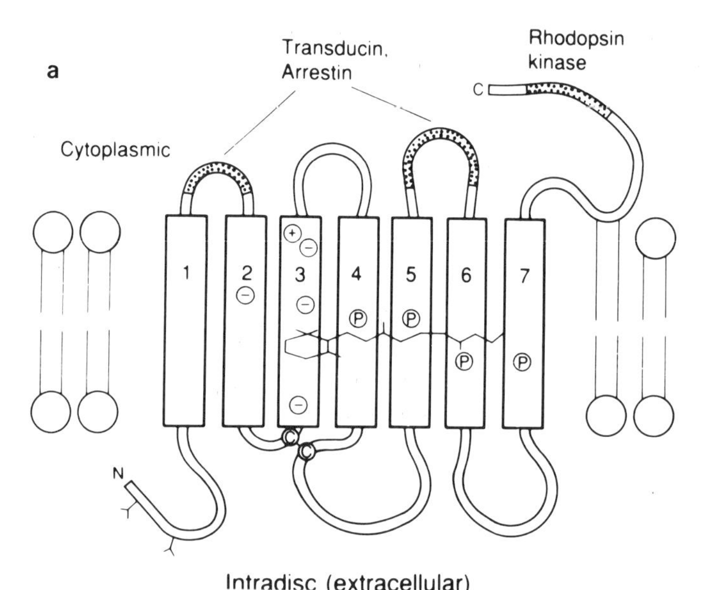

Figure 3. (a) Schematic view of the transmembrane topography of rhodopsin (from Saibil, 1990). The 11-*cis* retinal chromophore, which lies in the center of the ring of helices (see b), is linked to a lysine in helix 7. Sites of interaction with rhodopsin kinase, transducin and arrestin on the cytoplasmic loops are shaded. (b) Proposed packing of the helices within the membrane, viewed from the cytoplasmic surface, and the location of retinal between them.

Bleaching and Photoproducts

A visual pigment molecule that is not exposed to light contains retinal in the form of the 11-*cis* isomer (Fig. 2). Rhodopsin has a photosensitivity of 0.67 (Dartnall, 1957) so that two out of three photons absorbed lead to a photoisomerization of 11-*cis* to all-*trans* retinal. This is the only effect of photon absorption. All the subsequent changes to the molecule are thermal. The changes in the molecular structure fall into two phases; an initial very rapid sequence of photoproducts occurring in the ps (10^{-12} s) to ms (10^{-6} s) range, followed by a second much slower series in the ms to s range (Fig. 4). These can be followed spectrally either at very low temperatures, where the thermal changes occur relatively slowly, or by very rapid monitoring after brief (ps) intense flashes (for a detailed discussion, see Knowles and Dartnall, 1977).

The first detectable change occurs in less than 6 ps at 20°C with the formation of prelumirhodopsin (bathorhodopsin) that in the case of extracted bovine rhodopsin (λ_{max} = 498 nm) has an absorbance maximum at 543 nm (an even earlier photoproduct, hypsorhodopsin, has been described, see Knowles and Dartnall, 1977). The isomerization reaction is too rapid for any simultaneous large changes in the opsin-binding site to occur. Thermal reactions then can occur in the dark with prelumirhodopsin decaying to lumirhodopsin (λ_{max}= 497 nm) followed by metarhodopsin I whose absorbance maximum is displaced to shorter wavelengths at 478 nm. Metarhodopsin I can be stabilized at temperatures below –15°C. These early photoproducts are themselves photosensitive and can be photo-converted back to rhodopsin so that even intense flashes of light fail to bleach all the visual pigment.

At temperatures above –15°C, metarhodopsin I is converted to metarhodopsin II (λ_{max}= 380 nm), but the reaction is pH dependent and under acid conditions metarhodopsin II predominates. At 5°C metarhodopsin II is relatively stable and the dependence of the ratio of I/II on pH can be studied. The formation of the final photoproducts, metarhodopsin III (λ_{max}= 465 nm, sometimes termed pararhodopsin) and retinal + opsin (λ_{max}= 387 nm) is relatively slow: metarhodopsin II can decay either directly to retinal + opsin or via metarhodopsin III and again these reactions are pH sensitive. The exact reaction rates and absorbance maxima are dependent upon the molecular environment and differ somewhat between extracts of rhodopsin and rhodopsin *in situ* in the lipid environment of the intact membrane and also between different types of rod and cone (Bowmaker, 1973, 1977).

In the eye, all-*trans* retinal is enzymatically converted to 11-*cis* retinal, which spontaneously combines with opsin to regenerate rhodopsin. Under normal physiological light levels only relatively small percentages of visual pigment are in the bleached form and even under very bright sunlight perhaps only about 50% of the pigment in the cones in humans is "bleached".

Transduction

The absorption of a single photon and the subsequent isomerization of a single molecule of rhodopsin in the outer segment of a rod is sufficient to change the level of neuro-transmitter release at the synapse of the receptor cell. Since photoreceptor cells are long and thin, the synapse may be some distance (in the order of 10-100 μm) from the site of absorption so that the process of transduction must include considerable amplification.

In the dark, rods and cones have a resting membrane potential between the inside and outside of the cell of about –30 mV with a light-sensitive current ("dark" current) flowing from the inner segment to the outer segment. The current is carried primarily by Na^+ ions (for a recent review, see Lamb and Pugh, 1990). Excitation of the photoreceptor leads to a reduction of the current. A single absorbed photon decreases the dark current by about 3-5% and fewer than 100 photons saturate the response of the fully dark-adapted rod. The reduction

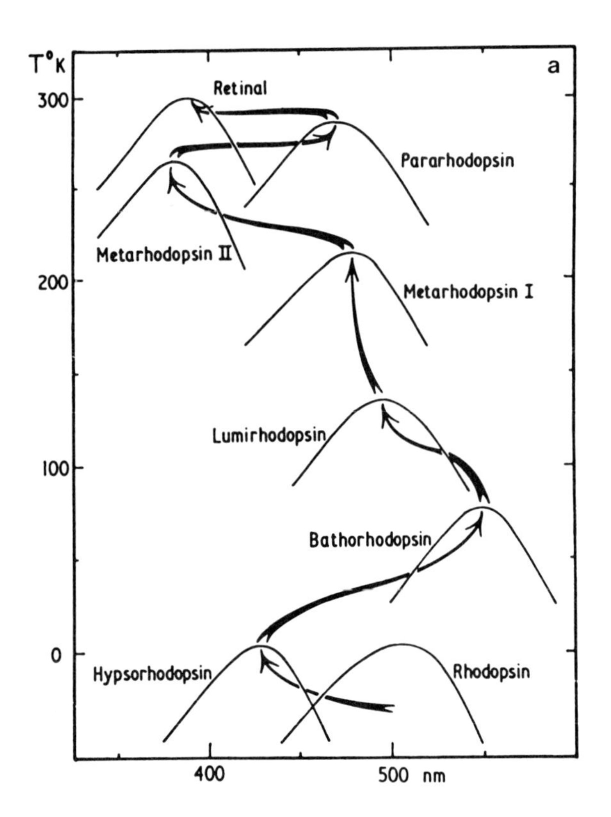

Figure 4. (a; left) Absorbance spectra of the intermediates in the bleaching of bovine rhodopsin as they appear by first bleaching at 4 K (–269°C) to yield hypsorhodopsin, and then allowing the temperature to rise in stages (from Knowles and Dartnall, 1977). Pararhodopsin is another term for metarhodopsin III. (b; below) A simplified scheme of the stages in the bleaching and regeneration of rhodopsin showing the absorbance maxima (in parentheses) of each photoproduct. The approximate half-times for interconversion of each product at 20°C are noted and the approximate temperatures above which interconversions occur are indicated by "T" (from Dratz, 1989).

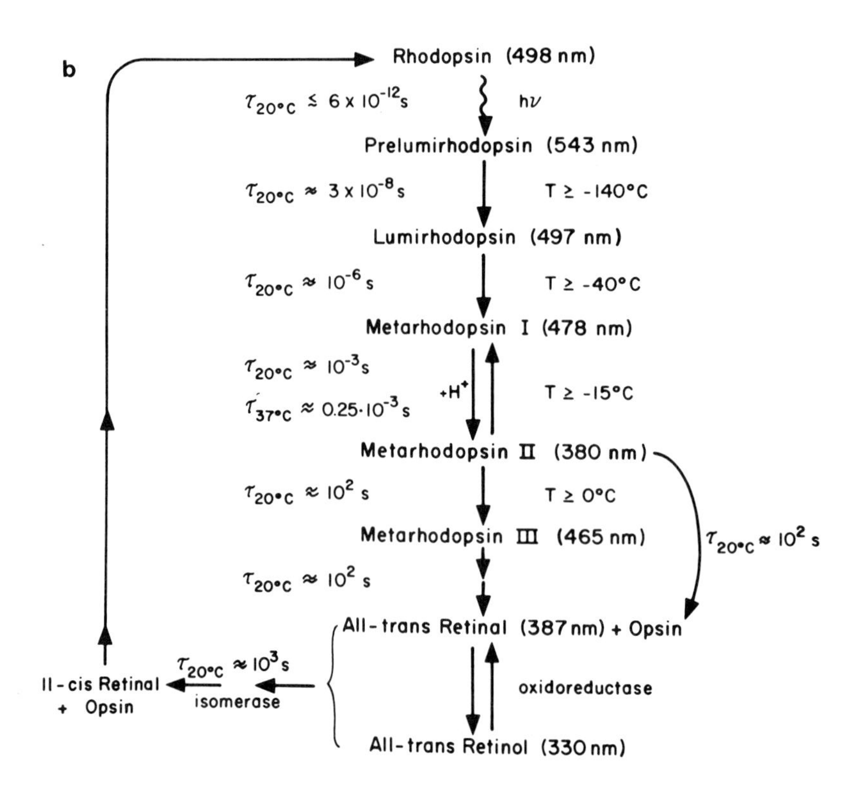

in the dark current is a consequence of a decrease in the permeability of the plasma membrane to ions (primarily the closing of Na^+ channels) and this can be recorded as a graded receptor potential that spreads electrotonically from the outer segment to the synapse. The receptor potential is a hyperpolarization, the membrane potential becoming more negative, and the amplitude of the hyperpolarization is graded with respect to the intensity of the light stimulus. The receptor potential is the major consitutuent of the a-wave of the electroretinogram as described in the following chapter.

Since, at least in rods, the activated rhodopsin will be contained within the membranes of the discs which are isolated within the outer segment, an internal transmitter is required to alter the ionic permeability of the plasma membrane. This is now known to be cGMP (cyclic guanosine monophosphate) and activation of rhodopsin leads to an enzyme cascade that regulates the level of cGMP within the cytoplasm. An enzyme cascade will act as an amplifying system since any one molecule at one stage of the cascade can activate many molecules in the next stage.

In rods, the activated rhodopsin interacts transiently with a GTP (guanosine triphosphate)-binding protein (transducin) on the cytoplasmic surface of the membrane of the disc (Fig. 3; for a recent review see Saibil, 1990). The activated GTP-binding protein exchanges GDP (guanosine diphosphate) for GTP and the GTP-binding protein + GTP in turn activates a phosphodiesterase that hydrolyzes cGMP. The reduced levels of cGMP in the cytoplasm close the Na^+ channels of the plasma membrane giving rise to the hyperpolarizing receptor potential of the cell. Amplification will occur at each stage in the cascade and the overall gain in rods is thought to be in the order of 10^6. The excited rhodopsin is inactivated by rapid phosphorylation by ATP and rhodopsin kinase and the binding of a 48 kDa protein (arrestin) (Fig. 3).

The levels of cGMP in the cytoplasm are also regulated by the intracellular concentration of Ca^{2+}. The membrane channels controlled by the levels of cGMP not only allow the passage of Na^+ but also Ca^{2+} and in turn the levels of Ca^{2+} in the cell are regulated by a Na^+/Ca^{2+} exchange mechanism in the plasma membrane in which Ca^{2+} is expelled in exchange for Na^+. Thus when the plasma membrane channels are closed by light stimulation, the levels of Ca^{2+} fall since Ca^{2+} continues to be pumped out. These lowered levels of Ca^{2+}, however, stimulate a further enzyme, guanylate cyclase, thereby raising the levels of cGMP that in turn opens the plasma membrane channels and switches off the light response. There are thus self-regulating processes that will both switch off activated rhodopsin and re-open the plasma membrane ionic channels.

Visual Pigments in the Frog

The frog retina has two classes of rods and two (or possibly three) classes of cones (Leibman and Entine, 1968; Hárosi, 1982). However, the predominant class of photoreceptors is composed of large rods with λ_{max} at about 502 nm. Although these cells absorb maximally in the green, they are often referred to as "red" rods, since they contain sufficient rhodopsin to look pinkish-red to the eye. The second class of rods, unique to certain amphibians, comprise about 8% of the total rod population and contains a visual pigment with λ_{max} at about 435 nm (Bowmaker, 1977). To add to the confusion of names, these rods, absorbing maximally in the blue, are normally referred to as "green" rods even though they would not look green to the eye!

Visual pigments, being membrane-bound proteins, are not water soluble and have to be extracted in a detergent-like solvent (most organic solvents destroy the pigments). Classically, the detergent used is digitonin that will extract all rod pigments and probably cone pigments, though these are usually in such low concentrations as to be undetectable in normal spectrophotometry. In the case of the frog, digitonin will extract the rhodopsins from both the red and the green rods. The absorbance spectrum of the extract will therefore contain a mixture of two visual pigments as well as any photostable compounds. Because the photosensitive visual pigments have different spectral sensitivities, it is possible to determine the true absorbance spectra of the rhodopsins by "partial" bleaching, a technique in which the extract is first exposed to long-wave light to selectively "bleach" the longer-wave-sensitive pigment and then to shorter-wave light (or white light) to "bleach" the second pigment. This technique has been fully described by Knowles and Dartnall (1977).

The "difference" spectrum derived by subtracting the first post-bleach absorbance spectrum from the pre-bleach spectrum will show a loss in absorbance (maximum change at about 500 nm) due to the loss of photosensitive rhodopsin from the "red" rods and a gain in absorbance at shorter wavelengths due to the formation of the late thermal photoproducts, primarily retinal + opsin and metarhodopsin III. Subsequent exposures to the same long-wave light will continue to bleach the P502 until it has all been converted to photoproducts. The difference spectrum derived from a final bleach with shorter-wave light will show the conversion of the "green" rod rhodopsin (P435) to its photoproducts with a maximum loss of optical density at about 440 nm.

These difference spectra will not give the true absorbance spectra of the visual pigments since the unbleached spectra and the photoproduct spectra will overlap. One way to obtain absorbance spectra that are less distorted by photoproducts, is to convert the photoproducts to a substance that absorbs at shorter wavelengths. This can be achieved by the addition of hydroxylamine because it competes with opsin for retinal and forms retinal oxime. This absorbs maximally at about 380 nm and has a narrower absorbance band than retinal + opsin. A problem with hydroxylamine is that it will react in the dark with some visual pigments, especially cone pigments, so that it needs to be used with care when dealing with extracts of visual pigments. Indeed, the "green" rod pigment is destroyed by hydroxylamine in the dark (Bowmaker and Loew, 1976).

EXPERIMENT 22: ABSORBANCE SPECTRA OF VISUAL PIGMENTS

OBJECTIVE

These experiments illustrate the measurement of the absorbance spectra of two spectrally distinct rhodopsins that have been extracted from the retina of the frog. Through the reaction of hydroxylamine, the late photoproducts of bleached rhodopsin will be examined.

The frog is used for extraction because it has relatively large eyes and the retina is dominated by rods with large outer segments (50 μm long and 5 μm in diameter). There is then a high yield of visual pigment per eye. An alternative is to obtain cattle eyes from an abattoir, but because of the blood circulation in mammalian retinas, the extraction process is more complex (for a full discussion of visual pigment extraction, see Knowles and Dartnall, 1977).

LIST OF MATERIALS

Frogs, either *Rana temporaria* or *Rana pipiens*
Digitonin (BDH "Analar"), 2% w/v solution in 25 mM phosphate buffer, pH 7.1
Hydroxylamine as hydroxylammonium chloride (BDH "Analar")
25 mM phosphate buffer, pH 7.1 ($Na_2HPO_4 \cdot 2H_2O$ - 3.12 $g \cdot l^{-1}$ + $NaH_2PO_4 \cdot 2H_2O$ - 1.15 $g \cdot l^{-1}$)
Bench top centrifuge (about 4000 rpm)
Recording spectrophotometer with micro-cuvettes
Tungsten light source
575 nm long-pass filter

PREPARATIONS

Digitonin Solution: Add 0.2 g of digitonin to 10 ml of phosphate buffer and heat until the solution is clear. As the solution cools, a flocculent precipitate will form. Ignore this since it will be removed when the visual pigment extract is centrifuged (see below).

Extraction of Visual Pigment: Dark adapt the animals (i.e., place them in a dark room) for at least 60 min, though for convenience, normally overnight. All the following operations should be carried out under dim red light (Kodak Safelight with Wratten No. 2 filter) to avoid bleaching the visual pigments. The pigments will not regenerate post mortem. Sacrifice the frogs in compliance with local regulations. If consistent with local regulations, cervical section and pithing are preferred. Extrude the retina from the eye by cutting across the cornea with a scalpel or razor blade and applying gentle pressure to the eye from inside the mouth. The pressure will first cause the lens to be extruded, followed by the retina. The retina will appear as a thin 'milky' sheet of tissue under red light. Lift off the retina and place it in ice-cold phosphate buffer at pH 7.1. When sufficient retinas have been obtained, spin them in a bench top centrifuge and discard the supernatant.

Add a small volume of 2% digitonin solution to the precipitated retinas and stir the suspension for 24 h at 5°C. Centrifuge the suspension and reserve the supernatant containing the extracted visual pigment. A second volume of digitonin can then be used for a second extraction of the tissue that may yield further visual pigment. The retinas from about four frogs, if extracted in 1.0 ml of digitonin, should yield sufficient rhodopsin to give an absorbance spectrum with a density greater than 0.1 at the λ_{max} of 502 nm (1 cm path length). The extracted pigment can be kept frozen almost indefinitely. However, before making absorbance measurements it is often necessary to recentrifuge since digitonin tends to precipitate.

EXPERIMENTAL PROCEDURE

Part A: Partial Bleaching Technique

1. Transfer the centrifuged solution of rhodopsin to a cuvette and place it in the spectrophotometer.
2. Record the absorbance spectrum of the unexposed solution from 350 to 700 nm.
3. Systematically expose the solution to long-wavelength light (575 nm and above) to bleach the visual pigment in stages. Record the absorbance spectra after each "bleach". A preliminary experiment will be necessary to determine the exact bleaching times since the exposure times required to achieve a satisfactory series of bleaches will be dependent on the brightness of the light source and the density of the rhodopsin.
4. When an exposure to red light no longer causes a significant reduction in absorbance, expose the solution to white light in order to bleach any other visual pigment that absorbs maximally at shorter wavelengths than the "red" rod rhodopsin. Again record the absorbance spectrum.

Part B: Hydroxylamine Technique

1. Repeat steps 1 and 2 of Part A with a fresh sample of extract.
2. Expose the solution to white light for 5 min. This should bleach all the visual pigments.
3. Add a single small crystal of hydroxylamine to the solution and record a final absorbance spectrum.

DISCUSSION

Typical Results – Part A

A family of absorbance spectra from such a series of bleaches (carried out at 15°C) is shown in Fig. 5. Curve 1 is the absorbance spectrum of the unexposed extract showing the main a-band with a λ_{max} at about 500 nm and rapidly rising absorbance at short wavelengths with a shoulder at about 400 nm. Curve 2 was obtained after an initial bleach of about 30 s with red light. Curves 3 to 8 were then obtained after repeated exposures (each of 60 s) to the red light. The increases in absorbance at shorter wavelengths due to photoproducts have been left off the graph for clarity. Curve 10 was obtained after a final 10 min exposure to the red light. The sample was then exposed to white light for 1 min and curve 11 recorded, showing the photoproducts formed at wavelengths below 400 nm.

Difference spectra, showing the effect of each bleach i.e., the loss of visual pigment and the gain of photoproducts, were then calculated. The difference spectra between curves 1 to 8 are shown in Fig. 6. Note that the differences 1–2 and 2–3 are significantly different from each other and from the remaining set of curves. Since each identical bleach should remove the same percentage of the pigment remaining at each stage, a graph of the loss in absorbance caused by each bleach should show an exponential decay. The total loss in absorbance as represented by the difference curve 1–8 is shown in Fig. 7 with a maximum loss between 500 and 505 nm. This can be compared with the difference spectrum 10–11,

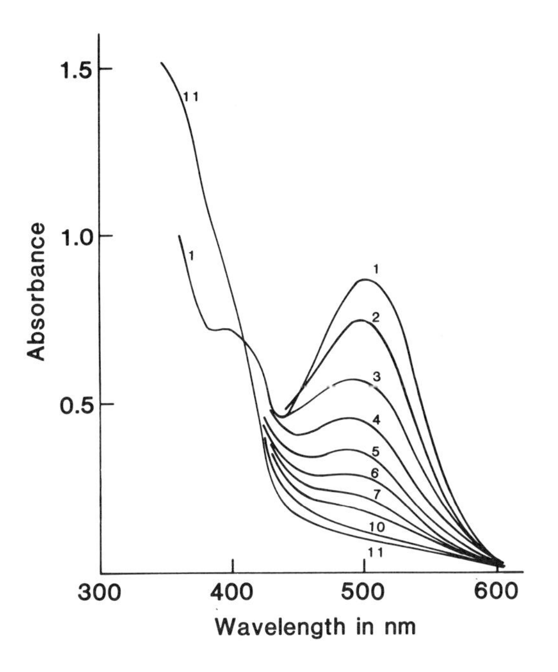

Figure 5. Absorbance spectra of a visual pigment extract from the retina of the frog (*Rana temporaria*). Curve 1, extract unexposed to light. Curve 2, after a 45 s exposure to red light (wavelengths longer than 575 nm). Curves 3 to 8, after a series of 60 s exposures to the red light. Curve 10, after a further 10 min exposure to the red light. Curve 11, after a 60 s exposure to white light.

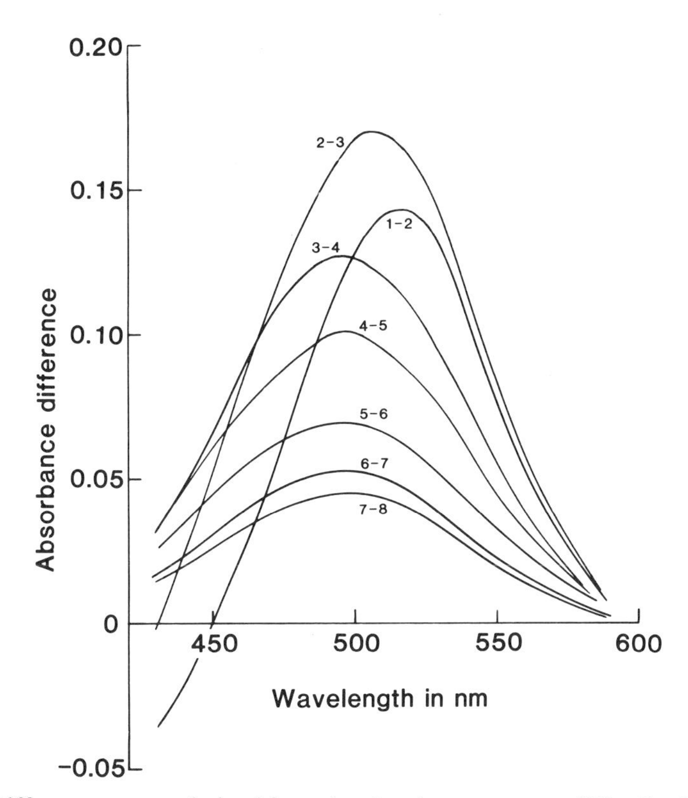

Figure 6. Difference spectra derived from the absorbance spectra of Fig. 5 as labelled. A loss in absorbance is shown as a positive value and a gain in absorbance as a negative value. Note the longer λ_{max} of spectra 1–2 and 2–3. All the difference spectra from 3–4 to 7–8 have λ_{max} at about 500 nm.

caused by the white light bleach, that has a maximum optical density loss at about 440 nm. This curve has been plotted on an expanded scale for clarity, because the decrease in density is only about 5% of the total loss (1–10) due to the long-wave light.

Typical Results – Part B

The pre- and post-bleach spectra are shown in Fig. 8 (curves 1 and 2) and the difference spectrum, 1–2 is shown in Fig. 9. Note the shoulder at about 470 nm in curve 2 and the photoproduct maximum at about 380 nm in curve 1–2. The final absorbance spectrum (after the addition of hydroxylamine) is shown as spectrum 3 in Fig. 8. The difference spectra representing the effect of the hydroxylamine (2–3) and the overall effect of the bleach and the hydroxylamine (1–3) are shown in Fig. 9. In curve 2–3 note the shoulders at about 470 and 420 nm and in curve 1–3 the photoproduct maximum at about 465 nm.

General Discussion

The absorbance spectrum of the unexposed extract from the frog retinas shows a broad absorbance band of rhodopsin in the visible region of the spectrum with λ_{max} at about

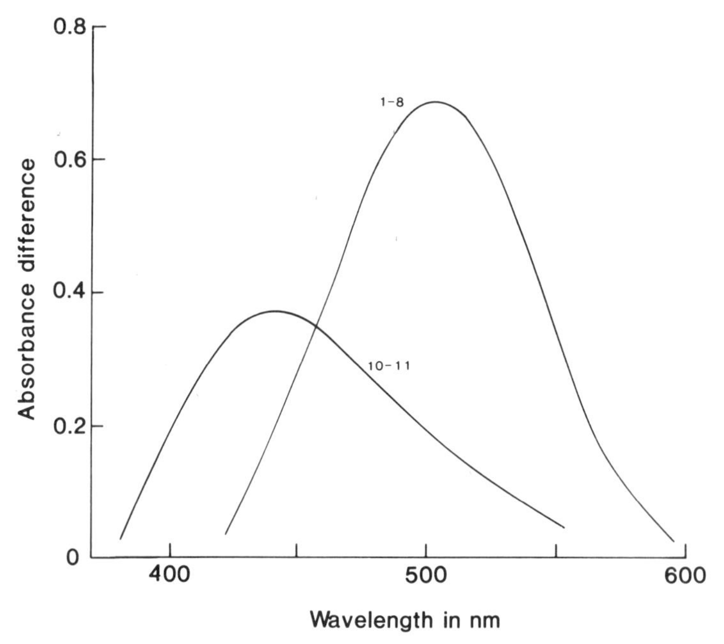

Figure 7. Difference spectra derived from the absorbance spectra of Fig. 5. Curve 1–8 is the total loss in absorbance after a series of 60 s bleaches with red light and has a λ_{max} at about 500–505 nm. The difference spectrum represents the loss of rhodopsin extracted from the "red" rods. Curve 10–11 is the loss in absorbance after exposure to white light and has a λ_{max} at about 440 nm. The curve illustrates the loss of rhodopsin extracted from the "green" rods. For clarity, this curve has been amplified by a factor of 10. The maximum loss in absorbance is about 0.04.

500 nm. The shoulder at about 400 nm is probably due either to retinal present in the dark adapted retinas or to the "Soret" bands of respiratory enzymes extracted primarily from the ellipsoid regions of the inner segments of the photoreceptors where there are high concentrations of mitochrondria (or probably a mixture of both). Since these substances are not photosensitive, they will have no effect on the difference spectra obtained from bleaching the extract.

The first difference spectrum (curve 1–2, Fig. 6) has a maximum loss in absorbance at about 530 nm with an increase in absorbance from below about 450 nm. The next two difference spectra show a shift in maximum optical density loss from about 510 nm (2–3) to about 495 nm (3–4). The subsequent difference spectra all have a maximum loss at about 495-500 nm. These displacements are due to the formation and decay of late thermal photoproducts, primarily metarhodopsin III. Metarhodopsin III absorbs maximally at about 465 nm and is shown more clearly in the difference spectra 2–3 in Fig. 9. Its formation will affect the initial difference spectra by displacing the maximum absorbance loss to longer wavelengths, away from the λ_{max} of rhodopsin. As the metarhodopsin III then decays, the difference spectra derived from subsequent bleaches (of smaller amounts of the visual pigment) will more truly represent the a-band absorbance spectrum of the rhodopsin.

The white light bleach (difference spectrum 10–11, Fig. 6) has a maximum absorbance loss at about 440 nm due to the bleaching of the "green" rod pigment. The difference spectrum is very broad at longer wavelengths because the white light will also remove any remaining "red" rod rhodopsin. The absorbance loss may also include some remaining decay

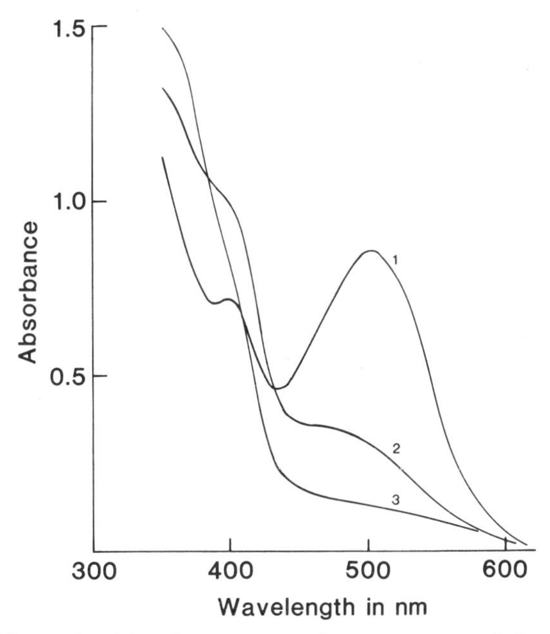

Figure 8. Absorbance spectra from an extract of rhodopsin from the retina of the frog (*Rana temporaria*). Curve 1, extract unexposed to light. Curve 2, after a 5 min exposure to white light. Curve 3, after the addition of a crystal of hydroxylamine.

of metarhodopsin III. The amount of "green" rod pigment is only about 5% of the total "red" rod pigment, but this is what would be expected since the "green" rods comprise about 8% of the rod population, but have shorter outer segments than the "red" rods.

The late thermal photoproducts, metarhodopsin III and retinal + opsin, are clearly seen in the difference spectra 1–2 and 2–3 in Fig. 9. The maximum gain in absorbance at about 380 nm in curve 1–2 shows the formation of retinal + opsin, whereas, after the addition of hydroxylamine, the loss of metarhodopsin III can be seen by the peak at about 465 nm in curve 2–3. Hydroxylamine converts the photoproducts to retinal oxime, which has a maximum absorbance at about 365 nm, as shown by the difference spectrum 1–3 (Fig. 9). The absorbance loss in curve 1–3 shows the relatively undistorted absorbance spectrum of rhodopsin, though it will, of course, be a mixture of both the "red" and "green" rod rhodopsins.

These experiments show that the digitonin extract of frog retinas contains at least two photosensitive (visual) pigments with λ_{max} at about 500 and 440 nm, as well as small amounts of non-photosensitive material. Bleaching of the rhodopsins reveal the presence of the late photoproducts, metarhodopsin III and retinal + opsin. Experiments that measure the scotopic spectral sensitivity of the frog's retina show that there is a direct correlation between the absorbance spectrum of the extracted pigment and the spectral sensitivity of the animal, demonstrating that the photosensitive pigment is indeed a "visual" pigment (see the next Chapter).

Further Experiments

The photoproducts of rhodopsin preceding metarhodopsin III can be visualized by bleaching a solution of rhodopsin at low temperature. If it is possible to regulate the temperature of the cuvette in the spectrophotometer and to maintain a low temperature during bleaching, then at about 5°C, the metarhodopsin II/III equilibrium can be observed. Bleach a solution of rhodopsin (pH 7.1) for 5 min at 5°C and record the absorbance spectrum. Raise the temperature slowly and monitor the changes in absorbance. Repeat the experiment, but

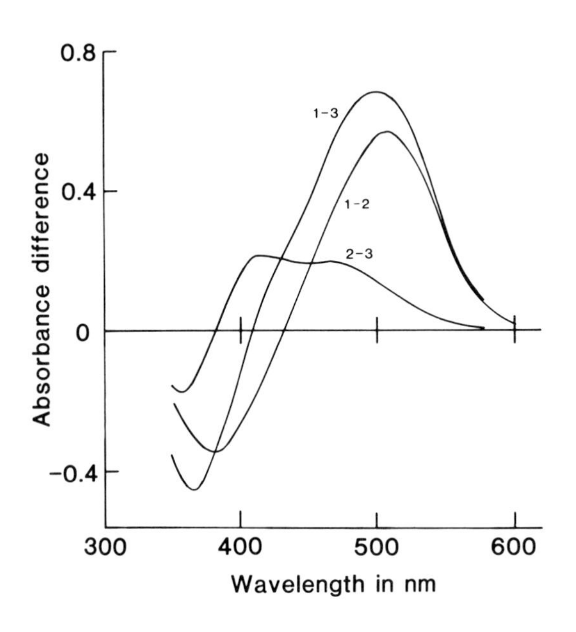

Figure 9. Difference spectra derived from the absorbance spectra of Fig. 8. Curve 1–2 represents the effect of a white light bleach showing the formation of retinal + opsin with λ_{max} at about 380 nm. Curve 2–3 illustrates the effect of hydroxylamine on the photoproducts showing the conversion of metarhodopsin III (λ_{max} at about 465 nm) and retinal + opsin to retinal oxime. Curve 1–2 shows the total effect of a white light bleach and the addition of hydroxylamine. Note the formation of retinal oxime with λ_{max} at about 365 nm.

at different pH, such as pH 8.5 and pH 5.0 (use a borate and phosphate buffer respectively). To observe metarhodopsin I it is necessary to record absorbance spectra at temperatures below 0°C. This can be achieved by making the rhodopsin extract in a glycerol and water mixture so that the solution does not freeze.

A porphyropsin can be readily extracted from the retinas of goldfish (*Carassius auratus*). The procedure is similar to that described for the frog, but the dissection of the retina is more difficult. Each eye must be hemisected and the retina gently pulled away from the back of the eye cup. The extracted porphyropsin has a λ_{max} at about 522 nm and has a broader absorbance spectrum than a rhodopsin.

REVIEW QUESTIONS

1. Describe the structure of a visual pigment molecule.
2. Discuss the technique of partial bleaching as used to separate mixtures of visual pigments.
3. Describe the sequence of changes in rhodopsin after exposure to light.

ANSWERS TO REVIEW QUESTIONS

1. The answer should include a description of the structure of opsin and retinal and their interrelationship. Brief details are given in the chapter under the heading "Visual Pigments" and extended information and further references can be obtained from Hargrave et al. (1984) and Saibil (1990).
2. A brief summary is given in the chapter in the section entitled "Visual Pigments in the Frog" and further details can be found in Knowles and Dartnall (1977), pp. 86-98.
3. A brief summary is given in the chapter under the heading "Bleaching and Photoproducts". More details can be obtained from Knowles and Dartnall (1977), pp. 289-340.

SUPPLEMENTARY READING

Dratz, E.A. (1989) Vision. In *The Science of Photobiology*, 2nd edition (Edited by K.C. Smith), pp. 231-271. Plenum Press, New York.

REFERENCES

Bok, D. (1985) Retinal photoreceptor-pigment epithelium interactions. *Invest. Ophthalmol. Vis. Sci.* **26**, 1659-1693.

Bowmaker, J.K. (1973) The photoproducts of retinal-based visual pigments *in situ*: a contrast between *Rana pipiens* and *Gekko gekko*. *Vision Res.* **13**, 1227-1240.

Bowmaker, J.K. (1977) Long-lived photoproducts of the green-rod pigment of the retina of the frog (*Rana temporaria*). *Vision Res.* **17**, 17-23.

Bowmaker, J.K. (1991) Photoreceptors, photopigments and oil droplets. In *Vision* and *Visual Dysfunction: Vol. 6, The Perception of Colour* (Edited by P. Gouras), MacMillan, London.

Bowmaker, J.K. and E.R. Loew (1976) The action of hydroxylamine on visual pigments in the retina of the frog (*Rana temporaria*). *Vision Res.* **16**, 811-818.

Dartnall, H.J.A. (1957) *The Visual Pigments.* Methuen, London.

Dartnall, H.J.A., J.K. Bowmaker and J.D. Mollon (1983) Human visual pigments: microspectrophotometric results from the eyes of seven persons. *Proc. Roy. Soc. B*, **220**, 115-130.

Dratz, E.A. (1989) Vision. In *The Science of Photobiology* (Edited by K.C. Smith), pp. 231-271. Plenum Press, New York.

Hargrave, P.A., J.H. McDowell, R.J. Feldman, P.H. Atkinson, J.K.M. Rao and P. Argos (1984) Rhodopsin's protein and carbohydrate structure: selected aspects. *Vision Res.* **24**, 1487-1499.

Hárosi, F.I. (1982) Recent results from single-cell microspectrophotometry: Cone pigments in frog, fish and monkey. *Color Res. Appl.* **7**, 135-141.

Knowles, A. and H.J.A. Dartnall (1977) The Photobiology of Vision. In *The Eye, 2nd Ed*, Vol. 2b (Edited by H. Davson). Academic Press, London and New York.

Lamb, T.D. and E.N. Pugh (1990) Physiology of transduction and adaptation in rod and cone photoreceptors. *Sem. Neurosci.* **2**, 3-13.

Liebman, P.A. and G. Entine (1968) Visual pigments of frogs and tadpoles. *Vision Res.* **8**, 761-775.

Nathans J., D. Thomas and D.S. Hogness (1986) Molecular genetics of human colour vision: the genes encoding blue, green and red pigments. *Science* **232**, 193-203.

Saibil, H. (1990) Cell and molecular biology of photoreceptors. *Sem. Neurosci.* **2**, 15-23.

Young, R.W. (1976) The renewal of photoreceptor cell outer segments. *J. Cell. Biol.* **33**, 61-72.

Measurement of Retinal Function Using the Electroretinogram

Ron DOUGLAS
City University
U.K.

INTRODUCTION

The Retina

In our waking hours, we are continually being subjected to a vast amount of information that needs to be analyzed and interpreted by the visual system. If we consider the simplest case of viewing a single object against a uniform background, we need to know, among other things, the brightness and color of both object and background, the object's speed and direction of movement, its distance and its size. If more than one object is involved, as is usually the case, one not only needs to know these parameters for each object, but also how they relate to one another. The interface between this complex visual environment and our nervous system is the retina; a thin translucent layer of nerve cells lining the posterior half of the eye. The optic media, primarily the cornea and, to a lesser extent, the lens, produce an inverted optical image of the world on our retina. As described in the previous chapter, the photons that make up this image are then absorbed by the visual pigment and converted into a series of electrical events in the photoreceptors through the processes of transduction. These electrical signals are then processed by the remainder of the visual system to extract all the relevant stimulus parameters. Although much of this processing occurs within the brain, a surprising amount is carried out within the retina itself. This is especially true for lower vertebrates such as amphibia and fish.

Rods and Cones

The ability of the visual system to respond to various aspects of a visual image is, however, not a constant, since it depends to a large extent on the intensity of the ambient illumination. In order to enable the visual system to remain active during the vastly different levels of illumination encountered during a normal 24 h period (a range of almost 12 log units), the eye of most vertebrates has evolved two distinct receptor systems based on separate photoreceptor types, the rods and cones. The former system is "designed" to maximize

---

RON DOUGLAS, Department of Optometry and Visual Science, City University
311-321 Goswell Road, London, EC1V 7DD, U.K.

sensitivity and is thus used primarily at low light levels (scotopic vision). However, in order to achieve such heightened sensitivity, the ability of the rod system to resolve both spatial and temporal detail and to process chromatic information has had to be severely diminished. The cone system, on the other hand, although less sensitive, is optimally adapted for processing detail and color, and is hence, utilized in conditions of high illumination (photopic vision) where optimizing sensitivity is not at a premium. Such functional differences between the rod and the cone systems are due to both differences in anatomy, physiology and biochemistry of the photoreceptors themselves and to the synaptic relationship between the receptors and the other neurones of the retina.

One of the most obvious demonstrations of the functional differences between the rod and the cone systems is to enter a darkened room from bright sunlight. Initially one is to all intents and purposes blind, but within a few minutes is able to make out certain details, and within half an hour, when the visual system has fully dark adapted, a surprising amount of detail is revealed. If one performs such a dark adaptation experiment under more controlled conditions and plots sensitivity as a function of time in darkness, the resulting curve often has two distinct limbs (Fig. 1), the first representing cone based vision and the other vision dependent on rods. (A number of excellent reviews of the mechanisms of light and dark adaptation are available and should be consulted for further details: Barlow, 1972; Dowling, 1987; Walraven et al., 1989).

The Electroretinogram

A convenient way to measure changes in various visual parameters during light/dark adaptation is through the use of the electroretinogram (ERG). The ERG, which is analogous to the heart's ECG and the EEG of the brain, represents the massed electrical activity of the retina and can be recorded from either intact animals, isolated eyes, eyecups or even isolated retinae using simple electrodes positioned on the cornea, in the vitreous or resting on the retina itself. It is a complex waveform consisting of several distinct components, and, not surprisingly, it differs in its exact configuration depending on the state of adaptation of the retina (Fig. 2). Both light and dark adapted ERGs display an initial corneal negative a-wave followed by a larger positive b-wave. The light adapted ERG then remains generally silent until a positive d-wave is observed at light offset. In the dark adapted condition, however, the b-wave is often followed by a much slower positive going c-wave while an off effect such as that observed in cones is absent. There are other minor components to the ERG, and major waves such as the a-wave can be subdivided into several separate components, but this need not concern us here.

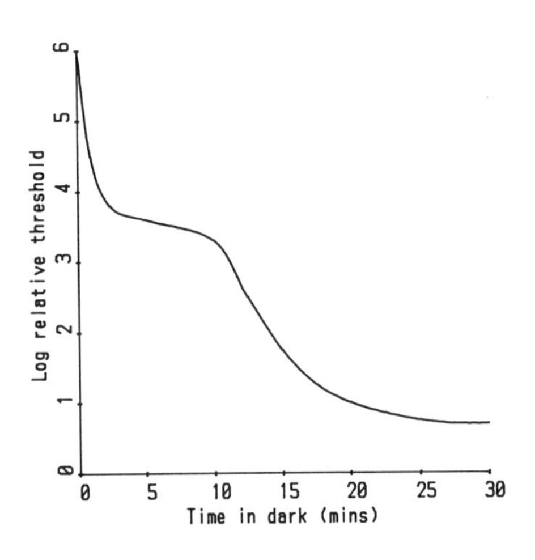

Figure 1. Schematic representation of a psychophysical dark adaptation experiment on humans. Relative threshold is plotted versus time in darkness on a semi-log scale. In such experiments the dark adaptation curve usually consists of 2 limbs, the first representing adaptation of the cone system and the lower limb representing rod dark adaptation. Although such sensitivity changes are primarily caused by a switch from cone to rod based vision, each of these 2 systems itself undergoes adaptive changes which depend on a variety of mechanisms, some occurring within the photoreceptors and some mediated by postreceptoral mechanisms.

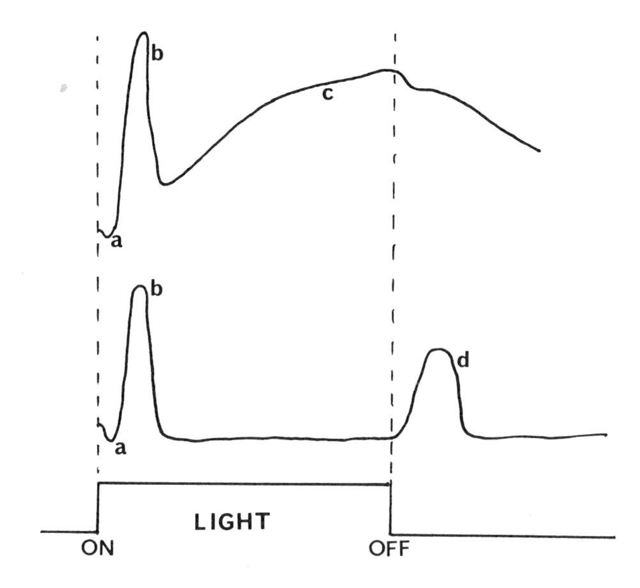

Figure 2. Schematic representation of "typical"; a) dark and, b) light adapted ERGs.

The major part of the a-wave is a consequence of photoreceptor activity, while the b-wave is thought to be generated by the Muller (glial) cells of the retina. The c-wave, which is usually, but not exclusively, observed in either rod dominated or dark adapted retinae, is largely a product of the pigment epithelium. The origin of the d-wave is uncertain, but not surprisingly since it is a characteristic of the light adapted retina, it is at least partially a reflection of cone activity (see Armington, 1974; Dowling, 1987 for reviews).

The ERG is a simple way to measure the retina's sensitivity to a variety of visual stimuli, its amplitude reflecting the retina's response. The b-wave is usually the most sensitive component of the ERG. Thus, as stimulus intensity is lowered the b-wave is usually the last component to disappear. For this reason, its amplitude is often used as a measure of retinal sensitivity. However, when studying photoreceptor mechanisms it seems best to use the amplitude of the a-wave as a measure of visual function, since it directly reflects the receptor output. However, as will be demonstrated below, both the a- and b-waves often produce comparable results.

EXPERIMENT 23: EFFECT OF LIGHT/DARK ADAPTATION ON THE FROG'S SPECTRAL SENSITIVITY (RETINAL ACTION SPECTRUM)

OBJECTIVE

The laboratory exercise outlined below is designed to demonstrate how retinal spectral sensitivity changes during adaptation from light to darkness. Spectral sensitivity is simply the relationship between stimulus wavelength and the preparation's sensitivity. Since, in this case, sensitivity will be measured using the ERG recorded from a frog eyecup, the resulting spectral sensitivity curve can be regarded as the retina's action spectrum. By examining its precise form and by comparing this to the animal's visual pigments, one can draw conclusions about the underlying photoreceptor mechanisms and the animal's color vision. By recording both light and dark adapted sensitivities, it should be possible to study inherent differences between the rod and cone systems.

LIST OF MATERIALS

Frogs
Dual beam storage oscilloscope
Preamplifier
Electrodes
Electrode leads
Light source (any source will do as long as it is bright enough - e.g., xenon arc or tungsten filament)
Optical components (lenses, etc.)
Faraday cage
Grounding leads
Diaphragm shutter (to present stimuli for 10 ms to 1 s)
Neutral density filters
Interference filters
Frog Ringer's (0.65 g NaC1, 0.025 g KC1, 0.03 g $CaC1_2$, 0.02 g $NaHCO_3$, 100 ml H_2O)
Petri dish

PREPARATIONS

Tissue: The following experiments describe how to record the ERG from an isolated frog eyecup measuring the a-wave amplitude to indicate sensitivity. However, comparable results can be obtained using intact animals, different species and by using the b-wave as an index of sensitivity (see above). Thus, although most data shown below will relate to frog eyecup a-wave measurements, some data using a-, b- and d-waves recorded from intact anesthetized rainbow trout (*Salmo gairdneri*) and goldfish (*Carassius auratus*) will be presented for comparative purposes.

Any species of frog can be utilized (*Rana temporaria* was used in the following examples). The animals must be killed in compliance with local regulations, preferably by decapitation and pithing, and the eyes isolated. An incision is then made around the limbal region dividing the eye into a posterior and anterior half. The posterior segment contains the retina and is referred to as an eyecup. This preparation, when placed on some Ringer soaked tissue paper in a Petri dish, should remain viable for several hours.

Experimental Apparatus and Recording: The eyecup preparation should be placed within an optical system that delivers a spot of uniform illumination covering the whole retina (Fig. 3). The intensity and wavelength of the stimulus should be adjustable through the insertion of neutral density and interference filters into the light beam. The eyecup and part of the optical system, as well as the preamplifier (see below), should, if possible, be enclosed in a Faraday cage to minimize electrical interference. A Faraday cage is essentially a metal box that surrounds the electrophysiological recording system. This can be most simply constructed of wire mesh, but for the experiments outlined below, it would be helpful if it were made of sheet metal, since it can then be made light-tight. The light stimulus can be delivered through a simple electronic shutter system (Fig. 3).

Two electrodes are required to record an ERG. An active electrode, which in this case rests on the retinal surface, and a reference electrode positioned on the outer coat of the eye. When intact animals are used, the active electrode can simply be inserted into the vitreous (a method used to obtain the goldfish data shown below) or even rest on the corneal surface (used to obtain rainbow trout data), while the reference electrode is positioned elsewhere on the animal, such as in the nostril or on the unstimulated eye. The electrodes used to obtain the frog eyecup ERGs shown here, were 0.13 mm diameter tungsten wire coated in epoxylite resin (both supplied by Clark Electromedical Instruments). A variety of alternative electrodes, such as a cotton wick surrounded by chloride coated silver wire (used to obtain the trout ERGs shown below), can also be used.

The output from these two electrodes should be differentially amplified by a preamplifier before being fed into an oscilloscope. Differential amplification eliminates any unwanted signals, such as those due to the heartbeat in an intact preparation, which will be picked up by both electrodes. To generate the data shown below, a Grass Instruments P16 preamplifier was used coupled with a Tektronix 5111 dual beam storage oscilloscope. The shutter signal also triggered the oscilloscope. Measurements of the ERG can be made directly from the screen, or a permanent record of the trace can be made either through the use of a computer aided digitizing system, a magnetic tape recorder or by simply photographing the stored image. In order to obtain a clear ERG as much background electrical interference as

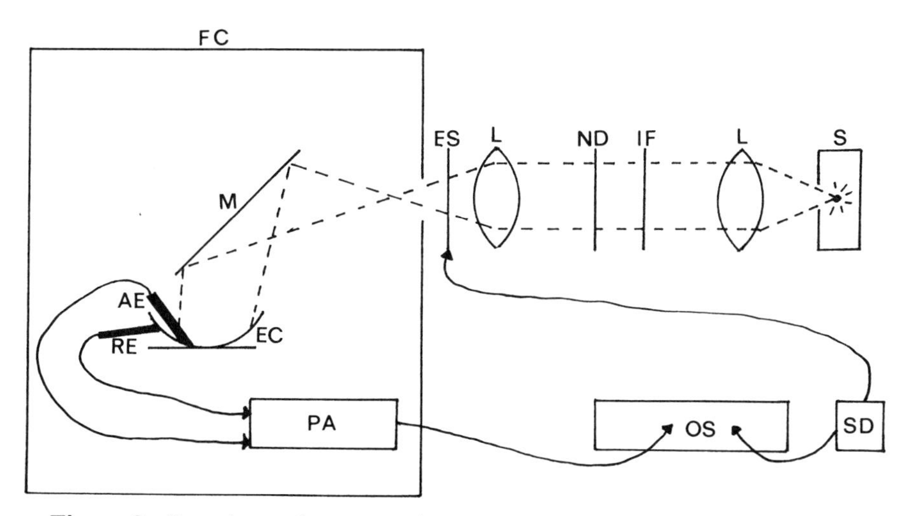

Figure 3. Experimental set-up used to record ERGs from frog eyecups. S–light source, L–lenses, IF–interference filter, ND–neutral density filters, ES–electronic shutter, FC–Faraday cage, M–mirror, EC–eyecup, AE–active electrode, RE–reference electrode, PA–preamplifier, OS–oscilloscope, SD–shutter drive.

possible should be eliminated. This is achieved primarily through the Faraday cage, but it is also advisable to earth all components and to shield any nearby electrical equipment. It is useful to use AC coupling, especially when using intact animals, as this tends to minimize drift. Such coupling will, however, also eliminate all slow potentials such as the c-wave. Since AC coupling was used to obtain the data shown here, the c-wave is not in evidence (e.g., Fig. 6).

EXPERIMENTAL PROCEDURE

1. Prepare the eyecup as described above.
2. Place eyecup in the optical system (see above) and leave it undisturbed to dark adapt for at least 1.5 h. It is important that during this time, absolutely no light reaches the retina.
3. Record ERGs in response to stimuli of varying intensity (e.g., Fig. 4). Plot a-wave amplitude against stimulus intensity to produce a response/intensity function. This should be done at a variety of wavelengths ranging from 400 nm to 650 nm (e.g., Figs. 5a and b). When working with a dark adapted preparation such as this, it is best to start at the lowest possible intensity and increase the level of illumination only until an ERG of intermediate size is obtained. Stimulating the retina with higher light levels will reduce the state of dark adaptation. It is also important, in the dark adapted condition, that the preparation is not stimulated too frequently (ideally about once a minute), as this will also reduce the level of dark adaptation. Wavelengths should be presented in a random order and it is advisable to redetermine one of the earlier wavelengths tested at the end of the experiment, to ensure that the sensitivity of the preparation has not altered during the course of the experiment.
4. Expose the preparation to room light for about 1 h.
5. Repeat Step 3 on the now light adapted preparation.

DISCUSSION

Form of the ERG – Effect of Adaptation and Stimulus Intensity

Figure 6 shows light and dark adapted ERGs recorded from a frog eyecup in response to stimuli of equal intensity, while Fig. 7, for comparative purposes, shows similar records for an intact goldfish. These ERGs show all the "classic" components described previously (Fig. 2). The initial response of both light and dark adapted preparations is a negative a-wave followed by a corneal positive b-wave. The light adapted ERGs also display a prominent off-

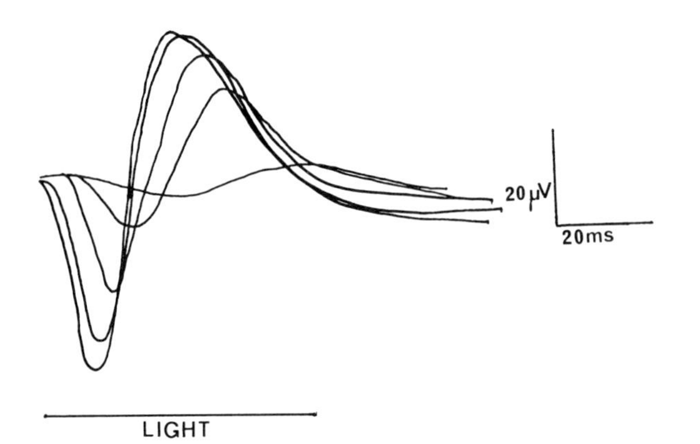

Figure 4. A series of 5 ERGs recorded from a frog eyecup following 15 min dark adaptation, in response to white light stimuli of varying intensity. The brighter the stimulus, the shorter the latency of the a- and b-waves and the greater their amplitude.

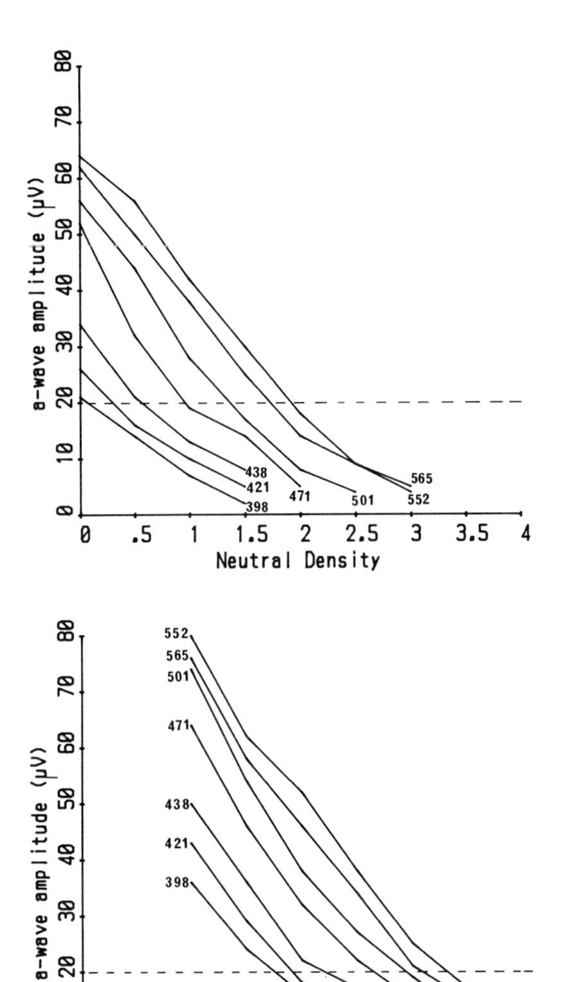

Figure 5. a-wave amplitude as a function of stimulus intensity (expressed as neutral density filters in the stimulus beam) for 7 wavelengths; top: in a light adapted frog eyecup and, bottom: following 40 min dark adaptation. Four other wavelengths were tested, but are not shown here to avoid overcrowding. The data for all 11 wavelengths are shown in Table 1.

effect, the d-wave, which is thought to be largely cone generated and is, therefore, absent from the dark adapted response. A c-wave, which is often observed under scotopic conditions, is not apparent since it is lost by the AC coupling employed in these studies.

For both light and dark adapted ERGs, the amplitude of all components generally increases with stimulus intensity while the response latencies decrease (Fig. 4). The response/intensity curves shown in Figs. 5, 8, 15 and 16 show this amplitude increase graphically and demonstrate a fundamental principle in the coding of sensory information; namely, that within certain limits, the response of a sensory system is proportional to the degree of stimulation. Such increases in response amplitude with stimulus intensity, however, only continue up to a certain point, after which any further increase in stimulus intensity will not result in an amplitude increase. In fact, very high levels of illumination often lead to an amplitude decrease. Such saturation responses are especially apparent for the b- and d-waves (Fig. 8).

Determination of Spectral Sensitivity

In the present experiment, spectral sensitivity is the relationship between stimulus wavelength and the retina's sensitivity. Usually, threshold (1/sensitivity) is taken as the

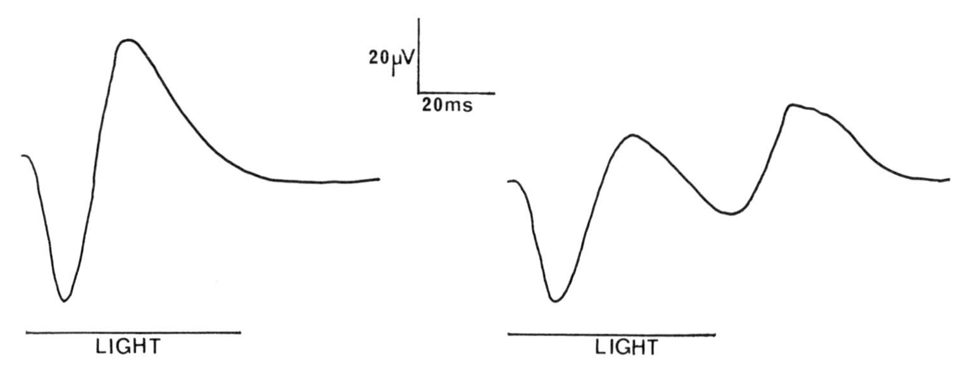

Figure 6. ERGs recorded from frog eyecups after; left: 15 min dark adaptation and, right: 30 min light adaptation, in response to white light stimuli of equal intensity.

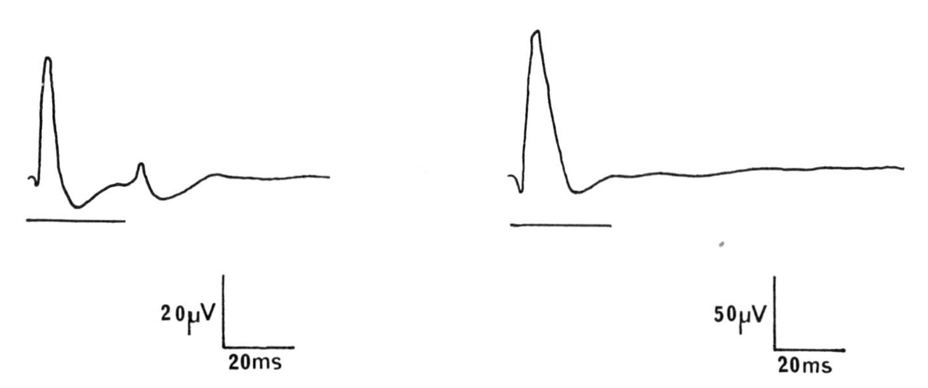

Figure 7. ERGs recorded from an intact goldfish after; left: 1 h light adaptation and, right: 2 h dark adaptation, in response to white light stimuli of equal intensity.

minimum intensity of stimulus that will elicit a response. However, in electrophysiological recordings such as the ERG, the point at which a response just disappears is hard to judge with absolute certainty given limited amplification and the inevitable background noise. For comparative purposes, it is, therefore, more usual to determine the level of stimulation required to elicit some criterion response. For example, Figs. 5A and B show response/intensity curves at 7 wavelengths for both a light and dark adapted frog retina. From such a series of curves, a relative threshold for each wavelength can be determined by reading off the intensity of illumination that produces a criterion response (in this case, 20 μV). As long as the response/intensity functions run parallel to one another, which they should at all but the highest and lowest levels of illumination, the criterion response chosen will not effect the relative sensitivity obtained. However, in order to use such thresholds to construct spectral sensitivity curves that can be compared to the animal's visual pigments, certain "corrective" procedures have to be performed.

In the first instance, thresholds determined as described above will be in terms of neutral density values (Table 1, Col. B). Such measurements by themselves, tell us nothing about the actual amount of light impinging on the retina at various wavelengths, since the output of the light source will vary with wavelength and different interference filters transmit different amounts of light. In order to convert such neutral density thresholds into radiometric units, an appropriate detector must be positioned in the optical system at the point usually occupied by the eyecup. Radiometric readings are then taken for all wavelength/intensity combinations. A plot of neutral density versus log relative irradiance for each wavelength (Fig. 9) can then be used to convert all electroretinographically determined neutral density thresholds into radiometric units (Table 1, Col. C). A further correction needs to be made to these figures since most radiometers have differing sensitivities at various wavelengths. The spectral sensitivity of the detector used here is shown in Fig. 10. All thresholds must thus be multiplied by a "correction factor" (Table1, Col. E) to account for this differential sensitivity, to produce true relative thresholds (Table 1, Col. F). The inverse of this is the retina's sensitivity to that wavelength (Table 1, Col. G).

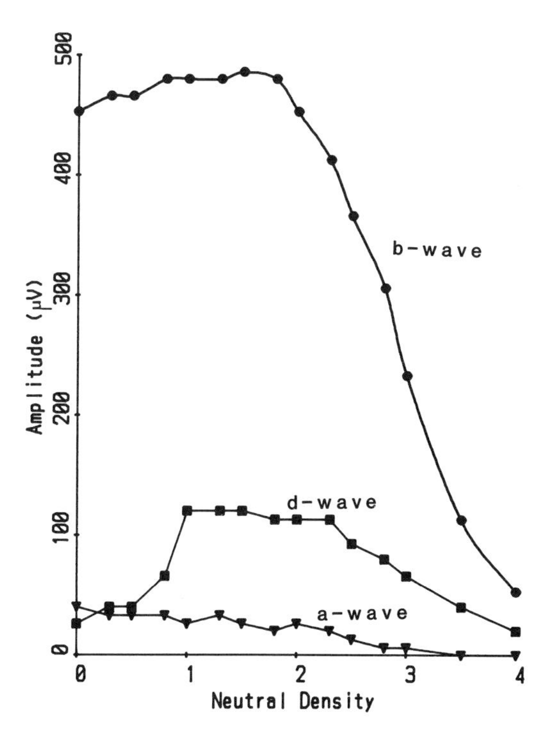

Figure 8. Relationship between stimulus intensity (expressed as neutral density filters in the stimulus and amplitude of ERG a-, b- and d-waves recorded from intact goldfish.

Table 1

Calculations required to convert ERG "thresholds" recorded from a light adapted frog eyecup at various wavelengths from initial values expressed in terms of neutral density into log relative quantal sensitivities.

A	B*	C§	D	E¶	F	G†	H	I∫
Wavelength	Neutral Density Required to Produce an a-Wave of 20 μV	B Expressed as Log Relative Irradiance	Antilog C	Detector "Correction Factor"	Relative Threshold = D × E	Relative Sensitivity = 1/F	Relative Quantal Sensitivity = G/A	Log Relative Quantal Sensitivity = Log H
398	0.06	3.16	1440	7.67	11040	906	2.276	0.357
421	0.32	3.13	1350	1.53	2060	4854	11.530	1.062
438	0.52	3.15	1410	1.10	1550	6452	14.731	1.168
471	0.95	2.98	955	1.15	1100	9091	19.301	1.286
501	1.35	2.75	562	1.05	590	16949	33.830	1.529
535	2.05	2.59	392	1.20	470	21293	39.800	1.600
552	1.90	2.75	562	1.15	640	15625	28.306	1.452
565	1.72	2.85	708	1.00	710	14085	24.929	1.397
576	1.77	2.90	794	1.05	830	12048	20.917	1.320
609	1.67	2.98	955	1.10	1050	9524	15.639	1.194
650	1.00	3.83	6740	1.35	9100	1099	1.691	0.228

* See Fig. 6 (left). § See Fig. 7. ¶ See Fig. 8.

† These values have been multiplied by a factor of 10^7 to simplify calculations. This is permissable since all values are expressed in relative terms.

∫ See Fig. 9.

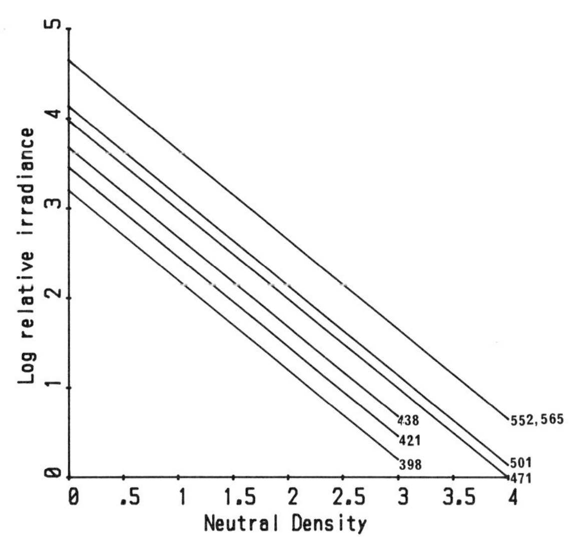

Figure 9. Relationship between the neutral density filters in the stimulus beam and the relative irradiance reaching the retina (determined using a UDT 10 DF detector) for the 7 wavelengths whose data are shown in Fig. 5.

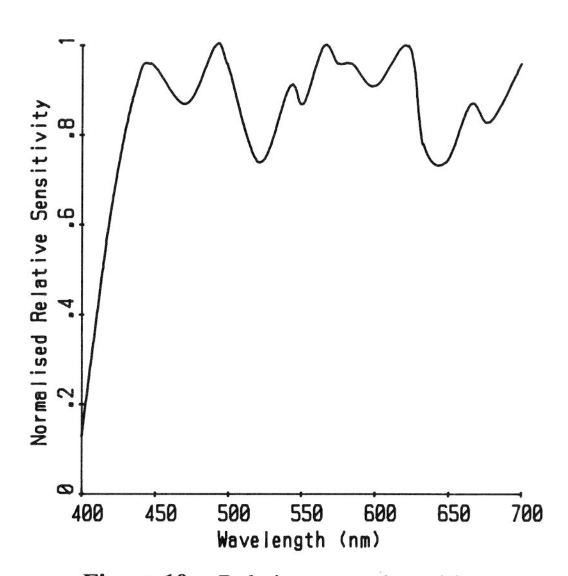

Figure 10. Relative spectral sensitivity of the detector used to obtain the data in Fig. 9.

Since the amount of visual pigment bleached depends on the number of quanta absorbed, and not on the energy of these quanta, such sensitivities must be expressed in quantal terms, if the resulting spectral sensitivity is to be related to the visual pigments of the animal. This is achieved by dividing the sensitivity by the stimulating wavelength (Table 1, Col. H). In the case of intact animals, a further correction is needed for selective absorption by the preretinal optic media. A plot of such "corrected" sensitivities against wavelength produces an action spectrum which can be compared to the animal's visual pigments (Fig. 11).

As described in the previous chapter, the first event of the visual process is the absorption of a quantum of light by visual pigment molecules situated within the photoreceptor outer segments. The probability of such a photon causing a visual pigment isomerization, depends primarily on its wavelength, since some parts of the spectrum are much more readily absorbed by the pigments than others. The sensitivity of a pigment to various wavelengths is described by its absorption spectrum. Consequently, the spectral sensitivity of the visual system depends mainly on the absorption spectra of the visual pigments its photoreceptors contain.

As outlined in an earlier section, the retinae of most vertebrates contain two types of photoreceptor, the rods and cones. In most cases, all the rods within a given species contain the same visual pigment, usually absorbing maximally between 480 and 520 nm. Thus, the spectral sensitivity of the dark adapted (scotopic) retina should resemble the absorption spectrum of this single pigment. Fig. 11 shows the scotopic action spectrum recorded from a single frog eyecup compared to the absorption spectrum of its rod visual pigment. The two spectra, as expected, correspond quite well, showing that the sensitivity of the dark adapted animal is indeed dependent on the absorption spectrum of the rod visual pigment. This correspondence is, however, not perfect (Fig. 11). Such a mismatch is caused partly by the fact that these are data from only a single experiment. An average spectral sensitivity constructed from several such spectra would give a closer fit to the pigment data. Fig. 12 shows such an average action spectrum for six intact rainbow trout, which gives a near perfect fit to that species' scotopic pigment. Another reason for the mismatch between the frog scotopic spectral sensitivity and the rod visual pigment, especially at long wavelengths,

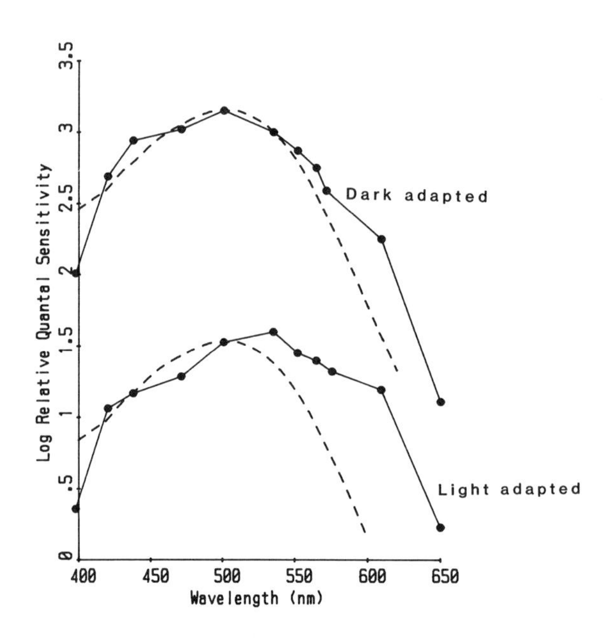

Figure 11. Relative spectral sensitivities of a light and dark (40 min) adapted frog eyecup at 11 wavelengths. The dashed curve represents the Dartnall (1953) nomogram for the extractable frog rod visual pigment fitted to the action spectra by eye.

might be that some of the cones are still active (see below). Finally, the frog is one of the relatively few animals where there is more than one spectral class of rod. Apart from containing "conventional" (red) rods absorbing maximally around 502 nm, frogs have also been shown to have a population of "green" rods with maximum absorption around 432 nm (Denton and Wyllie, 1955; Dartnall, 1957; Donner and Reuter, 1962; Muntz, 1966; Liebman and Entine, 1968; Bowmaker and Loew, 1976; Bowmaker, 1977). However, these green rods almost certainly do not function near threshold and, in fact, in many ways they behave more like cones than rods and might, therefore, be involved in color vision (Rushton, 1959; Muntz, 1963; Bowmaker, 1977, 1984). They are, therefore, unlikely to participate in a scotopic absorption spectrum such as that shown in Fig. 11.

The visual system in its scotopic state is unable to mediate color vision since, as described above, it generally only consists of one spectral class of photoreceptor. The reason why this is unsuitable for any form of wavelength discrimination can be most easily demonstrated using a theoretical example (Fig. 13a). From this figure, it would, at first glance, seem that a 500 nm stimulus would bleach about twice as much visual pigment as a stimulus of 550 nm. If the visual system were able to monitor the level of bleaching (receptor activity), one might, therefore, assume that it would have a way of judging stimulus wavelength. However, this would only be true if the two stimuli were of equal intensity since a 550 nm stimulus that was twice as bright as a 500 nm stimulus would result in the same photoreceptor activity as the dimmer 500 nm stimulus. A single photoreceptor type on its own is, therefore, unable to provide any information about stimulus wavelength, resulting in achromatic vision under scotopic conditions.

For wavelength discrimination, one needs at least two different photoreceptor types, each containing a spectrally distinct visual pigment (Fig. 13b). In this simplified example, the 500 nm stimulus will always bleach the two visual pigments in a ratio 100A/69B irrespective of its intensity, while a 550 nm stimulus will always result in a ratio of 48A/100B. Thus, each wavelength has its own unique "bleaching ratio" and the visual system

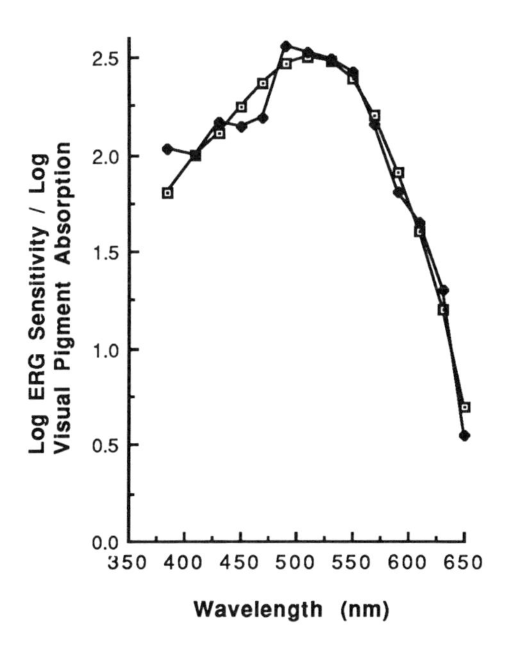

Figure 12. Average dark adapted spectral sensitivity determined using ERGs recorded from 6 intact rainbow trout. The dashed line represents the absorption spectrum of the scotopic visual pigment extracted from these 6 fish.

could gain information about the stimulus wavelength by comparing the output of these two receptor types. In fact, the situation we think of as "normal", primarily because it is the system adopted by man, is to have three distinct spectral classes of cone. Such a trichromatic system of color vision results in better wavelength discrimination than a system relying on only two visual pigments. It should be noted, however, that many vertebrates have color vision relying on two or even four spectral cone types.

Due to the presence of several different spectral cone types, photopic spectral sensitivity curves are usually more complex than those observed under scotopic conditions. Since frogs, like most vertebrates, are known to possess several distinct spectral cone types (Liebman and Entine, 1968; Liebman, 1972; Semple-Rowland and Goldstein, 1981; Harosi, 1982), it is not surprising that the photopic action spectrum determined here (Fig. 11) does not match the absorption spectrum of any single visual pigment (Muntz, 1977, for discussion). The spectrum is, for instance, significantly broader at long wavelengths, reflecting the participation in the response of long wavelength sensitive cones. Such cones also explain the shift in maximum sensitivity from 501 nm under scotopic conditions, to around 535 nm on light adaptation. Such a "Purkinje shift" is a very common phenomenon among vertebrates, and in frogs is usually thought to shift the maximal sensitivity of a fully light adapted individual to around 560 nm (e.g., Granit, 1955; Muntz, 1966, 1977). The precise form of photopic spectral sensitivity curves is, in fact, very variable and depends to a large extent on the precise experimental conditions. Under different circumstances, the various cone types make different contributions to the action spectrum. Fig. 14, for example, shows a photopic curve derived for rainbow trout, which is quite different to that obtained for the frog here, and shows clear contributions from three different spectral classes of cone. You will, therefore, find the precise form of the spectral sensitivity curves you will determine in these experiments to be quite variable, depending on your precise adaptation and experimental conditions and on the species you use.

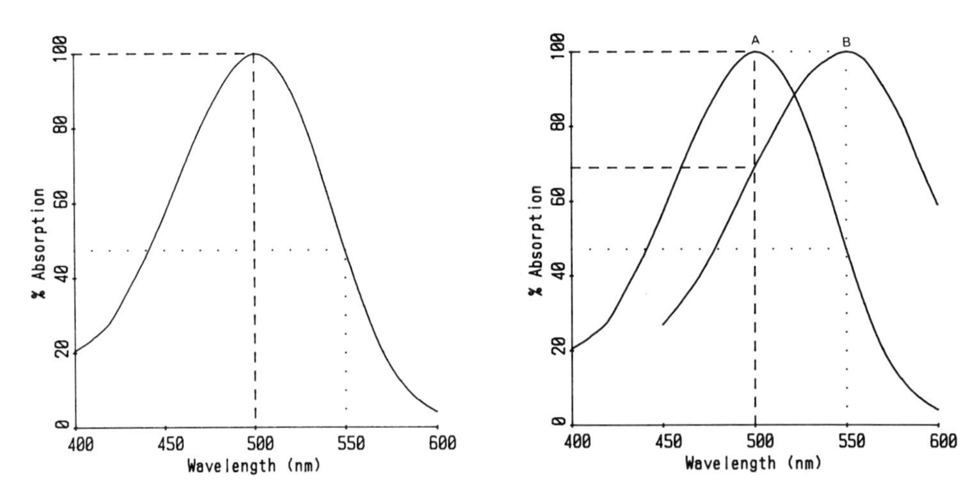

Figure 13. Left: Absorption spectrum of a visual pigment absorbing maximally at 502 nm (from Dartnall, 1953). The dotted and dashed lines indicate the relative amounts of visual pigment bleached by a 550 nm and a 500 nm stimulus of equal intensity. Right: Absorption spectra of two visual pigments absorbing maximally at 502 nm and 550 nm (from Dartnall, 1953). The dotted lines indicate the relative amounts of each visual pigment bleached by a 550 nm stimulus and the dashed line, the relative amounts bleached by a 500 nm stimulus.

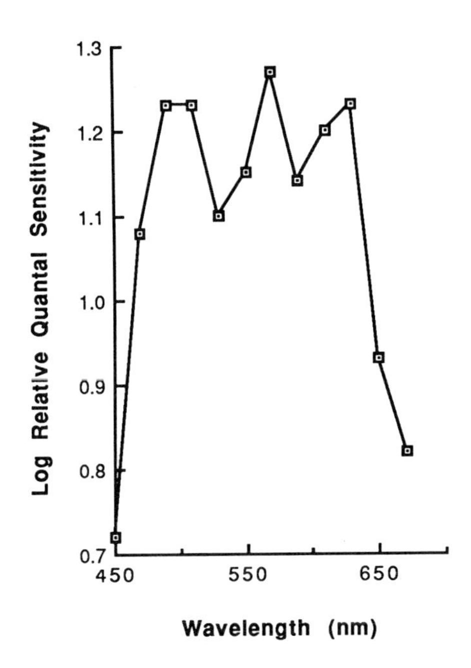

Figure 14. Average light adapted spectral sensitivity determined using ERG b-waves from 5 intact rainbow trout.

Other Retinal Parameters that can be Measured Using the ERG

The above examples have shown that the ERG is a useful tool for studying the relative changes in retinal spectral sensitivity during dark and light adaptation. However, as outlined in an earlier section, this is not the only parameter that changes during adaptation. Most obviously perhaps, retinal *absolute sensitivity* also changes during light adaptation. This change in absolute sensitivity can be determined quite simply by comparing the threshold obtained in response to any one stimulus wavelength before and after light exposure. Fig. 15 shows the response/intensity curves for a single retina both when light and dark adapted in response to a 501 nm stimulus. The relative thresholds for the light and dark adapted retina can then be determined by reading off the intensity of stimulation required to produce a criterion response. A criterion response of 30 μV used in this example, gives a shift in sensitivity of 1.5 log units following 40 min dark adaptation. Similarly, Fig. 16 shows response/intensity curves obtained from goldfish at various times during dark adaptation indicative of a change in sensitivity of around 4.4 log units after 2.5 h in darkness. The degree of this change in sensitivity is, among other things, highly dependent on the intensity of the light the preparation was exposed to prior to dark adaptation, the length of this exposure and of course, the duration of dark adaptation. In general, the more light the retina is exposed to before being put in darkness, the longer it will take to fully dark adapt.

Apart from changes in absolute and spectral sensitivity, the ability of the visual system to resolve, for example, temporal detail is also greater under photopic conditions. This can also be demonstrated using the ERG. An index of the retina's ability to resolve temporal detail is its *flicker fusion frequency*. ERGs can easily be recorded in response to a flickering stimulus of various frequencies (Fig. 17). The flicker fusion frequency is that frequency at which individual ERGs are no longer apparent. Invariably, photopic flicker fusion occurs at higher frequencies than fusion under conditions of dark adaptation (Fig. 18), indicating the higher temporal resolving power of the cone system.

Thus, the ERG is a very versatile tool for studying a variety of functional changes in the retina during the processes of light and dark adaptation.

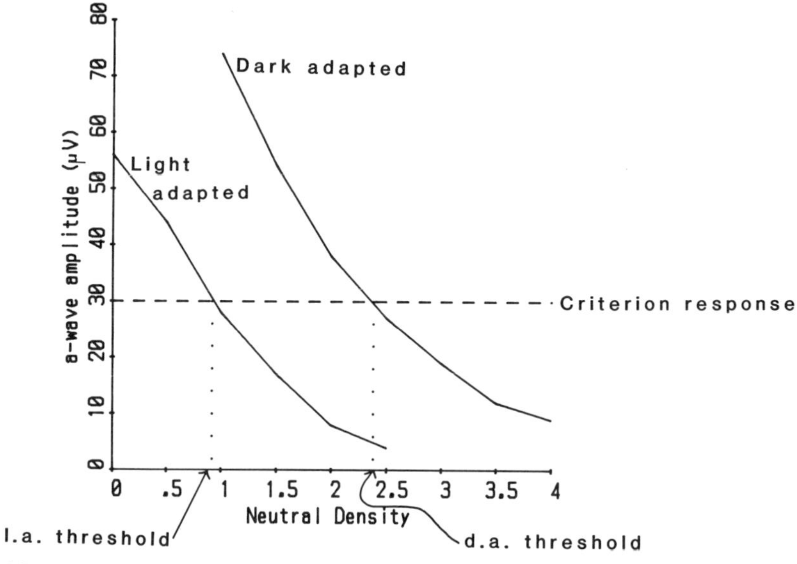

Figure 15. a-wave amplitudes of ERGs recorded from frog eyecups, in response to varying intensities of a 501 nm stimulus (expressed as neutral density filters in the stimulus beam) in both the light adapted retina and following 40 min dark adaptation. The dotted line indicates a response amplitude of 30 μV which was taken as the threshold in this example. The dashed lines indicate the light intensity required to produce this criterion response (light adapted 0.92, dark adapted 2.38) demonstrating an increase in sensitivity during dark adaptation of 1.46 log units.

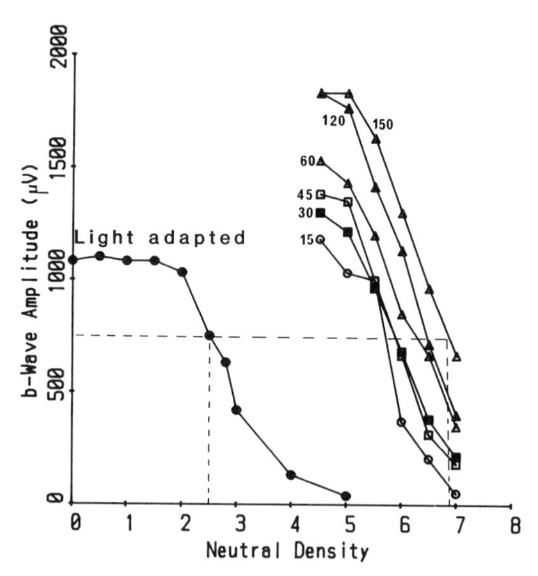

Figure 16. Relationship between b-wave amplitude and stimulus intensity (expressed as neutral density filters in the stimulus beam) at various times during dark adaptation in an intact goldfish. The numbers indicate the length of time (min) that the fish has been in darkness. The dashed lines indicate the stimulus intensity required to produce a response amplitude of 750 μV in the light adapted condition (2.5) and following 150 min dark adaptation (6.9), representing an increase in sensitivity of 4.5 log units during dark adaptation.

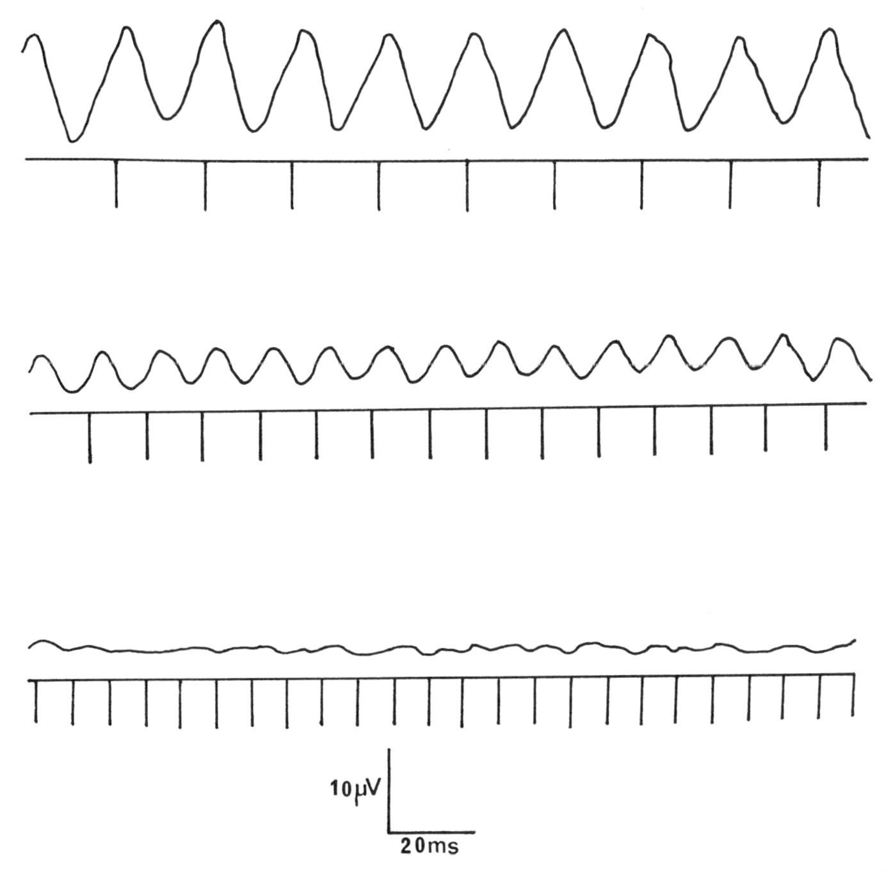

Figure 17. ERGs recorded from a light adapted frog eyecup, in response to flickering white light stimuli of different frequencies; top: 4 Hz, middle: 7 Hz, bottom: 12 Hz.

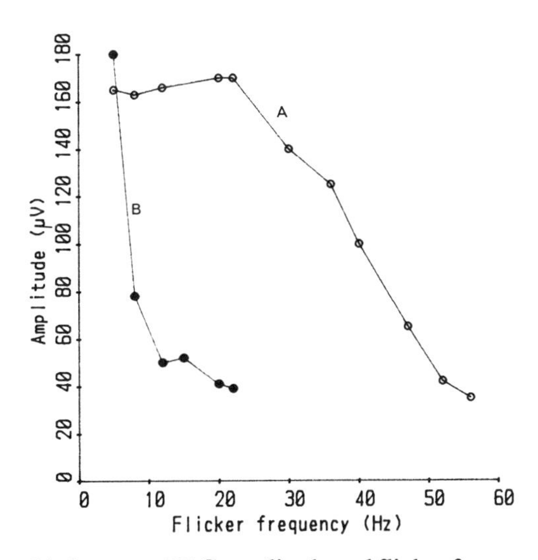

Figure 18. Relationship between ERG amplitude and flicker frequency in; A) light and, B) dark adapted rainbow trout.

REVIEW QUESTIONS

1. How does visual perception change when switching from cone based vision to vision mediated by rods?
2. How does the form and size of the ERG change during dark adaptation?
3. Why does one need more than one spectral type of visual pigment to perceive color? In what way must the receptor systems containing these different pigments interact to produce color vision?
4. What anatomical and physiological factors result in the rod system being more sensitive than the cone system?
5. What causes cones to have a higher spatial acuity than rods?
6. What can the shape of a spectral sensitivity curve tell us about the underlying receptor mechanisms?

SUPPLEMENTARY READING

Dratz, E.A. (1989) Vision. In *The Science of Photobiology*, 2nd edition (Edited by K.C. Smith), pp. 231-271. Plenum Press, New York.

REFERENCES

Armington, J.C. (1974) *The Electroretinogram.* Academic Press, New York.

Barlow, H.B. (1972) Dark and light adaptation: psychophysics. In *Handbook of Sensory Physiology VII/4* (Edited by D. Jameson and L.M. Hurvich). Springer Verlag, Berlin.

Bowmaker, J.K. (1977) Long-lived photoproducts of the green-rod pigment of the frog, *Rana temporaria. Vision Res.* **17**, 17-23.

Bowmaker, J.K. (1984) Microspectrophotometry of vertebrate photoreceptors. *Vision Res.* **24**, 1641-1650.

Bowmaker, J.K. and E.R. Loew (1976) The action of hydroxylamine on visual pigments in the intact retina of the frog (*Rana temporaria*). *Vision Res.* **16**, 811-818.

Dartnall, H.J. (1957) *The Visual Pigments.* Methuen, London.

Denton, E.J. and J.H. Wyllie (1955) Study of photosensitive pigments in the pink and green rods of the frog. *J. Physiol.* ***127***, 81-89.

Donner, K.O. and T. Reuter (1962) The spectral sensitivity and photopigment of the green rods in the frog's retina. *Vision Res.* ***2***, 357-372.

Dowling, J.E. (1987) *The Retina: An Approachable Part of the Brain.* Belknap Press, Cambridge, Mass.

Granit, R. (1955) *Receptors and Sensory Perception.* Yale University Press, New Haven.

Harosi, F.I. (1982) Recent results from single cell microspectrophotometry: cone pigments in frog, fish and monkey. *Colour Res. Applic. 7*, 135-141.

Liebman, P.A. (1972) Microspectrophotometry of photoreceptors. In *Handbook of Sensory Physiology VII/1* (Edited by H.J.A. Dartnall). Springer Verlag, Berlin.

Liebman, P.A. and G. Entine (1968) Visual pigments of frog and tadpole (*Rana pipiens*). *J. Exp. Biol.* ***40***, 371-379.

Muntz, W.R.A. (1966) Visual pigments and spectral sensitivity in *Rana temporaria* and other European tadpoles. *Vision Res.* ***6***, 601-618.

Muntz, W.R.A. (1977) The visual world of amphibia. In *Handbook of Sensory Physiology, VII/5* (Edited by F. Crescitelli). Springer Verlag, Berlin.

Rushton, W.A.H. (1959) Visual pigments in man and animals and their relation to seeing. *Prog. Biophys. Chem.* ***9***, 239-283.

Semple-Rowland, S.L. and E.B. Goldstein (1981) Segregation of Vitamin A_1 and Vitamin A_2 cone pigments in the bullfrog retina. *Vision Res.* ***21***, 825-828.

Walraven, J., C. Enroth-Cugell, D.I. Hood, A. MacLeod and J.L. Schnapf (1989) The control of visual sensitivity: receptoral and postreceptoral processes. In *Visual Perception: The Neurophysiological Foundations* (Edited by L. Spillman and J.S. Werner). Academic Press, San Diego.

Photomorphogenic Responses to UV Light: Involvement of Phytochrome and UV Photoreceptors

Bartolomeo LERCARI
Universitá di Pisa
Italy

INTRODUCTION

Photomorphogenesis

Plant life depends upon light as an energy source. In photosynthesis, photon energy is transformed into chemical energy. However, in addition to the conversion of light energy to chemical energy in the biosphere, plants have evolved mechanisms regulating their growth and development according to the light environment (photomorphogenesis). Photomorphogenesis can be defined as the control by light of growth and differentiation (and therewith development) of a plant independently of photosynthesis (Mohr, 1972). Therefore, in photomorphogenesis light is a signal which determines the pattern of differentiation, but does not satisfy its energy requirements. Indeed the light energy required to induce a photomorphogenic response (e.g., seed germination, flowering) is many orders of magnitude lower than the amount of energy required for the expression of the phenomenon itself. Photomorphogenic responses embrace many processes in plants from the macroscopic-morphological level (germination, flowering, elongation growth, leaf expansion) down to the molecular level (gene expression) (Table 1, Fig. 1). By sensing changes in light conditions, i.e., quality and quantity, direction and duration of light, plants can adapt for optimal growth.

The separation between photomorphogenesis and photosynthesis is basic. Indeed the photoreceptors involved in photomorphogenesis differ from the photoreceptors of photosynthesis from many points of view. Photosynthetic pigments are generally present in great amounts in cells, and are concentrated in specialized membranes. By contrast the photomorphogenic pigments are sensor pigments; they are present in very low concentrations and their biological function does not necessarily depend on membranes.

Plants are capable of sensing light accurately throughout the whole spectrum of sunlight (290-800 nm). At least three different sensor pigments are present in higher plants: the red/far red light absorbing phytochrome, the blue/UV-A light absorbing photoreceptor(s) and UV-B photoreceptor(s), as indicated below.

BARTOLOMEO LERCARI, Dipartimento di Biologia delle Piante Agrarie, Sezione di Orticoltura e Floricoltura, Universitá di Pisa, Viale delle Piagge 23, I 56124 Pisa, Italy

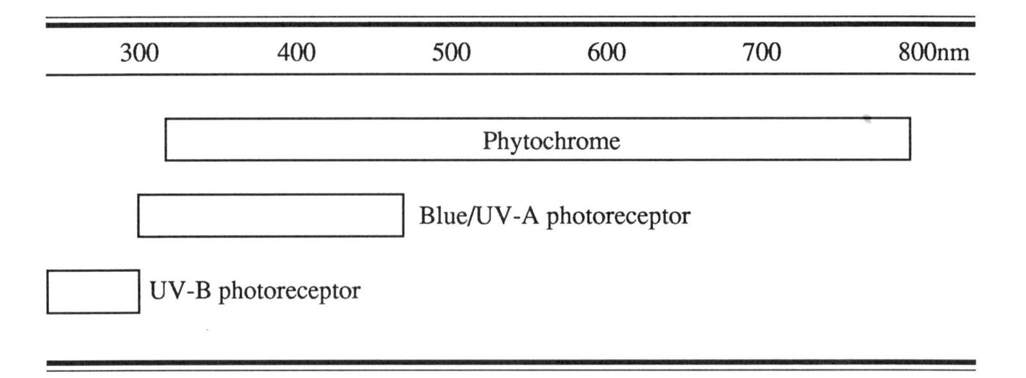

Phytochrome

The isolation and characterization of phytochrome is a classic example of how photobiological techniques can predict the nature of an unknown photoreceptor. Indeed the existence of phytochrome and its photochromic properties were predicted by Borthwick et al. (1952) on the basis of physiological experiments on seed germination and photoperiodic flower induction.

The basic features of the phytochrome system are clearly illustrated by describing the photocontrol of seed germination. To germinate, seeds have to be moistened at 20-25°C on an inert medium (sand, filter paper). As a further requisite for germination, many soaked seeds have to be illuminated with white light for a few seconds. Placed in darkness after the

Table 1

Some phytochrome-mediated photoresponses of mustard seedling, *Sinapis alba L.* (Mohr, 1972)

Inhibition of hypocotyl lengthening
Inhibition of translocation from the cotyledons
Enlargement of cotyledons
Unfolding of the lamina of the cotyledons
Hair formation along the hypocotyl
Opening of the hypocotylar ("plumular") hook
Formation of leaf primordia
Differentiation of primary leaves
Increase of negative geotropic reactivity of the hypocotyl
Formation of tracheary elements
Differentiation of stomata in the epidermis of the cotyledons
Formation of plastids in the mesophyll of the cotyledons
Changes in the rate of cell respiration
Synthesis of anthocyanin
Increase in the rate of ascorbic acid synthesis
Increase in the rate of carotenoid synthesis
Increase in the rate of protochlorophyll formation
Increase of RNA synthesis in the cotyledons
Decrease of RNA content in the hypocotyl
Increase of protein synthesis in the cotyledons
Changes in the rate of degradation of storage fat
Changes in the rate of degradation of storage protein

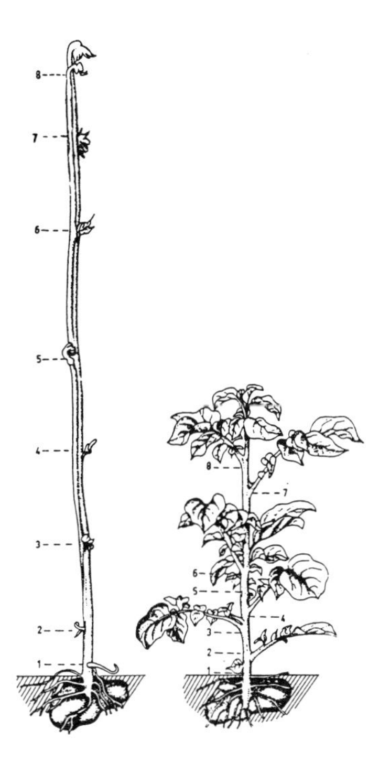

Figure 1. Photomorphogenesis in potato plants. Left: grown in dark. Right: grown in light.

light flash, the seeds germinate completely (positive photoblastic seeds). The spectral responsivity of positively photoblastic seeds determined by action spectroscopy, shows that it is the radiation between 550 and 700 nm (red light) which induces germination. According to the action spectra, the radiation around 660 nm is the most effective. Far red light ($700<\lambda<800$ nm) does not stimulate germination.

By 1952, the red/far red antagonism had been discovered; it was found that the induction of germination by red light can be nullified by a subsequent irradiation with far red light. The action spectrum for the reversing effect shows that light of wavelength between 700 and 800 nm (far red, peak at 730 nm) is particularly effective.

The phenomenon of red/far red reversibility is repeatable. The decisive light quality is that which has been given last. This property has been found in a large number of systems and is involved in a great number of photoresponses (Table 2, Fig. 2). From this fact, it was concluded that the photoreaction itself is reversible, which means that the pigment (phyto-

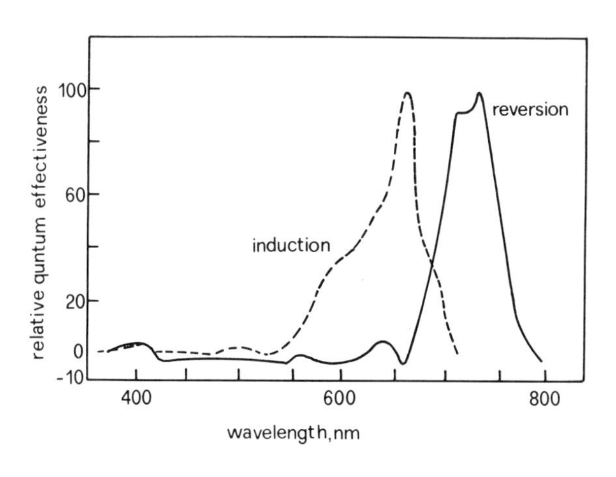

Figure 2. Action spectra of phytochrome-mediated, red light/far red light reversible, photoresponses induced by light treatments of short duration. These spectra are representative of many low fluence responses (LFR; see below).

Table 2

Some examples of photomorphogenic (red/far red light, reversible) responses (from Kendrick and Kronenberg, 1986)

Response	Delay Between Irradiation and Response
Flower induction or suppression	days to weeks
Induction of germination	hours to days
Enzyme synthesis	hours
Chloroplast movement	minutes
Nyctinastic leaf movement	minutes
Elongation growth	minutes
Transmembrane potential	seconds

chrome) must exist in two forms, interconvertible by radiation. Phytochrome is, by far, the most studied and best characterized photomorphogenic photoreceptor in higher plants. It is a bluish chromoprotein having two forms: P_r, which absorbs maximally in the red light spectral range, with a peak around 660 nm, and P_{fr}, which absorbs maximally in the far red spectral range with a peak near 730 nm. Phytochrome exists as a dimer of two polypeptides (monomer has a mol. wt. of about 120 kDa depending on the species), each with a covalently attached linear tetrapyrrole chromophore. Phytochrome is synthesized in the dark as P_r, the red absorbing form.

Absorption of light by P_r converts the molecule to P_{fr}, the far red absorbing form, which initiates the biological responses to light. The two forms, P_r and P_{fr}, are reversibly interconvertible by light.

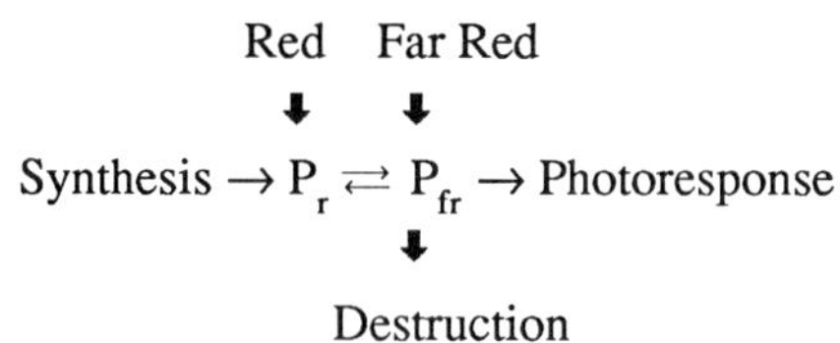

The amount of phytochrome present in light grown plants is only 1% of the dark value because P_{fr} is proteolytically degraded.

As previously suggested on the basis of physiological experiments which were found rather puzzling and which were described as "phytochrome paradoxes" (Hillman, 1967), it has been shown that at least two physically distinct forms of phytochrome exist in green tissues of pea and oat. Type I or "unstable" phytochrome is abundant in etiolated tissue and declines upon illumination. Type II is poorly characterized, but is relatively stable in light grown plants.

Since the first studies, the far red absorbing form of phytochrome, P_{fr}, has been proposed to be the physiologically active one. On the other hand, it has also been suggested that it is the loss of P_r (red absorbing form) which is the active process. Mutants lacking phytochrome, such as the *aurea* mutant of tomato, provide unequivocal evidence that P_{fr} is the active form. Indeed the dark grown *aurea* mutant, which has a very low level of P_r in the dark, is elongated like the wild type. If the loss of P_r was the active process in phytochrome action, the mutant would be expected to be short.

Absorption spectra of highly purified solutions of phytochrome under the P_{fr} form (after irradiation with red light) or under the P_r form (after irradiation with far red light) are shown in Fig. 3. The absorption bands of the two forms, P_r and P_{fr}, are broad and overlap considerably (except λ >700 nm). Both P_r and P_{fr} absorb throughout the solar spectrum of sunlight, 290-800 nm, so that upon irradiation, a mixture of P_r and P_{fr} is always established.

The wavelength dependent photoequilibrium is defined as follows:

$$\varphi(\lambda) = \frac{P_{fr}}{P_r + P_{fr}}$$

P_r does not absorb appreciably in the far red region (λ>700 nm), therefore, if the photoequilibrium is established with far red light (e.g., 720 nm) the ratio φ is about 0.03. Red light drives phytochrome mainly to the P_{fr} form and the ratio φ is of the order of 0.8 (Fig. 4). The measurements of the *in vivo* photoequilibrium in blue and UV light is complicated either by strong scattering and light attenuation by flavin and carotenoid absorption, as well as by the low quantum yield and extinction coefficient values of the phytochrome system in this spectral range.

The phytochrome system is further complicated by the formation of intermediates for both photoconversions $P_r \rightarrow P_{fr}$ and $P_{fr} \rightarrow P_r$.

Blue–UV Photoreceptors

Though Julius Sachs demonstrated as early as 1864 that plant orientation with respect to light direction (phototropism) is only induced by the blue region of the spectrum, the nature of the UV-blue absorbing pigment(s) has yet to be elucitated. Many action spectra have been determined, having maxima or shoulders at 370, 420, 450 and 480 nm (Fig. 5). These data have been interpreted as being indicative of either a carotenoid or a flavin type pigment. The blue-UV absorbing receptor(s) appears to be involved in a number of other responses, i.e., chloroplast orientation, stomatal aperture, elongation growth, etc.

In many higher plants, specific UV-B responses have been found, but no action spectrum or other evidence is available which permits definitive conclusions on the nature of the UV-B absorbing pigment(s).

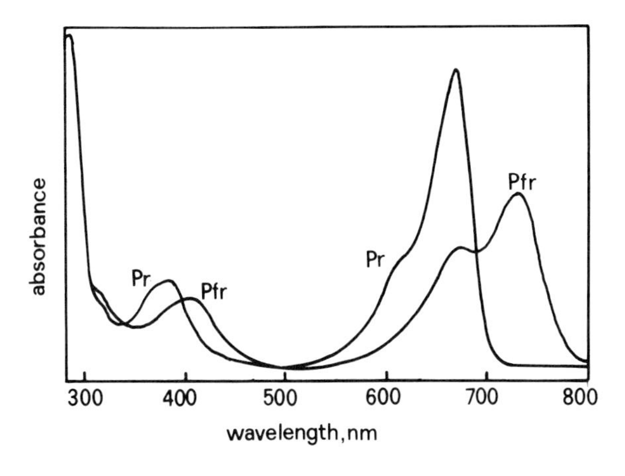

Figure 3. Absorption spectra of purified oat phytochrome. The absorption spectrum of P_{fr} is actually the absorption spectrum of the mixture of P_r and P_{fr} established by a saturating red light pulse. (after Smith, 1986)

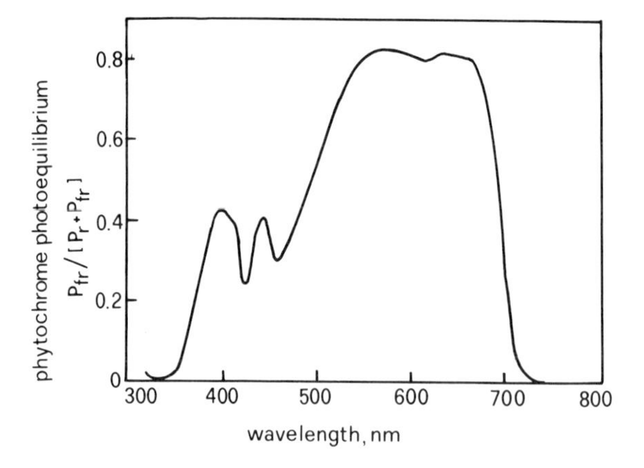

Figure 4. Photostationary photoequilibrium established by different wavelengths of light in mustard hypocotyls. (after Mohr, 1972)

Phytochrome Mediated Responses

The photomorphogenic responses resulting from the interaction between light and the phytochrome system can be classified as low fluence responses (LFR), high irradiance responses (HIR), and very low fluence responses (VLFR). General properties of these photomorphogenic responses are summarized in Table 3.

The classical red light-induced, far red light-reversible responses are LFR or inductive responses. For LFR to occur, a short exposure with low fluence is required, 1-1000 $\mu mol \cdot m^{-2}$ of red light, and the response correlates well with the photoconversion of P_r to P_{fr}. Inductive responses require a high amount of P_{fr} but only for a relatively short time. Therefore, the light pulses act as inductive signals with photomorphogenesis continuing in the darkness.

VLFR, such as the induction of rapid chlorophyll synthesis and growth modification of completely dark-grown maize coleoptiles, are induced by only 10^{-4}- 10^{-2} $\mu mol \cdot m^{-2}$ of red light which does not cause measurable conversion of P_r to P_{fr} (such fluences produce φ = 10^{-6} - 10^{-3}). Obviously, VLFR are not reversible by far red light, which establishes a lower phytochrome photoequilibrium, and are fully induced by low fluence of far red light.

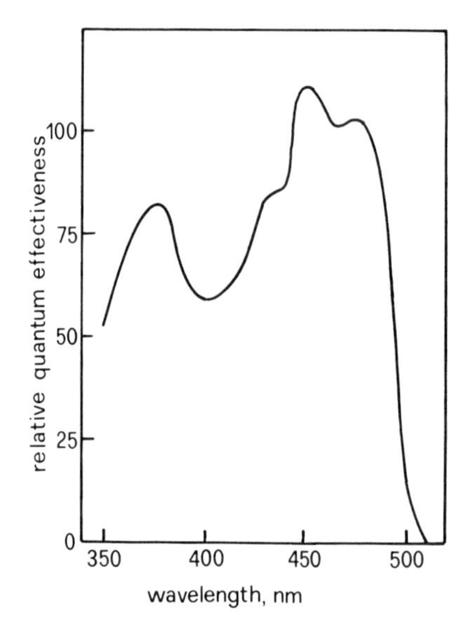

Figure 5. Action spectrum for carotenoid synthesis in *Fusarium acquaeductum* as an example of a blue light-induced response.

Table 3

General properties of phytochrome-mediated processes in plant photomorphogenesis (Modified after Mancinelli, 1975).

Low Fluence Responses (LFR)	High Irradiance Responses (HIR)	Very Low Fluence Responses (VLFR)
Inductive Responses	Steady-State Responses	Inductive Responses
Low Fluence Requirement (0.2×10^{-7} to 0.2 $J{\cdot}cm^{-2}$) Response saturated by a single irradiation at a low fluence rate	High Fluence Requirement (Higher than 1.0 $J{\cdot}cm^{-2}$) Full expression of the response requires long irradiations (hours or days) at high fluence rates. Responses strongly dependent upon the fluence rate.	Very Low Fluence Requirement (10^{-11} to 10^{-9} $J{\cdot}cm^{-2}$) Response saturated by a single irradiation
Red/Far Red Reversibility	No Red/Far Red Reversibility	No Red/Far Red Reversibility
Action spectra: peak of action for the R effect at about 660 nm; peak of action for the FR effect at about 730 nm	Action spectra: peaks of action in the near UV (around 370 nm), in the blue (420 to 480 nm), in the red (around 650 nm), and in the far red (710 to 730 nm)	Action spectra: peaks of action in the red (673 nm) and in the blue
The extent of the response is not a function of the fluence and obeys the Bunsen-Roscoe reciprocity law	The extent of the response is a function of fluence, but does not obey the Bunsen-Roscoe reciprocity law	
Photoreceptor: Phytochrome	Photoreceptors: Phytochrome Photosynthetic pigments UV-Blue light photoreceptors	Photoreceptor: Phytochrome

Many photomorphogenic responses are of the HIR type. They require, or are greatly enhanced by, prolonged exposures (many hours) to high fluence rates of far red and blue-UV light (Fig. 6). These responses are fluence-rate dependent at fluence rates above those required to establish phytochrome photoequilibrium and do not show the classical red/far red reversibility. An increasing amount of evidence suggests that phytochrome is the functional pigment involved in the photoresponses of etiolated seedlings but the effectiveness of far red light almost disappears when such seedlings, are grown in the light. On the other hand, the wavelength and irradiance dependency of photoperiodic responses in long-day green plants suggest a role for phytochrome in the promoting effect of far red light (Schneider *et al.*, 1967, Deitzer *et al.*, 1979; Lercari, 1983).

The same peak at 714-718 nm was reported in action spectra determined for long-day responses (Schneider *et al.*, 1967), as well as in many other high-irradiance action spectra measured in dark-grown seedlings. It is well known that action spectra determined under prolonged irradiations do not represent the absorption spectrum of the photoreceptor involved. This is due primarily to the fact that these action spectra are the result of the kinetic properties (cycling rate and dark reactions) of the functional photoreceptor. Phytochrome cycling rate (i.e., the rate of P_r/P_{fr} interconversions) is fluence rate dependent and therefore, it has been suggested as a possible factor for phytochrome-mediated responses which are fluence rate dependent.

According to Kazarinova-Fukshansky *et al.* (1985), the action spectra for the effects of continuous irradiations in green plants are affected by the presence and distribution of other pigments that can shift the apparent maximal cycling rate of phytochrome between 695 and 715 nm. There is therefore, a good agreement between the wavelength dependence for HIR in green plants, and the calculated values of cycling rate at different wavelengths in the green tissues (Kazarinova-Fukshansky *et al.*, 1985). But there remains a discrepancy between photoperiodic and photomorphogenic responses to prolonged far red irradiations of green plants. This different effectiveness of far red light may be due either to the presence of different species of phytochrome in dark- and light-grown plants (Shimazaki and Pratt, 1985; Tokuhisha *et al.*, 1985) or to the presence of alternative receptors for different phytochrome-dependent photoresponses.

The high efficiency shown by far red light is explained on the basis of a saving mechanism for P_{fr}. When continuous red light is applied, it converts most of phytochrome to the P_{fr} form. This results in a rapid disappearance of the phytochrome pool, since P_{fr} is unstable and undergoes a rapid destruction. Far red light, on the contrary, establishes a low φ ratio which results in a higher P_{fr} concentration averaged over a long experimental period.

Photomorphogenesis in Nature

In the natural environment, plants are not exposed for short periods to monochromatic light, but to sunlight whose quantity (i.e., fluence) and quality (i.e., the spectral distribution of radiation) vary with time of day, season, climatic conditions and local attenuation.

The photosynthetic pigments, the chlorophylls and carotenoids, present in the assimilatory organs (leaves), strongly absorb radiation below 700 nm, but hardly absorb any radiation between 710 and 800 nm. Thus light below a dense green canopy is equivalent to

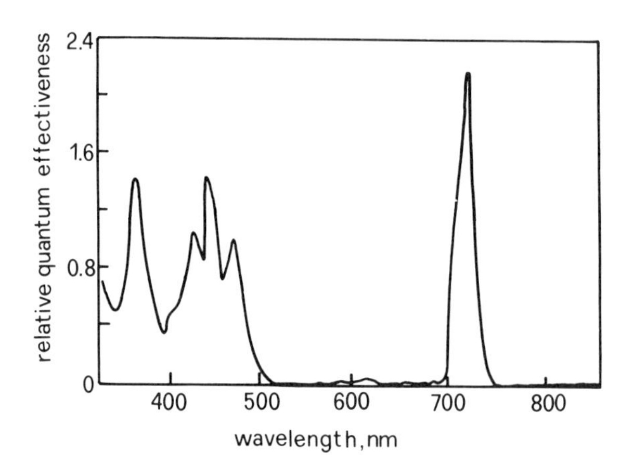

Figure 6. Action spectrum for inhibition of hypocotyl lengthening in lettuce seedlings by continuous light (18 h). This spectrum is representative of HIR. (after Hartmann, 1966)

far red radiation around 740 nm. Consequently, the photoequilibrium of the phytochrome system under direct sunlight is $\varphi = 0.56 \pm 0.02$, while under dense canopies of corn, wheat or sugar beet, this value is reduced to $<0.10 \pm 0.01$. The perception by phytochrome of changes in the red light/far red light ratio, allows the plants to respond by changes in metabolism, in leaf morphology, and elongation of the stem to the light conditions present inside or outside a dense canopy (Fig. 7).

Plants can be divided into shade tolerant species and shade intolerant species, characterized as follows:

(i) *Shade Tolerant:* slow growth rates, leaf structure which allows efficient photosynthesis at low photon fluence rates, increase of area of leaf per unit dry mass of leaf under shade.

(ii) *Shade Intolerant:* high growth rate, relatively inefficient photosynthesis at low photon fluence rates, fast response to shade by changing pattern of growth leading to shoot extension until direct sunshine is received.

The ecological significance of phytochrome in relation to green shade inside the canopy is demonstrated even in the control of seed germination. It is possible to demonstrate the red/far red reversibility by alternatively irradiating seeds directly with white light, or after filtering it through a leaf. This mechanism ensures seed germination only in gaps in the vegetation, but prevents it under leaf shade.

The possible increase of UV radiation reaching the earth's surface, as a consequence of pollutant-induced destruction of atmospheric ozone, has led to a growing interest in the effect of UV on living systems. During evolution, higher plants have optimized their ability to collect solar radiation. They are, therefore, the primary target, and obviously, the living organisms most affected by changes in the level of UV radiation in the biosphere. The morphological differences between high and low altitude plants (shorter stem, thicker leaves and brightly colored flowers in mountain plants) have been ascribed to the increased UV portion of sunlight in mountain environment. So far, there is no convincing evidence supporting this hypothesis. On the other hand, it is well known that plants grown in greenhouses covered with UV-absorbing glass, are strongly damaged after transferring them

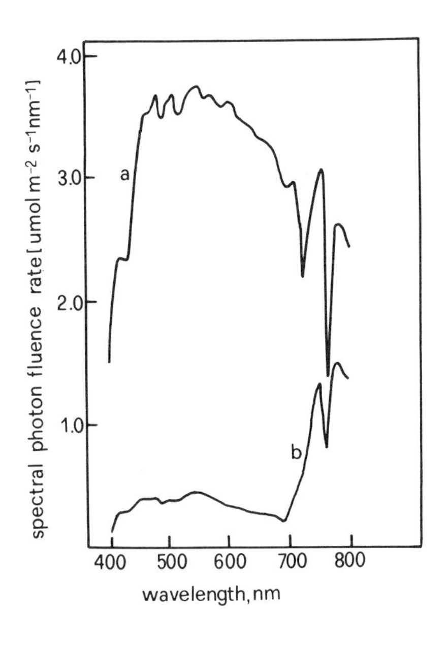

Figure 7. Spectral photon distribution of light a.) in the open and, b.) under a leaf canopy. (after Holmes and Smith, 1977)

to full sunlight, whereas plants grown under UV-enriched radiation are well protected against UV damage. The role of UV itself in inducing such adaptation is suggested, but it remains to be demonstrated.

A considerable number of studies have been reported on the effects of UV radiation on higher plants. Nevertheless, the nature and the mode of interaction between the photoreceptors involved in the control of the photomorphogenic responses induced by UV irradiation are still open questions. UV may induce both negative (damaging) and positive (non-damaging) effects in higher plants, and the same response can be differently affected (positively or negatively), depending on the fluence rate. UV light can be absorbed by phytochrome, the red/far red reversible photoreceptor of plants, causing photoconversion of P_r to P_{fr} and *vice versa*, both *in vitro* and *in vivo*. A specific blue-UV receptor with one of its peaks at about 370 nm has been characterized by action spectroscopy. In addition, a little studied UV-B photoreceptor has been reported. Yatsuhashi et al. (1987), using short irradiation periods have been able to separate, in sorghum plants, a UV response to 290 nm radiation that is independent of phytochrome, and another at 385 nm that is mediated exclusively through phytochrome. The mechanism of the UV response is, possibly, more complicated when the effects of prolonged irradiations are studied. The results of Lercari et al. (1989), suggest that phytochrome alone or phytochrome and a separate UV-absorbing receptor are involved in mediating the action of prolonged UV irradiations on both anthocyanin production and hypocotyl elongation.

The presence of many photosensors, some of them with multiple mechanisms of action (which depend on experimental conditions such as fluence rate, duration of irradiation, temporal sequence of irradiation, etc.) makes the study of photomorphogenesis very complicated. The overlapping absorption spectra of different photoreceptors, which control the same response in the blue-UV regions of the daylight spectrum, do not allow adequate conclusions about the nature, role and interactions among the absorbing pigments (Gaba and Black, 1987; Lercari *et al.*, 1989; Mohr, 1986). Morphogenic responses to blue-UV radiation may be mediated by phytochrome, UV photoreceptors, or both. Since any UV treatment causes phytochrome photoconversion, it is impossible to exclude a role for P_{fr} in a given response. Therefore, the nature and the interaction between the photoreceptors absorbing in the UV part of the spectrum, which are involved in plant photomorphogenesis are still open questions.

EXPERIMENT 24: PHOTOCONTROL OF SEED GERMINATION BY SHORT LIGHT IRRADIATIONS (RED/FAR RED REVERSIBILITY)

OBJECTIVE

The aim of this experiment is to illustrate a technique for separating the action of phytochrome and UV-photoreceptors in higher plants. To determine whether phytochrome alone or phytochrome and a separate UV-absorbing receptor are involved in mediating the action of UV irradiations on plant photomorphogenesis, this experiment demonstrates the basic red/far red photoreversibility test for phytochrome.

LIST OF MATERIALS

Seeds of lettuce, *Lactuca sativa L*, cv. Grand Rapids
Petri dishes, glass, 5.5 cm diameter
Disk filter papers, Whatman #41, 5 cm diameter
Cardboard boxes *(Empty boxes from photographic film work well)*
Black cloths
Forceps
Distilled water
Beaker, 250 ml
Pipette, 10 ml
Redlight: Red fluorescent tubes, 40 W (2) with red filter (Rosco Laboratories Inc. #823)
Far Red light: Incandescent bulbs, 100 W (6) with far red filters (Rank Strand Electric), with "Cinemoid" filter types, cin 5 a + cin 20
UV Radiation: Black light (UVA + UVB) fluorescent tubes, 40 W (4) Sylvania GTE type BLB or Westinghouse type BLB
Green Light: Green fluorescent tubes, 40 W (2) with green filters (Rosco Laboratories Inc. #874 + 877)
Spectroradiometer and/or a thermopile

PREPARATIONS

Seed Selection: Carefully select seeds in order to eliminate extreme sizes and seeds with bacterial or fungus infections.

EXPERIMENTAL PROCEDURE

PERFORM THESE STEPS USING NO LIGHT OTHER THAN A DIM GREEN SAFELIGHT.

Day 1

1. Sow fifty (50) seeds at 25 ± 1° C in Petri dishes on two layers of filter paper soaked with 3 ml of distilled water.
2. Wrap the Petri dishes in black cloth and place them in a black cardboard box. For each irradiation program, use a separate black box and mark it with the time of sowing and the irradiation program. Use at least 4 Petri dishes for each irradiation program, i.e. at least 32 dishes in total. Store them at 25° ± 1° C.

3. Six h after sowing the seeds, irradiate sets of 4 dishes as follows:

   a) 5 min red light (20 μmol·m^{-2}·s^{-1})
   b) 15 min far red light (20 μmol·m^{-2}·s^{-1})
   c) 15 min UV (50 μmol·m^{-2}·s^{-1})
   d) 5 min red + 15 min far red
   e) 15 min UV + 15 min far red
   f) 5 min red + 15 min far red + 5 min red
   g) 15 min UV + 15 min far red + 15 min UV
   h) dark control (no irradiation)

4. After irradiation wrap the Petri dishes in black cloth and place them in a black cardboard box at 25° ± 1°C.

Day 2

5. Twenty-four h after sowing, count the number of germinated seeds with root length > 1mm for each irradiation condition.

DISCUSSION

In darkness the percentage of germinated seeds should be low, about 15-20%. The saturating red pulse and UV pulse should bring about a very high percentage of germination. The saturating far red pulse should induce a very low percentage of germinations (0 - 5%). Furthermore, a far red light pulse immediately following the red light pulse should nullify the increase of germination induced by the red light.

It is largely accepted that if induction of a photoresponse by a red light pulse ($\varphi = 0.86$) can be fully reversed by immediately following it with a saturating far red light pulse ($\varphi = 0.03$), the response is controlled by phytochrome. On the other hand, UV radiation is also absorbed by phytochrome, causing photoconversion of P_r to P_{fr} and *vice versa*, both *in vitro* and *in vivo,* and establishing $\varphi = 0.75$. It is, therefore, possible that if the promoting effect of UV irradiation is reversed by immediately following it with a saturating far red light pulse, phytochrome is involved in this photoresponse. However, if the photoresponse induced by UV radiation is not reversed by an immediately following far red light pulse, this action must be attributed to another specific UV-absorbing photoreceptor.

The far red pulse must be given immediately after the red or UV pulse because the initial action of P_{fr} is fast. Irreversible effects become detectable, in some instances, after only a few seconds of red light irradiation.

The UV radiation pulse in this experiment is given with a high photon fluence rate (50 μmol·m^{-2}·s^{-1}) and for a relatively long time, 15 min. This is due to the low molar absorption coefficient and quantum efficiency values for phytochrome in this waveband, and also to the high scattering and light attenuation by flavin and carotenoid absorption.

EXPERIMENT 25: PHYTOCHROME, UV PHOTORECEPTORS AND THE PHOTOREGULATION OF HYPOCOTYL ELONGATION UNDER PROLONGED UV RADIATION

OBJECTIVE

The aim of this experiment is to illustrate a technique for separating the action of phytochrome and UV-photoreceptors in higher plants. To determine whether phytochrome alone or phytochrome and a separate UV-absorbing receptor are involved in mediating the action of UV irradiations on plant photomorphogenesis, this experiment demonstrates the use of the light equivalence principle for photochromic pigments (Schäfer *et al.*, 1983).

LIST OF MATERIALS

Seeds of tomato, *Lycopersicon esculentum* Mill, cv Cal J
Petri dishes, glas, 10 cm diameter, 8 cm height
Disk filter papers, Whatman #41, 9 cm diameter
Cardboard boxes *(Empty boxes from photographic film work well)*
Black cloth
Forceps
Scalpels
Ruler
Distilled water
Beakers, 250 ml (3)
Pipettes, 10 ml (3)
Norflurazon (4 - chloro - 5 - (methylamino) 2 - (2-2-2, trifluoro-m-tolyl) - 3(2H) - pyridarizone (Sandoz, A.G., Basel, Switzerland)
Streptomycin sulfate
Magnetic stirrer
Red light, green light and UV radiation sources as in the previous experiment
Spectroradiometer and/or thermopile
Aluminum gauze (1.4 × 1.4 mm grids, 0.28 mm wire diameter) as neutral density filters

PREPARATIONS

Norflurazon Solution: Dissolve 3.80 mg of Norflurazon in 250 ml of distilled water to produce a 10^{-5} M solution.

Streptomycin Sulfate Solution: Dissolve 62.50 mg of streptomycin sulfate in 250 ml of distilled water to produce a 250 mg·l^{-1} solution.

Seed Selection: Carefully select seeds in order to eliminate extreme sizes and seeds with bacterial or fungus infections.

Sowing: Four days before illumination sow seeds and store them in the dark, as follows.

PERFORM THESE STEPS USING NO LIGHT OTHER THAN A DIM GREEN SAFELIGHT.

(i) Sow fifty (50) seeds at 25 ± 1° C in Petri dishes on four layers of filter paper soaked with 10 ml distilled water or streptomycin (250 mg·l^{-1}) or Norflurazon (10^{-5} M) solutions.

Table 4

Photochemical parameters of phytochrome at selected wavelengths (Mancinelli, 1989)

Wavelength (nm)	Photoequilibrium	N ($\mu mol \cdot m^{-2} \cdot s^{-1}$)
300	0.66	2.1
350	0.75	4.1
400	0.57	2.7
450	0.42	11.9
500	0.48	28.9
550	0.80	17.4
600	0.89	5.9
660	0.87	1.6
700	0.30	2.8
730	0.02	28.7
760	0.02	60.5

N is the photon flux required to maintain the photoconversion rate (Pr $\rightleftarrows$ Pfr).
$k = 10^{-3}\ s^{-1}$ at photoequilibrium.

(ii) Wrap the Petri dishes in black cloth and place them in a black cardboard box. For each irradiation program, use a separate black box and mark it with the time of sowing and the irradiation program. Use at least 3 Petri dishes for each irradiation program, i.e., at least 36 dishes in total. Store them at 25° ± 1° C.

Photon Fluence Rates: Different photon fluence rates can be obtained from the UV and red light sources by filtering the light with layers of aluminum gauze (usually used as mosquito netting). The UV and red light photon fluence rate are chosen according to Table 4. The desired fluence rates, which should be determined using a spectroradiometer or a calibrated thermopile, are the following.

UV radiation: 0.04; 0.4; 4.1; 41; 63 $\mu mol \cdot m^{-2} \cdot s^{-1}$
Red light: 0.02; 0.2; 1.6; 16; 24 $\mu mol \cdot m^{-2} \cdot s^{-1}$

EXPERIMENTAL PROCEDURE

Day 1

1. Measure the length of the 96 h dark-grown hypocotyls (3 Petri dishes for each solution). This represents CONTROL A. When measuring hypocotyl lengths, always discard the measurements of the 10 tallest and the 10 shortest seedlings in each Petri dish. Thus you should record the lengths of 30 seedlings in each dish.

2. Place the 3 Petri dishes for each irradiation program (i.e., water, streptomycin and Norflurazon) under the appropriate UV or red source. You will have at least 3 dishes for each of the 10 irradiation conditions prepared above (see Photon Fluence Rates in Preparations).

Day 3

3. Measure the length of the hypocotyls that have been grown in continuous darkness

(96 h + 48 h = 144 h dark). This is CONTROL B. Remember to discard the measurements of the 10 longest and shortest.

4. Measure the length of the hypocotyls in each of the 10 different irradiation conditions (96 h dark + 48 h light). These are the experimental (EXP) results.

5. Calculate the percentage inhibition of hypocotyl growth elongation according to the following formula:

$$\frac{\text{CONTROL B} - \text{EXP}}{\text{CONTROL B} - \text{CONTROL A}} \times 100$$

This represents the suppression of elongation growth caused by 48 h of UV or red irradiation relative to growth in darkness.

6. Compare the results obtained for seedlings grown in the presence of water, Norflurazon, and streptomycin.

DISCUSSION

The overlapping absorption spectra of different photoreceptors, which control the same response in the blue-UV regions of the daylight spectrum, do not allow adequate conclusions about the nature, role and interactions among the absorbing pigments. Here, in order to determine whether inhibition of hypocotyl elongation brought about by prolonged exposures to UV radiation is mediated through phytochrome or through a specific UV receptor, experiments based on the principle of "light equivalence for photochromic pigment" (Schäfer *et al.*, 1983) are performed.

According to the light equivalence principle, two light treatments are equivalent with respect to the phytochrome system when both the photoconversion rate k, ($k = k_1 + k_2$)

$$P_r \underset{k_2}{\overset{k_1}{\rightleftarrows}} P_{fr}$$

and the photoequilibrium, φ, are the same at two different wavelengths. According to Table 4, red light (660 mm) produces a photoequilibrium φ(red) = 0.86 similar to the photoequilibrium established by UV radiation φ(UV) = 0.75. According to the values presented in Table 4, the same rate constant for phytochrome photoconversion, $k = 10^{-3}\ s^{-1}$, is obtained when the photon fluence rate at 660 nm (red) is 1.6 $\mu mol \cdot m^{-2} \cdot s^{-1}$ (and the photon fluence rate at 350 nm (UV) is 4.1 $\mu mol \cdot m^{-2} \cdot s^{-1}$). Therefore, equal rates of phytochrome cycling are obtained under UV and red irradiations when the ratio between the photon fluence rates UV/red = 4.1/1.6 = 2.6 is kept constant.

It is thus possible to test the effect of increasing photon fluence rates of red light and UV radiation, with identical phytochrome conditions (same φ and k) under the two light sources. Therefore, if phytochrome is the only photoreceptor, the two light sources should induce similar effects. If the physiological response induced by red and UV are quantitatively different, the differences must be attributed to the action of separate photoreceptors working at the two wavelengths. In this experiment, when seedlings are irradiated with photon fluence rates below 2 $\mu mol \cdot m^{-2} \cdot s^{-1}$ for red light and below 4 $\mu mol \cdot m^{-2} \cdot s^{-1}$ for UV radiation, the growth inhibition is low (10-15%) and very similar for both the irradiations, suggesting that only phytochrome is the functional photoreceptor. At higher photon fluence rates, however, the comparison between the action of UV radiation (absorbed by phytochrome and UV absorbing receptors) and the action of red light (absorbed by phytochrome and not by UV absorbing receptors) should show a greater response to UV radiation, suggesting that UV radiation effects are mediated by both the phytochrome and UV-absorbing receptors.

In some experiments, plants are grown on Norflurazon (10^{-5} M), an inhibitor of carotenoid synthesis, thereby allowing chlorophyll photodestruction, and on streptomycin (250 mg·l^{-1}), an inhibitor of chloroplast development and chlorophyll synthesis. Seedlings grown with both streptomycin and Norflurazon are referred to as "chlorophyll-free" seedlings. In these "chlorophyll-free" seedlings, the wavelength-dependent differences in light attenuation due to chlorophyll as well as any other photosynthesis-dependent phenomenon should be considerably reduced.

Compare the reponses of seedlings grown without inhibitor and of chlorophyll-free seedlings. They should respond similarly to red light and UV irradiations. This kind of result supports the conclusion that UV- and red light-induced inhibition of growth is independent of photosynthesis.

Obviously, it is not possible with only one experimental approach to clarify the action and interaction of phytochrome and specific UV- absorbing photoreceptors. Several other techniques can be employed, including the following:

(a) Measurement of Action Spectra;

(b) Dichromatic Irradiations (Hartman, 1966). Prerequisites for dichromatic irradiations are: 1) the two wavelengths under investigation must establish very different phytochrome photoequilibria; 2) one of the two wavelengths is ineffective if applied alone;

(c) Cyclic Irradiations. Cycles of different lengths are used instead of continuous irradiations, the UV radiations being applied for 25% of the time, independent of cycle length. The plants are thus given the same total fluence during the experimental periods (12-48 h). Common cycles are as follows: 2 s UV/6 s dark; 10 s UV/30 s dark; 1 min UV/3 min dark; 15 min UV/45 min dark. Continuous UV radiation with the same total energy fluence is used as a control; i.e., with 25% of the photon fluence rate used in cyclic irradiations. Cyclic treatments are very useful because they allow alternate treatments in different wavebands (i.e., 1 min UV + 1 min far red + 1 min dark; 15 min UV + 15 min far red + 30 min dark); thus it is possible to investigate the far red reversibility (phytochrome involvement) under prolonged irradiations.

The availability of photoreceptor mutants, such as the *aurea* mutant of tomato which is lacking the unstable pool of phytochrome, represents a new and important tool for the study of higher plant photomorphogenesis.

REVIEW QUESTIONS

1. Why do plants have several kinds of photoreceptors?
2. Why is a red light-induced response nullified by an immediately following far red pulse and not nullified by a far red pulse given some hours later?
3. Why is the red light/far red light photon ratio important for plants?
4. Why is far red light more efficient than red light under prolonged irradiations?
5. Why is it impossible to exclude a role for phytochrome in UV-induced responses?

6. In lettuce, far red light almost completely inhibits seed germination, while in darkness 20% of seeds germinate. Can you explain this result?

ANSWERS TO REVIEW QUESTIONS

1. Plants need several photoreceptors in order to appreciate properly the quantity, quality, direction, and duration of light environment. Different wavebands, i.e., red, blue, UV, can specifically affect plant growth and development, requiring therefore specific sensor pigments in order to cover the range between 280 and 800 nm.

2. When there is a long dark period between red irradiation and a subsequent far red irradiation, there is loss of reversibility. During this dark period, a series of cell reactions are induced by P_{fr}, the active form of phytochrome. If given too late, far red light cannot reverse these effects.

3. The daylight spectrum presents a nearly constant red/far red ratio of about 1. Inside a vegetation canopy, however, this ratio is lower, in the range of 0.1-0.5. The perception by phytochrome of changes in red/far-red ratio allows the plants to respond by changes in metabolism, in leaf morphology, in elongation of the stem and in seed germination, to the light conditions present inside or outside a dense vegetation canopy.

4. Prolonged far-red irradiations are very efficient because they establish a very low photoequilibrium of the phytochrome system, $P_{fr} = 0.03\%$. Therefore only a few P_{fr} molecules are present and undergo degradation.

5. Phytochrome has an absorption peak at 280 nm. Under UV radiation, phytochrome is photoconverted and a large amount of P_{fr} is formed. This effect of UV light on phytochrome makes it difficult to separate the phytochrome reactions from purely UV-induced reactions.

6. When seeds are kept in darkness, some of the phytochrome in dry seeds is in the P_{fr} form. This P_{fr} is hydrated during imbibition and induces some germination (about 20%). Under far red light, however, nearly all the phytochrome is converted to the P_r form and germination cannot start.

SUPPLEMENTARY READING

Pratt, L.H. and M.-M. Cordonnier (1989) Photomorphogenesis. In *The Science of Photobiology*, 2nd edition (Edited by K.C. Smith), pp. 273-304. Plenum Press, New York.

REFERENCES

Adamse, P., P. Jaspers, J.A. Bakker, J.C. Wesselius, G.H. Heeringa, R.E. Kendrick and M. Koornneef (1988) Photophysiology of a tomato mutant deficient in labile phytochrome. *J. Plant Physiol.* **133**, 436-440.

Adamse, P., R.E. Kendrick and M. Koornneef (1988) Photomorphogenetic mutants of higher plants. *Photochem. Photobiol* **48**, 833-841.

Bartels, P.G. and A. Hyde (1970) Chloroplast development in 4-chloro-(dimethylamino)2-(2,2,2-trifluoro-m-tolyl)-3(2H) pyridazinone (Sandoz 6708)-treated wheat seedlings. *Plant Physiol.* **54**, 807-810.

Beggs, C.S., M.G. Holmes, M. Jabben and E. Schäfer (1980) Action spectra for the inhibition of hypocotyl growth by continuous irradiation in light- and dark-grown *Sinapis alba L.* seedlings. *Plant Physiol.* **66**, 615-618.

Bortwick, H.A., S.B. Hendricks, M.W. Parker, E.M. Toole and V.K. Toole (1952) A reversible photoreaction controlling seed germination. *Proc. Natl. Acad. Sci. USA* **38**, 662-666.

Deitzer, G.F., R. Hayes and M. Jabben (1979) Kinetic and time dependence of the effect of far red light on the photoperiodic induction of flowering in Winter barley. *Plant Physiol.* **64**, 1015-1021.

Frankland, B. (1986) The perception of light quantity. In *Photomorphogenesis in Plants.* (Edited by R.E. Kendrick and G.H.M. Kronenberg), pp. 219-236. Martinus Nijhoff/Dr. W. Junk Publ., Dordrecht.

Gaba, V. and M. Black (1987) Photoreceptor interaction in plant photomorphogenesis: The limits of exprimental techniques and their interpretations. *Photochem. Photobiol.* **45**, 151-156.

Hartman, K.M. (1966) A general hypothesis to interpret "high energy phenomena" of photomorphogenesis on the basis of phytochrome. *Photochem. Photobiol.* **5**, 349-366.

Holmes, M.G. and H. Smith (1977) The function of phytochrome in the natural environment. II. The influence of vegetation canopies on the spectral energy distribution of natural daylight. *Photochem. Photobiol.* **25**, 539-545.

Holmes, M.G. and E. Wagner (1982) The influence of chlorophyll on the spectral control of elongation growth in *Chenopodium rubrum L.* hypocotyls. *Plant Cell Physiol.* **23**, 745-750.

Hillman, W.S. (1967) The physiology of phytochrome. *Ann. Rev. Plant. Physiol.* **18**, 301-330.

Kazarinova-Fukshansky, N., M. Seyfried and E. Schäfer (1985) Distortion of action spectra in photomorphogenesis by light gradients within plant tissue. *Photochem. Photobiol.* **41**, 689-702.

Koornneef, M., J.W. Cone, R.G. Dekens, E.G. O'Herne-Robers, C.J.P. Spruit and R.E. Kendrick (1985) Photomorphogenic responses of long hypocotyl mutants of tomato. *J. Plant Physiol.* **120**, 153-165.

Koornneef, M. and R.E. Kendrick (1986) A genetic approach to photomorphogenesis. In *Photomorphogenesis in Plants.* (Edited by R.E. Kendrick and G.H.M. Kronenberg), pp. 547-563. Martinus Nijhoff/Dr. W. Junk Publ., Dordrecht.

Koornneef, G.H.M. and R.E. Kendrick (1986) Phytochrome. The physiology of action. In *Photomorphogenesis in Plants.* (Edited by R.E. Kendrick and G.H.M. Kronenberg), pp. 99-114. Martinus Nijhoff/Dr. W. Junk Publ., Dordrecht.

Lercari, B. (1983) Action spectrum for the photoperiodic induction of bulb formation in *Allium cepa L. Photochem. Photobiol.* **38**, 219-222.

Lercari, B., F. Sodi and M. Biagioni (1986) An analysis of UV-A effects on phytochrome-mediated induction of phenylalanine ammonia-lyase in the cotyledons of *Lycopersicon esculentum Mill. Environ. Exp. Bot.* **26**, 153-157.

Lercari, B., F. Sodi and C. Sbrana (1989) Comparison of photomorphogenic responses to UV light in red and white cabbage (*Brassica oleracea L.*). *Plant. Physiol.* **90**, 345-350.

Lipucci Di Paola, M., F. Collina Grenci, L. Caltavuturo, F. Tognoni and B. Lercari (1988) A phytochrome mutant from tissue culture of tomato. *Adv. Hort. Sci.* **2**, 30-32.

Mancinelli, A.L. (1989) Interaction between cryptochrome and phytochrome in higher plants photomorphogenesis. *Am. J. Bot.* **76**, 143-154.

Mohr, H. (1972) *Lectures on Photomorphogenesis.* Springer, Berlin, Heidelberg, Tokyo.

Mohr, H. (1986) Coaction between pigment systems. In *Photomorphogenesis in Plants.* (Edited by R.E. Kendrick and G.H.M. Kronenberg), pp. 521-546. Martinus Nijhoff/Dr. Junk Publ, Dordrecht.

Parks, B.M., A.M. Jones, P. Adamse, M. Koorneef, R.E. Kendrick and P.H. Quail (1987) The *aurea* mutant of tomato is deficient in spectrophotometrically and immnunochemically detectable phytochrome. *Plant Molec. Biol.* **9**, 97-107.

Sachs, J. (1864) Wirkungen des farbigen lichts auf pflanzen. *Bot. Z.* **22**, 353-358.

Schäfer, E., L. Fukshanky and L. Shropshire Jr. (1983) Action spectroscopy of photoreversible pigment systems. In *Photomorphogenesis.* Encyclopedia of Plant Physiology, Vol. 16 A, (Edited by W. Shropshire Jr. and H. Mohr), pp. 39-68. Springer Verlag, Berlin.

Schneider, M.J., H.A. Borthwick and S.B. Hendricks (1967) Effects of radiation on flowering of *Hyoscyamus niger. Am. J. Bot.* **54**, 1241-1249.

Shimazaki, Y. and L.H. Pratt (1985) Immunochemical detection with rabbit polyclonal and mouse monoclonal antibodies of different pools of phytochrome from etiolated and green *Avena* shoots. *Planta* **164**, 333-344.

Smith, H. and G. Holmes (1984) *Techniques in Photomorphogenesis.* Academic Press.

Smith, H. (1986) The perception of light quality. In *Photomorphogenesis in Plants* (Edited by R.E. Kendrick and G.H.M. Kronenberg), pp. 187-218. Martinus Nijhoff/Dr. W. Junk Publ, Dordrecht.

Tevini, M. and A.H. Teramura (1989) UV-effects on terrestrial plants. *Photochem. Photobiol.* **50**, 479-487.

Tokuhisa, J.G., S.M. Daniels and P.H. Quail (1985) Phytochrome in green tissues: Spectral and immunochemical evidence for two distinct molecular species of phytochrome in light-grown *Avena sativa. L. Planta* **164**, 321-332.

Urbach, F. (1989) The biological effects of increased ultraviolet radiation: An update. *Photochem. Photobiol.* **50**, 439-441.

Yatsuhashi, H., T. Hashimoto and S. Shimizu (1982) Ultraviolet action spectrum for anthocyanin formation in Broom sorghum first internodes. *Plant Physiol.* **70**, 735-741.

Photomovement

Wilhelm NULTSCH and Jürgen PFAU
Philipps-Universität Marburg
Germany

INTRODUCTION

Basic Terminology

Many microorganisms, as well as several higher plants respond to light with a movement response. Photomovement may be broadly defined as any light-induced motility or behavioral response involving the spatial displacement of all or part of an organism. Such movements vary in their basic phenomena, their mechanisms, their photoreceptors, the mode of photoresponse and, as far as is known, their sensory transduction. The overall process of photomovement can be seen as the consecutive series of a stimulus (photons in this case), a stimulus perception, a transduction, and finally, motile and behavioral responses. Here it seems to be useful to clearly define some of the basic terms used in the study of photomovement:

Phenomena: Movements of organs or parts of organs, cells, cell-aggregates or filaments, as well as movements of organelles inside the cells such as chloroplasts and cytoplasm.

Mechanisms: Differential growth or changes in turgor-pressure, both resulting in either bending or opening and closing by a hinge mechanism; movements caused by the microtubule system such as flagellar and ciliary movements; movements due to activities of the acto-myosin system such as ameboid movements and intracellular movements; gliding movements caused by other contractile protein-filaments or by unilateral secretion of mucilage. It is to be noted that the use of the term "mechanism" in this sense does not necessarily imply a detailed knowledge on a molecular basis.

Photoreceptors: Molecular systems such as flavoproteins, rhodopsin, stentorin, phytochrome, photosynthetic pigments that absorb the photons (stimulus).

Sensory Transduction: The entire process between the perception of the light stimulus and the response (movement). This process is, in general, poorly understood, and is often referred to as a "black box".

Modes of Photoresponses: These may be subdivided into four major categories (Fig. 1): *Light-oriented movements; transient changes in movement; effects on the steady state velocity of movement; setting the clock of circadian movements (Zeitgeber).*

WILHELM NULTSCH AND JÜRGEN PFAU, Fachbereich Biologie der Philipps-Universität Marburg, Karl-von-Frisch-Straβe, D-3550 Marburg/Lahn, Germany

Classification of Photomovement

Since the detailed mechanisms of the sensory transduction have not been elucidated, there is no general agreement among the workers in this field for an acceptable set of definitions to describe a particular photomovement (Song and Poff, 1989). A set of terms, however, describing the various movement responses in photosensitive microorganisms have been proposed for general use in photobiology (Diehn et al.,1977; Nultsch and Hader, 1988). Based on the definitions given by the latter authors, the various types of photomovements can be summarized as shown in Fig. 1.

The various types of photomovements are briefly described below.

Phototaxis: Oriented movement of locomotive microorganisms with respect to the light direction, resulting in a movement either toward the light source (positive phototaxis, Fig. 2), or away from it (negative phototaxis). A special case is a movement direction perpendicular to the light direction (transversal or dia-phototaxis). Mostly light of shorter wavelengths (violet, blue, blue-green) is active, but in some cases the photosynthetic pigments, e.g., phycobiliproteins and chlorophyll *a*, are also involved in the perception of the phototactically active light (Fig.3; Nultsch and Häder, 1979, 1988).

Phototropism: Oriented movement of sessile organisms or their organs with respect to the light direction. This movement is usually brought about by differential growth or, in some cases, by differential changes in turgor pressure in the two opposite flanks of an organ. The result is a bending of the organs toward the light source (positive phototropism) or away from it (negative phototropism). In green higher plants the stem reacts positively, the root either negatively (Fig.4) or, more frequently, not at all (phototropic indifference), whereas the leaves orient themselves perpendicularly to the light direction (transversal or dia-phototropism), in order to capture a maximum number of quanta for photosynthesis. Mostly light of shorter wavelengths (UV-A, violet and blue) is active

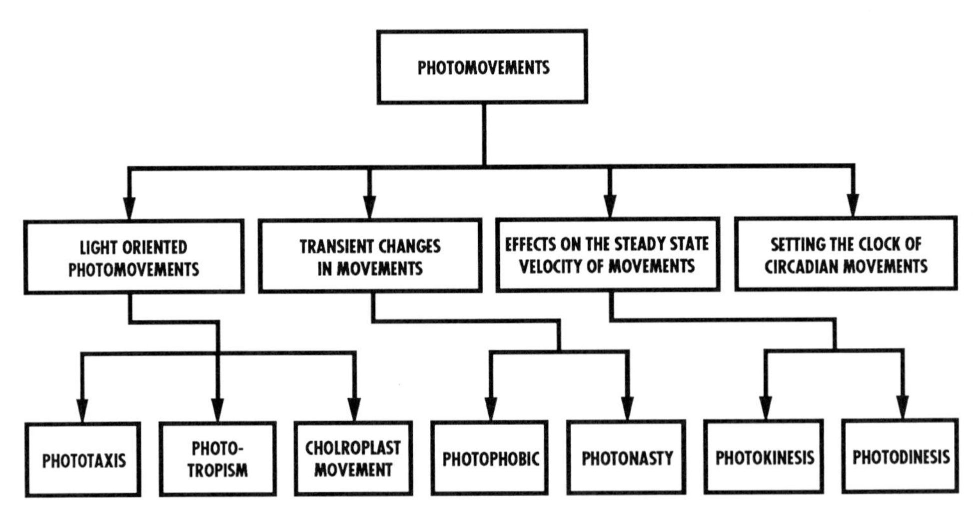

Figure 1. Classification of the various types of photomovements based on the concept of a photoresponse. In *Light Oriented Movement*, it is the *direction* of light that is perceived. In *Transient Changes in Movements*, either a sudden decrease or an increase in the *photon fluence rate* are perceived. These types of photomovements occur independently of the light direction. In *Effects of Light on the Steady State Velocity of Movements*, the light can function as a trigger, but in photosynthetic organisms it serves mainly as an energy source for movement.

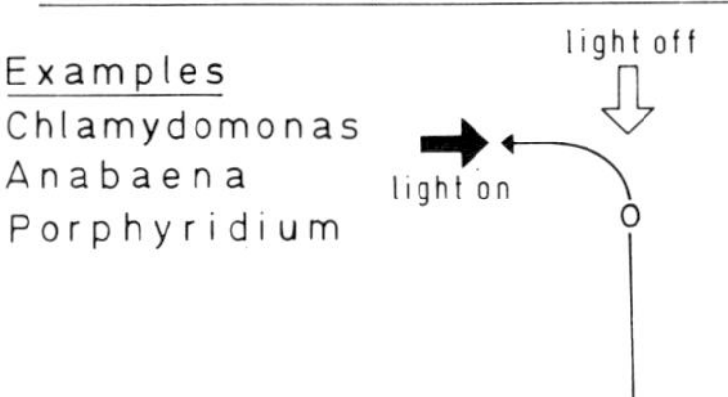

Figure 2. Schematic representation of phototactic orientation by true steering.

(Fig.5). The blue light receptors, frequently called cryptochrome, are most probably flavoproteins (Dennison, 1979).

Chloroplast Movements: Chloroplasts (chromatophores of higher plants, mosses, some green and many brown algae) can occupy two different positions with respect to the light direction, called "high intensity" and "low intensity arrangement". These arrangements depend on the morphology of the cells as well as on the number and the shape of the chloroplasts. In the freshwater alga *Mougeotia* the chloroplast has the shape of a plate which either faces the light (low intensity arrangement) or turns by an angle of 90° (high intensity arrangement). In many plants that contain a larger number of chromatophores, such as the moss *Funaria* or the brown alga *Dictyota*, the chromatophores occupy the periclinal cell walls which are perpendicular to the light direction in dim light but migrate to the anticlinal walls in strong light. As a general rule, one can say that they move to the least irradiated parts of the cells in strong light and *vice versa* (Britz,1979). This phenomenon has also been called "chloroplast phototaxis".

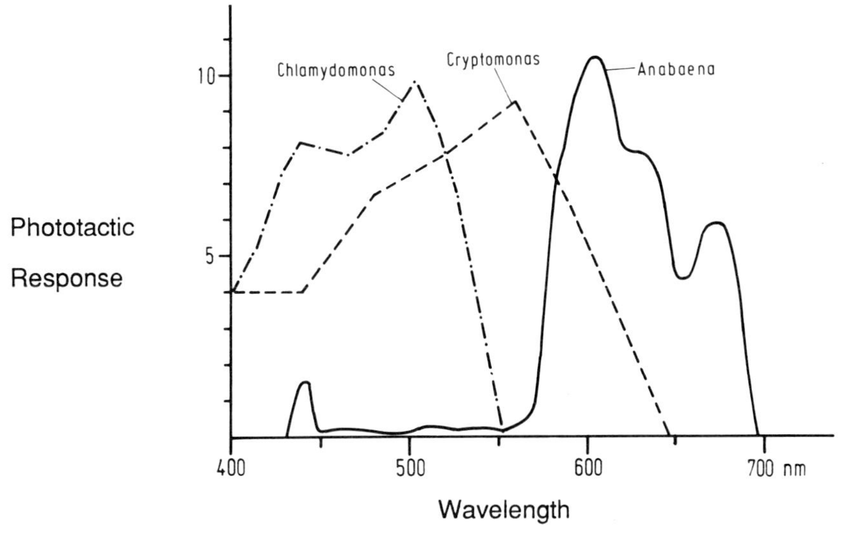

Figure 3. Phototactic action spectra of the flagellates *Chlamydomonas* (green algae), *Cryptomonas* (cryptphyceae) and the cyanobacterium *Anabaena variabilis*. Abscissa: wavelength in nm. Ordinate: phototactic response in relative units.

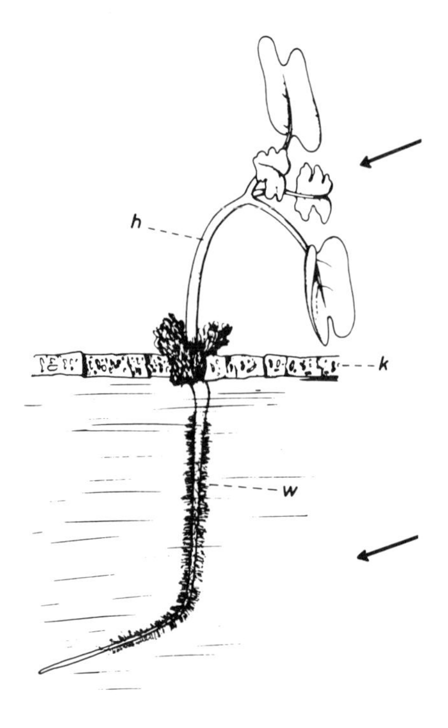

Figure 4. Phototropism of the mustard seedling (*Sinapis alba*). Arrows indicate the light direction; k = cork-disk; h = hypocotyl; w = root. The stem (hypocotyl) reacts positively, the root negatively. The leaves show diaphototropism (after Noll).

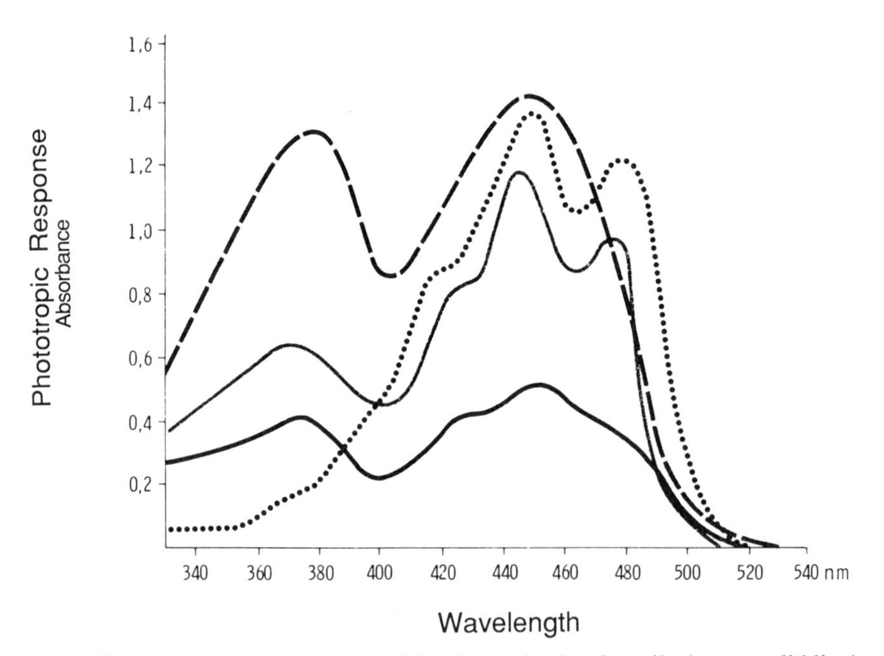

Figure 5. Phototropic action spectrum of the *Avena* (oat) coleoptile (upper solid line) and absorption spectra of a flavoprotein (lower solid line), riboflavin (broken line) and β-carotene (dotted line). Abscissa: wavelength in nm. Ordinate: phototropic sensitivity in relative units and absorbance (after Curry and Thimann, Karrer and Jucker, Bellin and Holstrom, Williams).

Photophobic Responses: Transient changes in locomotion of microorganisms caused by a sudden increase (step up) or decrease (step down) of the fluence rate, resulting in either a tumbling or turning of the cells (Fig.6), in a reversal of the movement direction or simply in a stop. Such responses can be brought about either by a temporal change, e.g., when the fluence rate is suddenly changed, or by a spatial gradient, e.g., when an organism moves from an area of given fluence rate to another with a higher or lower one. This is perceived by moving organisms as a temporal change as well. Thus, they accumulate in a light field or leave it (Nultsch and Häder, 1979, 1988), as a result of step down or step up responses, respectively.

Photonasty: Photomovement of organs of sessile plants caused by changes in the photon fluence rate. Since these movements are brought about by a hinge mechanism, they are independent of the light direction, e.g., flowers can open if the fluence rate is increased and close if it is decreased (by clouding). The contrary is true for plants flowering during the night. Also green leaves whose petioles (stalks) possess a pulvinus (leaf joint) show photonastic movements.

Photokinesis: Effect of light on steady state velocity of locomotive (micro)organisms. If movement is initiated or accelerated, photokinesis is called positive, if it is decelerated or stopped, it is called negative. Photokinesis is independent of the light direction. The signal perceived is the fluence rate. As far as is known, light acts via the proton motive force or by photophosphorylation (Nultsch and Häder, 1979, 1988).

Photodinesis: Initiation of protoplasmic streaming in plant cells by light or its effect on the speed of streaming. When the cells contain chloroplasts, they are passively transported by the streaming protoplasm (Seitz, 1979). This transport must not be mistaken with the above mentioned light induced chloroplast movement.

Zeitgeber: In plants, several different types of autonomous movements exist which are governed by the circadian clock. In this case light acts only as a "Zeitgeber", i.e., it sets the clock (see Chronobiology experiment).

Experimental Considerations

In many cases it is difficult to quantify photomovement phenomena. Because of their great diversity, specialized measuring methods have to be devised for each type of photomotion. Furthermore, instrumental details often have to be modified when different species are studied. A further complication is that the necessary equipment often is not commercially available. In order to build up a photomovement measurement system, essential parts may be purchased, but they usually have to be modified to meet special requirements. For these reasons, two movement phenomena have been selected for which measuring systems are described. These include: 1) light induced chloroplast movements in the moss *Funaria*

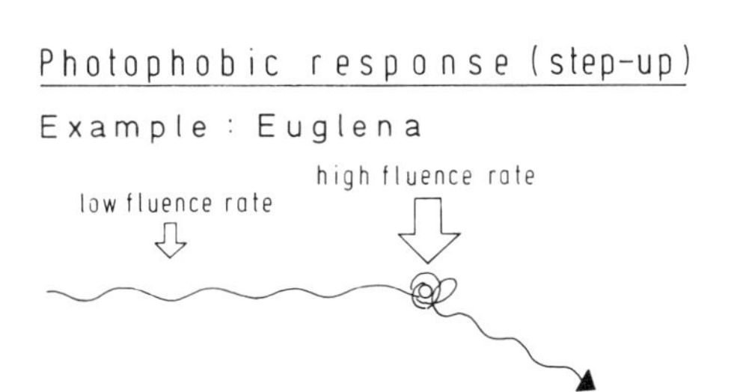

Figure 6. Schematic representation of a step-up photophobic response.

hygrometrica or the brown alga *Dictoyota dichotoma*, 2) phototaxis of the green alga *Chlamydomonas reinhardtii*, coupled with measurements of motility.

Light Induced Chromatophore Movements in *Funaria hygrometrica* or *Dictyota dichotoma*

In the moss *Funaria hygrometrica*, the chloroplasts occupy the periclinal walls facing the light in the so-called low intensity arrangement and the anticlinal walls in the high intensity arrangement (Fig. 7). The same applies to the phaeoplasts of *Dictyota dichotoma*. The so-called dark arrangement is disregarded here. At intermediate levels of light, the chromatophores take positions in between these two extremes, i.e., they occupy both anticlinal and periclinal walls to different extents depending on the fluence rate. These movements are brought about by an actomyosin system. In *Funaria*, as in most plants, only light of shorter wavelengths (UV-A, violet, blue and blue-green light (up to 520 nm)) is effective. The photoreceptor is supposed to be a flavoprotein, contrary to the green alga *Mougeotia* where phytochrome is the photoreceptor. For reviews see Britz (1980), Haupt and Wagner (1984) or Zurzycki (1980).

Chromatophore Movements: Measurement Principle

For quantitative measurements one could count the number of the chromatophores in the low intensity arrangement and compare this to the number in a high intensity arrangement. However, this is a time-consuming and cumbersome task. Even the use of an image analysis system presents considerable difficulties with respect to those chromatophores, which are partly in the face and partly in the profile position. Since the absolute transmittance values increase by 20% when the chromatophores move from the low to the high intensity arrangement, such optical transmittance changes can be used as an indirect, but reliable measure of the light induced chromatophore displacement (Pfau et al., 1974; Nultsch and Benedetti, 1978).

The extent of the transmittance changes may be demonstrated on an idealized model cell with six chromatophores in face position (Fig. 8A) or in profile position (Fig. 8B). The following assumptions are made: 1) a chromatophore in face position transmits 50%. 2) a "profile chromatophore" transmits 25%. 3) a "face chromatophore" takes double the area of a "profile chromatophore".

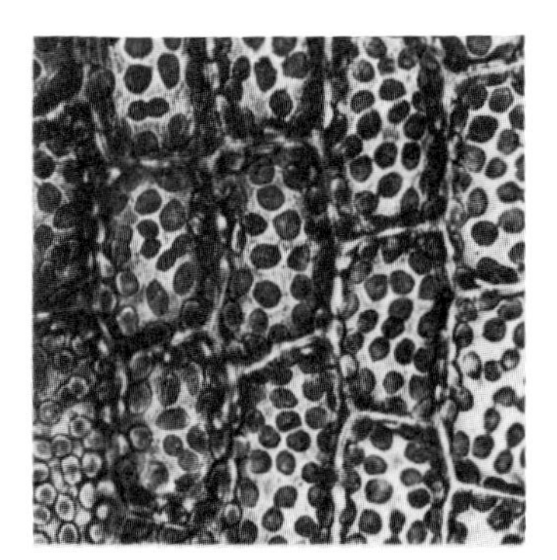
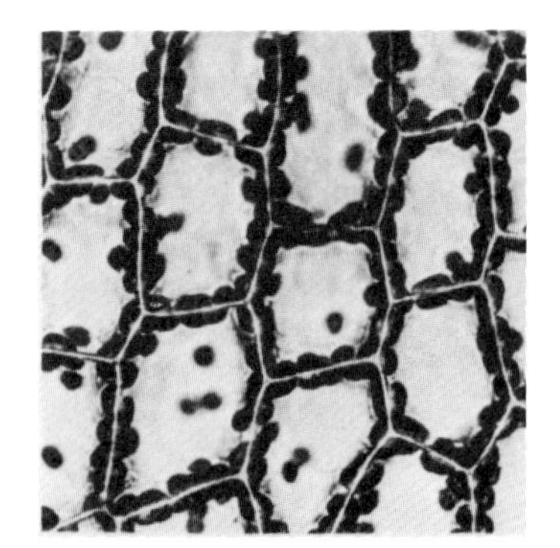

Figure 7. Light induced chloroplast movement of the moss *Funaria hygrometrica*. Left: low intensity arrangement. Right: high intensity arrangement.

A calculation of such a model cell results in a 50% transmittance in the face position, whereas the cell with chromatophores in profile position transmits about 73% of the incident radiation. Thus, transmittance measurements with a photometer should give meaningful results when the extent and the kinetics of chromatophore movements have to be determined.

If transmittance measurements are performed with *small* leaves or even filamentous algae, a microphotometer is needed. Two types of microphotometers will be described: a simple single beam microphotometer system, and a more sophisticated dual beam system. Moreover, if marine or freshwater algae are used for the experiments, a flow-through system of nutrient medium will be necessary in order to simulate natural conditions. By maintaining a steady gas exchange and nutrient supply for the algae, the experiments can be extended over a period of a month or more.

Phototaxis of *Chlamydomonas reinhardtii*

Chlamydomonas reinhardtii is a green unicellular alga with two flagella at its anterior end (Fig. 9). The cells have one chloroplast which contains an orange spot, the stigma (Nultsch, 1983). The flagella beat in the so-called breast-stroke style, (Fig. 10), but in a 3D pattern rather than in a plane (Rüffer and Nultsch, 1985, 1987). This causes a rotation and a helical swimming path (Fig. 11). Under unilateral irradiation *Chlamydomonas* swims toward the stimulus light source (+) at lower or away from it (–) at higher fluence rates. If the light direction is changed, the swimming direction is correspondingly altered. The photoreceptor is supposed to be located in those areas of the cytoplasmic membrane overlying the stigma. The light direction is apparently perceived by modulation of the stimulus light during rotation (irradiated-shaded-irradiated-shaded-etc.).

The phototactic response of single cells could be quantified by image analysis. Another possibility is a population method in which the phototactic movement of a whole population is measured.

Phototaxis Assay: Measurement Principle

Phototaxis can simply be demonstrated by pouring a cell suspension into a Petri dish or into a small clear plastic box and exposing it to unilateral light. In the case of positive phototaxis the cells will move toward the light source and accumulate at the light-facing edge

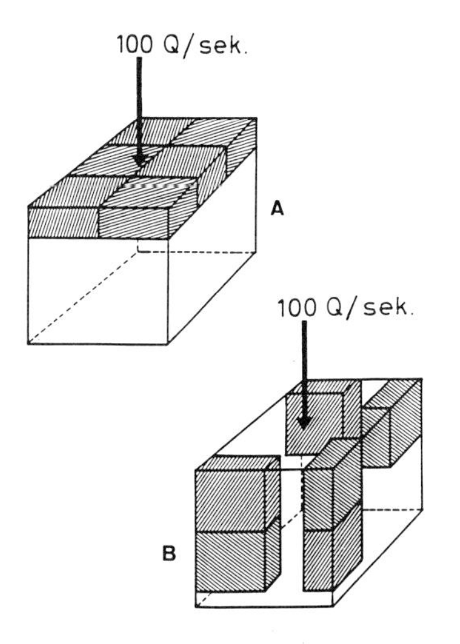

Figure 8. Schematic model of a plant cell with A) chromatophores in low intensity arrangement (face position) and B) chromatophores in high intensity arrangement (profile position). Of 100 quanta incident on the top of the cell, 50 quanta are transmitted in face position and about 73 quanta are transmitted in profile position of the chromatophores.

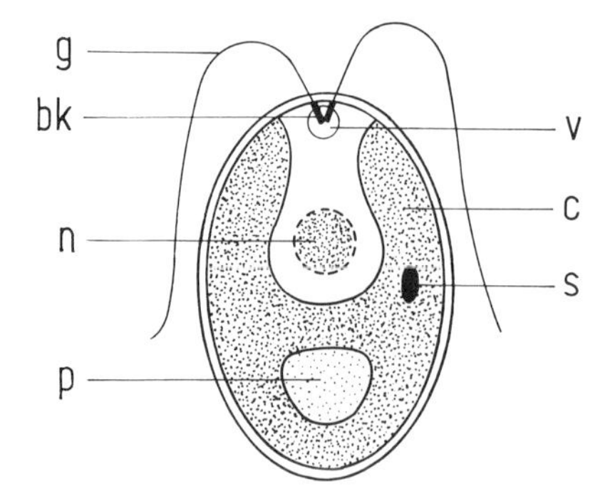

Figure 9. Cell of *Chlamydomonas reinhardtii*. bk = basal body; c = chloroplast; g = flagellum, n = nucleus; p = pyrenoid; s = stigma; v = vacuole.

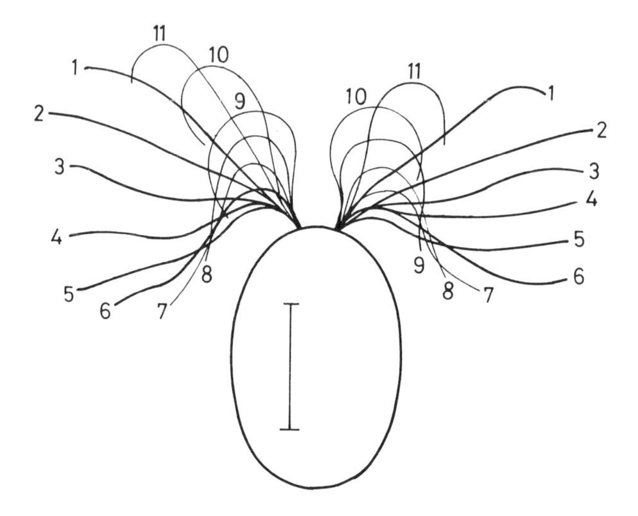

Figure 10. Flagellar beat of *Chlamydomonas reinhardtii* (after Rüffer and Nultsch, 1985). Numbers refer to sequential phases of the flagellar beat. No. 1 - early phase of the effective (back) stroke; No. 7 - start of the recovery (forward) stroke; No. 11 - end of the recovery stroke. Bar = 5 µm.

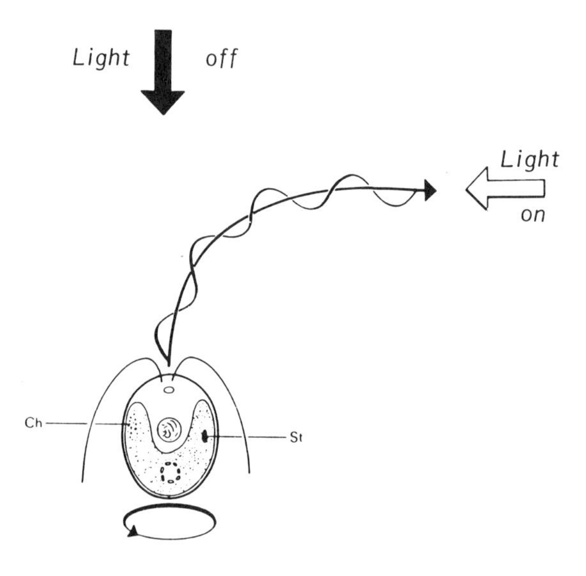

Figure 11. Schematic representation of movement and phototactic orientation of *Chlamydomonas reinhardtii*. Ch = chloroplast, St = stigma.

of the dish. In the case of negative phototaxis the cells eventually accumulate at the opposite end of the dish or cuvette. For quantitative measurements, however, a more sophisticated instrumentation is necessary.

For such a system (Fig. 12) a clear rectangular glass or Pyrex cuvette is needed as a basic element. We recommend a cuvette equipped with an inlet and an outlet tube on both sides, 8 mm in width, 3-5 mm in height and with a path length of 25-40 mm. The cuvette, filled with a *Chlamydomonas* cell suspension of sufficient optical density, is evenly irradiated from above with the measuring light (normally red, but in special cases, blue). The even distribution of cells along the cuvette path is disturbed by the stimulus light causing phototactic movement and an optical asymmetry. Two matched photoresistors or photodiodes of appropriate size, mounted below the cuvette, sense this asymmetry. For amplification the photosensors should be connected to an operational amplifier in an "active bridge" configuration (Sheingold, 1980). The bridge circuit will be unbalanced by the disturbed optical equilibrium in the cuvette and will give a corresponding electrical signal at the amplifier output.

Motility Assay: Measurement Principle

The asymmetric density caused by the phototactic movement does not only depend on the directedness of the movement, but also on the number and speed of moving cells. Therefore, measurements of phototaxis should be combined with motility measurements. In principle this could be done microscopically, measuring the velocity with an ocular micrometer and a stop watch. This "manual" method, however, is not only tedious, but also has some pitfalls. Thus the use of a motility assay is strongly recommended.

For the motility assay a similar principle to that suggested by Rikmenspoel and van Herpen (1957) and Ojakian and Katz (1973) is proposed. Cells swimming through a defined small area produce minute transmittance changes. Each cell entering the field reduces the

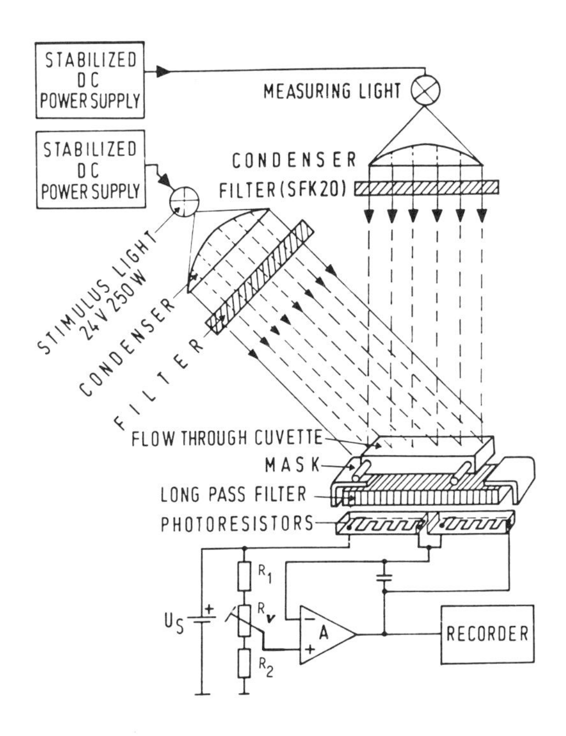

Figure 12. Schematic diagram and optical set-up of the phototaxis measuring assay. The flow-through cuvette is irradiated with a measuring and stimulus light beam. Two photoresistors connected to an operational amplifier in an active bridge configuration sense phototactically induced transmittance differences in both halves of the cuvette. The resulting analog signalmay be recorded for control purposes and digitized for off-line evaluation.

transmittance but increases it when leaving. To detect these transmittance changes, a microscope photometer rather similar to the single beam instrument described in the chromatophore movement section can be used. It is recommended to replace the blue filter with a red light filter transmitting at about 670 nm (e.g., SFK 16). If irradiance levels of 0.1 to 1 $W \cdot m^{-2}$ are used, no phototactic or photophobic responses will be detected in this wavelength band (Nultsch et al., 1971). Either a photomultiplier or a photodiode could be used to sense these changes in transmittance and to convert them into electrical current signals. The number of transmittance changes per unit time are counted and this provides a measure of the motility.

Simultaneous Measurement of Phototaxis and Motility

If it is desired to set up a fully automated system, a simultaneous measurement of phototaxis and motility is suggested. Figure 13, which is taken from the literature (Pfau et al., 1983), will provide you with some guidelines. A peristaltic pump (instead of a centrifugal pump) is recommended for connecting the culture vessel to two different flow-through cuvettes, being parts of the phototaxis/motility measuring system. The motility photomicroscope can be additionally equipped with a TV-camera and a time-lapse recorder (e.g., National WV 1350 AE (6)) to record and analyze the swimming behavior of the organisms.

At the beginning of a measurement cycle (every 15 min) the algal suspension must be circulated from the culture tube through the measuring cuvettes. Thus two solenoid valves are activated and open. Simultaneously a peristaltic pump is switched on to pump the cell suspension from the culture tube into a separator vessel. There the suspension is freed from suspended air bubbles. The main outlet of the separator vessel feeds into the phototaxis and motility cuvettes, whereas an overflow outlet is connected back to the main culture tube. Thus, a constant level of culture medium in the separator can be maintained, resulting in a defined hydrostatic pressure and a smooth flow of the cell suspension through both the measuring cuvettes. During the cuvette filling phase it is advisable to interrupt the pumping action several times. This interruption initiates big air bubbles which, by producing turbulence, "brush off" micro bubbles and *Chlamydomonas* organisms settling at the inner cuvette surfaces. The solenoid valves should be deactivated immediately after the pumping

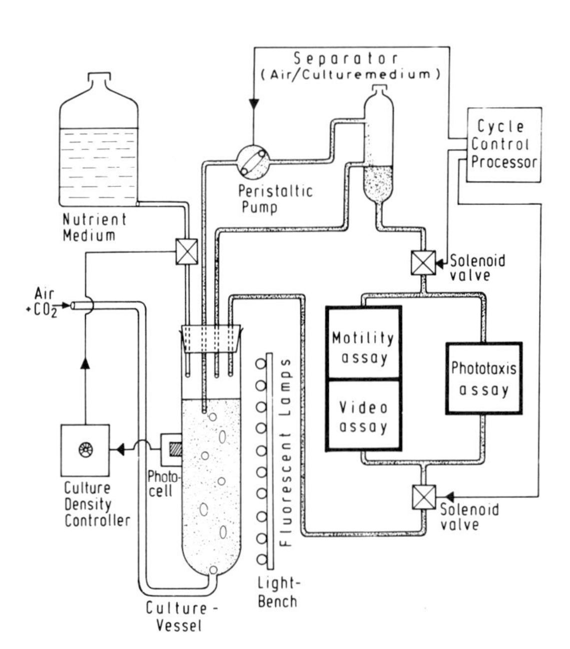

Figure 13. Schematic view of a complete system. A density controlled homocontinuous culture is coupled to a motility and phototaxis assay. The cycle control processor gives exact timing for the opening of the valves, pumping the cell suspension into the measuring cuvettes and the starting of data acquisition.

cycle, and simultaneous measurements of both motility and phototaxis may be started. Though analog data may be recorded on strip chart recorders, a computer system with a multichannel digitizer board is strongly recommended for further data processing.

EXPERIMENT 26: MEASUREMENT OF CHROMATOPHORE MOVEMENTS

OBJECTIVE

This experiment demonstrates how the chromatophore movement within a thin leaf or an algal thallus can be determined by a method of transmittance measurements using a microscope-photometer.

LIST OF MATERIALS

Plants of *Funaria* or a culture of *Dictyota*
Nutrient medium for *Funaria* or *Dictyota*
Monocular or binocular microscope, type "Biomed" (Leitz), type "Standard" (Zeiss), or type FM-31, (SWIFT, San Jose)
Low voltage DC power supply: choose voltage/current rating according to the specified lamp rating in the microscope. For the dual beam system a second stabilized power supply is needed. A dual power supply, e.g., type 6227B (Hewlett-Packard) is a convenient choice.
LED, type HLMP 4101 (Hewlett-Packard) or LED type TLDR 5101 (AEG) (measuring light in the dual beam system)
Photodiode, such as a BPW 20 (Siemens)
Photomultiplier, side-on broadband (200-800 nm), type R508 with housing and integrated high voltage power supply (PTI)
Photodiode, small area, type PIN-5 (UDT)
Photodiode, small area, eye response, type BPW21, (Siemens)
Current amplifier with current suppression circuit and electrical low-pass filter, type 427 (Keithley)
General purpose type luxmeter (Lambrecht) and/or thermopile, e.g., type CA-1 (Kipp & Zonen)
Strip chart recorder, e.g., type SE110, sensitivity ranges between 0.1 to 20 V and chart speeds in the order of 60 mm/h (BBC)
Glass filter, blue light transmittance in the range between 350 and 550 nm; peak at about 400 to 450 nm (e.g., BG-28 or BG-12, Schott)
Neutral density glass filters, or at least those having 5%, 20%, and 25% transmittance
Cut off filter, such as RG-645 (Schott)
Glass compound filter, peak transmittance at 670 nm, with a half band width of approximately 30 nm (e.g., SFK 16, Schott)
Set of interference filters (Hoya, Oriel, Schott)
Glass vessel, approximately 1000 ml, with a 4 mm outlet at the bottom
Glass vessel, approximately 300 ml, with a 4 mm outlet at the bottom and an additional overflow outlet positioned near the upper section
Pyrex cuvette with a 4 mm inlet and outlet piece
Flexible silicon tubing
Tubing clamps
Air pump, such as used to aerate aquaria, or a peristaltic pump (approximately 300 ml/min, e.g., type UNI-V, Verder)
Transparent paper

PREPARATIONS

Nutrient Medium for *Funaria hygrometrica*:: The following are dissolved per liter of solution (solvent is doubly distilled water): 2 g of KH_2PO_4; 0.59 g of $Na_2HPO_4 \cdot 2H_2O$; 0.3 g of $CaCl_2 \cdot 2H_2O$; 1 g of $MgSO_4 \cdot 7H_2O$; 0.01 g of $FeSO_4 \cdot 7H_2O$; 0.02 g of Na_2 EDTA Titriplex III and 3 ml of P_{IV} metal solution. The P_{IV} metal solution is prepared by dissolving the following per liter of distilled water: 0.097 g of $FeCl_3 \cdot 6H_2O$; 0.041 g of $MnCl_2 \cdot 4H_2O$; 0.005 g of $ZnCl_2$; 0.002 g of $CoCl_2 \cdot 6H_2O$; 0.004 g of Na_2MoO_4 and 0.750 g of Na_2 EDTA.

Nutrient Medium for *Dictyota dichotoma* (Laub-Schreiber Solution): *Ulva lactuca* may be cultivated in the same seawatermedium as *Dictyota*. In some cases (depending on where *Ulva* was collected), dilution of the *Dictyota* medium with one quarter distilled water will give better results. The following stock solutions are first prepared (using doubly distilled water as the solvent), then the corresponding volumes in parentheses are added per liter of filtered seawater to give the equivalent of 1 liter of enriched seawater solution: 9.89 mg of $MnCl_2 \cdot 4H_2O$/500 ml (1 ml); 27.8 mg of $FeSO_4 \cdot 7H_2O$/100 ml (1 ml) + concentrated H_2SO_4 (0.01 ml); 1.075 g of $Na_2HPO_4 \cdot 12H_2O$/100 ml (1 ml); 4.25 g of $NaNO_3$/100 ml (1 ml); 37.2 mg of Na_2EDTA/100 ml (10 ml) and vitamin mixture (5 ml). 100 ml of vitamin mixture solution is prepared by adding the respective amounts, in parentheses, from the following stock solutions (using doubly distilled water as a solvent): 9.0 mg of Ionosit/100 ml (1 ml); 4.4 mg of folic acid/100 ml (1 ml) + 1.68 g of $NaHCO_3$/100 ml (2 ml); 9.5 mg of calcium pantothenate/100 ml (1 ml); 1.4 mg of vitamin B_{12}/100 ml (1 ml); 17.3 mg of vitamin B_1/100 ml (1 ml); 6.1 mg of Biotin/50 ml (1 ml); 13.7 mg of p-aminobenzoic acid/100 ml (1 ml); 20.7 mg of pyridoxinehydrochloride/100 ml (1 ml); 12.2 mg of niacineamide/500 ml (1 ml); 12.2 mg of Rutin crystal/40 ml (4 ml) + 0.1 N NaOH (0.5 ml); 3.8 mg of vitamin B_2/100 ml (2 ml) and 8.8 mg of ascorbic acid/50 ml (2 ml).

Single Beam Microscope System: The simplest way to perform photometric transmittance measurements with a microscope is to set up a single beam system, as shown in Fig. 14. Since such a system essentially depends on the microscope used, the design rules given below are more or less general, but hopefully, sufficient details are provided to enable one to properly set up a successful system. Each major component is thus described and discussed.

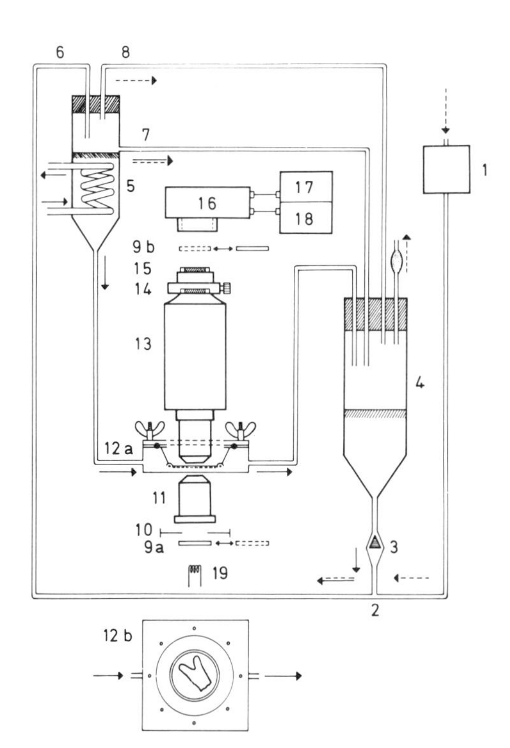

Figure 14. Scheme of a single beam microphotometer coupled to a nutrient medium recycling system. 1. membrane air pump, 2. mixing point, 3. back pressure valve, 4. lower storage bottle, 5. upper storage bottle with heating/cooling coil, 6. medium inlet upper storage tube, 7. overflow level, 8. pressure equalizing connection, 9a. blue stimulus filter and neutral glass density filters, 9b. adapter box and neutral glass density filters, 10. field diaphragm, 11. condenser, 12a/12b. cuvette sights, 13. microscope, 14. polarizer, 15. ocular and analyzer, 16/17. photo-multiplier with power supply, 18. amplifier and recorder, 19. halogen cycle lamp.

Illumination System: It is advisable to use a microscope equipped with a tungsten filament halogen cycle lamp for illumination. These lamps have good emission characteristics in the blue region of the visible spectrum and in general maintain these characteristics for the lifetime of the bulb. In cases where the lamp is fed from a separate transformer, detach the lamp from the transformer and make a connection to a DC power supply of sufficient power rating and with good regulation of either the DC current or the DC voltage. A regulation of 0.1% or better for line or load variations is required.

Modern microscopes often have a transformer integrated into the microscope stand. If such a system is used, the (mostly soldered) connections between the lamp and the transformer *must* be disconnected and a new connection between the lamp and an external regulated DC power supply has to be made with isolated wires of sufficient thickness (min. 0.75 mm^2).

An alternative light source can be made from a slide projector beam focused into the microscope condenser. (Many microscopes allow the illumination system to be removed in order to gain access to the light path). The slide projector has to be powered by a constant voltage or current supply as well.

The Detector: Place a microscope slide preparation with an object of your choice on the microscope object holder and prepare your microscope by making all necessary adjustments: adjusting the field brightness, focusing on the preparation, imaging and centering the field diaphragm into the object plane by vertical and horizontal adjustments of the microscope condenser. Then remove the slide preparation and determine the focal plane by holding a sheet of transparent paper above the eyepiece of the microscope (you should find a small bright spot; the distance of the spot is approximately 20 mm above the lens of the eyepiece). Attach a light-tight adapter to the eyepiece (can easily be made in a workshop) which positions the photomultiplier or, even better, a photodiode, coaxially in the focal plane of the eyepiece. This adapter should be constructed as a light-tight box with enough space so that the neutral density glass filters can be inserted between the eyepiece and the photomultiplier.

Using a photomultiplier, the usual output currents will range from a few tenths of a microampere to about 10 μA. A stabilized high voltage power supply will be required if such a system is not integrated in the multiplier housing. Photodiodes, however, give photocurrents in the nanoampere region; thus, the corresponding current amplifier must be much more sensitive.

The Current Amplifier: It is important that the current amplifier be equipped with an internal current suppression feature in order for the basic input signal to be compensated so that only the *changes* of the current, due to the changes in transmittance, appear at the output. This allows one to switch the recorder to an increased sensitivity. Furthermore, a low-pass filter at the output is needed to reduce any residual noise. The type 427 amplifier is nearly ideal for this purpose.

The Recorder: The strip chart recorder can be of a general purpose type. Its speed and sensitivity are not critical, as the amplification and signal conditioning is performed by the current amplifier. Data acquisition systems in combination with a computer should be considered for processing as well as for permanent data storage.

The Dual Beam System: The above single beam microphotometer system can be used conveniently as long as the measuring/stimulus wavelength is not changed. If, however, measurements are made using different stimulus or measuring wavelengths (e.g., action spectra), the resulting transmittance changes are not comparable, as the amount of change is dependent on the absorbance spectrum of the leaf. In such a case a dual beam microphotometer is necessary. The basic idea of a dual beam system is to use a stimulus light beam of variable wavelength and fluence rate along with a measuring beam of fixed wavelength and fluence rate. Both beams are mixed and focused on the same area of the object, e.g., a

Funaria leaf or a *Dictyota* thallus. Details for the construction of a simple dual beam system are thus provided below (see Fig. 15), whereas the description of a more sophisticated system can be found in the literature (Pfau et al., 1988).

Microscope: A small, rugged and rather inexpensive inverted field microscope (such as an FM-31) can be used. The object holder is removed and replaced by a specially built unit that contains the stimulus lamp housing, the measuring light housing, the filter compartment, the condensers, the beam mixer and the cuvette holder.

Measuring Beam: The measuring beam is now a source separate from the stimulus beam. Evaluation samples of red Light-Emitting-Diodes (LEDs) with beam widths of either 4 degrees (type TLDR 5101) or 8 degrees (type HLMP 4101) replace the formerly used lamps. The wavelength of the measuring beam should be as far as possible from 550 nm (the longest wavelength which could cause chromatophore movements in *Funaria* or *Dictyota*) but close to 670 nm, the red absorption band of chlorophyll *a*. The power supply for the diodes must be highly regulated. The output of the HP-6227 DC power supply is set at 21.5 V in the constant voltage mode. The LED is connected via a current-limiting resistor of 1000 ohms in series with the power supply.

Stimulus Light Source: For this, a light compartment for microscopes, rated for 100 VA halogen cycle lamps may be modified in order to use either the original lamps or low power halogen lamps rated at 20 W and 12 V. Regulation of the stimulus light source is much less demanding. When using the 20 W lamp, the second output of the HP-6227 supply might be used. If a projector serves as the light stimulus, a magnetic stabilizer may serve well for most of the experiments.

A simple and inexpensive microscope cover slip is used to mix the stimulus and the measuring beam. More than 90% of the stimulus light is transmitted whereas approximately 8% of the measuring light beam is reflected to the cuvette.

Detection: A photodiode may be used as a replacement for the photomultiplier, in which case the high voltage power supply is not necessary, but the current amplifier has to be more sensitive. In the dual beam system, photocurrents in the order of 10^{-8} to 10^{-10} amperes have

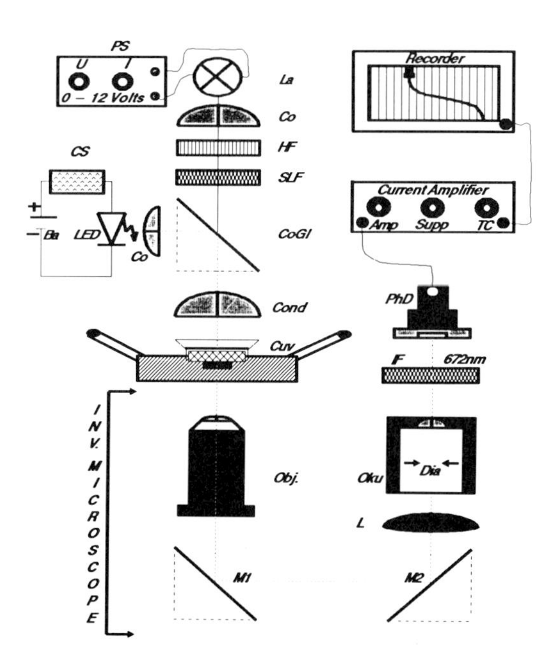

Figure 15. Schematic drawing of a dual beam microphotometer. PS: power supply for the stimulus light; La: halogen lamp; Co: collector; HF: heat filter; SLF: stimulus light filter; CoGl: cover glass; CS: current source; Ba: battery; LED: light emitting diode; Cond: condenser; Cuv: cuvette; Obj: objective; M1, M2: mirrors; L: lens; Oku: ocular; Dia: diaphragm; F: glass compound filter (672 nm); PhD: photodiode (light receiver).

to be compensated in order to switch the recorder to a higher sensitivity. The stimulus light, still present in the plane of the eyepiece, would normally drive the photodiode into saturation. A glass compound filter of 30 nm half bandwidth and peak transmittance at 672 nm (e.g., SFK 16) is positioned just below the detector photodiode, passing the measuring light to the diode but removing the shorter stimulus wavelengths.

The Flow-Through System for Object Maintenance: For preliminary testing, a microscope slide preparation of a *Funaria* leaf is sufficient. For long-term experiments, however, a flow-through system is necessary. Figure14 illustrates such a flow-through system.

EXPERIMENTAL PROCEDURE

A. Measurements with the Single Beam Instrument

Equipment Preparation

1. Make the basic alignments of the microscope (focusing, imaging of the field diaphragm and centering (see above).
2. Insert the blue glass filter (equivalent to a Schott BG 12 or, better, a BG 28) together with a neutral density glass filter of 5% transmittance into the beam below the object. A suitable position would be the space between the field diaphragm and the condenser. (For white light experiments, insert only the neutral density glass filter).
3. Increase the microscopic light to near full power. The irradiation with blue light should be about 1 $W{\cdot}m^{-2}$ measured in the object plane or have an illumination of approximately 1000 lux. To determine the (blue light) fluence rates in the object plane, photodiodes with a small active area are suggested after they have been calibrated against a thermopile standard together with the blue filter. Illumination levels (white light) may be determined in the same way by the use of a photodiode with eye-response characteristics which have been calibrated with a luxmeter.
4. Transfer a *Funaria* leaf onto a microscope slide or mount it into the flow-through cuvette. Be sure that the rims of the small leaves do not curl up and fold over.
5. Check the preparation for low intensity arrangement (face position). Nearly all chloroplasts should be spread over the periclinal cell walls leaving the anticlinal walls clearly visible.
6. Connect the photodiode to the current amplifier and turn down the suppression current completely. Set the amplifier gain in order to obtain a full scale deflection (1 V range) on the recorder. Set the low-pass filter to a time constant of 0.3 s. Now switch the suppression current to the level of the current of the photodiode. The diode current will then be compensated and the output signal at the recorder will be approximately zero. Switch the recorder to a higher sensitivity (e.g., 0.2 V full scale). In such cases where a photomultiplier is used: connect the photomultiplier output to the current amplifier and turn down the suppression current. Switch the amplifier gain to 10^7 volts/ampere. Increase the photomultiplier voltage (approximately 450 to 700 V) until full scale deflection is obtained on the 1 V recorder range (corresponding to a photomultiplier current of 100 nA). If you are working with a very sensitive photomultiplier, the amplifier gain might be reduced to 10^6 volts/ampere. Proceed as described above if using photodiodes.

Chromatophore Movement–Baseline Recording

7. Set the recording pen in a mid-scale position. Record the baseline, i.e., the transmittance in the low intensity arrangement over a few minutes. Variations are due to rearrangements of the chloroplasts under low intensities.

Recording of the Movement to the High Intensity Arrangement

8. Shut off the light beam to prevent overloading of the photomultiplier or photodiode. Do *not* switch off the power supply, but insert a solid piece of cardboard under the two filters on the field diaphragm. Then remove the 5% neutral density filter and insert it into the adapter box between the eyepiece and the photodiode (photomultiplier). Remove the cardboard. The recorder pen should move to approximately the same position as before, since the amount of light has not changed at the photodetector plane. However, the amount of light impinging onto the object has now increased 20-fold. The chloroplasts should now start to move to the anticlinal walls. A resulting increase in transmittance will be recorded until all chloroplasts are positioned at the anticlinal walls, and a steady state in transmittance is reached.

Recording of the Movement to the Low Intensity Arrangement

9. Shut off the light beam again. Reposition the neutral density glass filter over the field diaphragm and open the beam. The transmittance will decrease and a reverse recording will be obtained.

Setting an Intermediate Level of Stimulus Light

10. Replace the neutral density glass filter of 5% transmittance by two stacked neutral density filters of 20% and 25% transmittance. Total transmittance is now 5% again. Proceed as in step 8 but take only *one* of the filters and place it into the adapter box above the eyepiece. The stimulus light level increases either 5-fold (taking the 20% filter) or 4-fold (25% filter). Typical movement plots are presented in Fig. 16.

B. Measurements With the Dual Beam System:

11. Separately adjust the stimulus and the measuring beam concentrically and coaxially in the object plane. For eye protection during the alignment procedure, reduce the level of the stimulus light to safe, convenient levels. The output at the detector diode (SFK 16 filter removed during the adjustment) should be a maximum at correct alignment.
12. Replace the SFK filter and set the stimulus power supply for the highest needed stimulus irradiation in the object plane (e.g., 20 $W \cdot m^{-2}$). Measure the amount of light with a calibrated small-area photodiode (see step 3). To reduce the stimulus light to lower levels use appropriate neutral density glass filters (e.g., 5%).
13. Place a *Funaria* preparation into the cuvette, attach the cuvette to the flow through system, and place the cuvette under the microscope.
14. Switch the amplifier sensitivity in order to obtain a full range deflection of the recorder. Switch on the suppression current, set the output filter time constant, and increase the recorder sensitivity.
15. Make recordings of the baseline.
16. Test the system by recording the chloroplast movement to the high intensity arrangement and back to the low intensity arrangement. Do this by removing/reinserting appropriate neutral density glass filters from or into the stimulus beam.
17. If the tests are satisfactory, try obtaining fluence-response curves by inserting neutral density filters of increasing transmittance.

18. Check different stimulus wavelengths for their effectiveness on the chloroplast movement by using interference filters in the stimulus beam. Note that for UV stimulations, either a Xe arc lamp or a mercury lamp has to be used as a source.

DISCUSSION

Two systems for determining chromatophore movements have been described. The advantage of the single beam system is its simplicity. It can be easily constructed from parts available in most laboratories. The alignment is not too critical. Such an instrument serves ideally for white light studies as well as for studies on circadian rhythms. This latter type of experiment may be easily performed since the stimulus wavelength does not have to be changed frequently. It is recommended, however, that the experiment be set up in a temperature stabilized room with a smooth temperature regulation; (ON/OFF controllers normally cause large, sudden temperature changes). The dual beam system requires some experience and experimental skill in mechano-optical systems for precisely aligning, mixing and focusing the beams into the object plane. Chopping of the measuring beam, which would require an AC detection system such as a lock-in amplifier, generally is not needed. The system may, however, be prone to stray light leaking into the measuring beam path, which could then influence the results of experiments being carried out at low light levels. If one prefers to work with a lock-in system, one should consider modulating the LEDs by current-pulsing them instead of using a mechanical chopper. LEDs permit very rapid on and off switching. If action spectra are to be measured, the experimenter has to be aware of the dark reaction which may be caused by ineffective stimulus wavelengths. The dark arrangement of chloroplasts in *Funaria* and *Dictyota* is very similar to the high intensity arrangement. Note that action spectra reflect the effectiveness of an equal amount of quanta at the different wavelengths instead of an equal amount of energy.

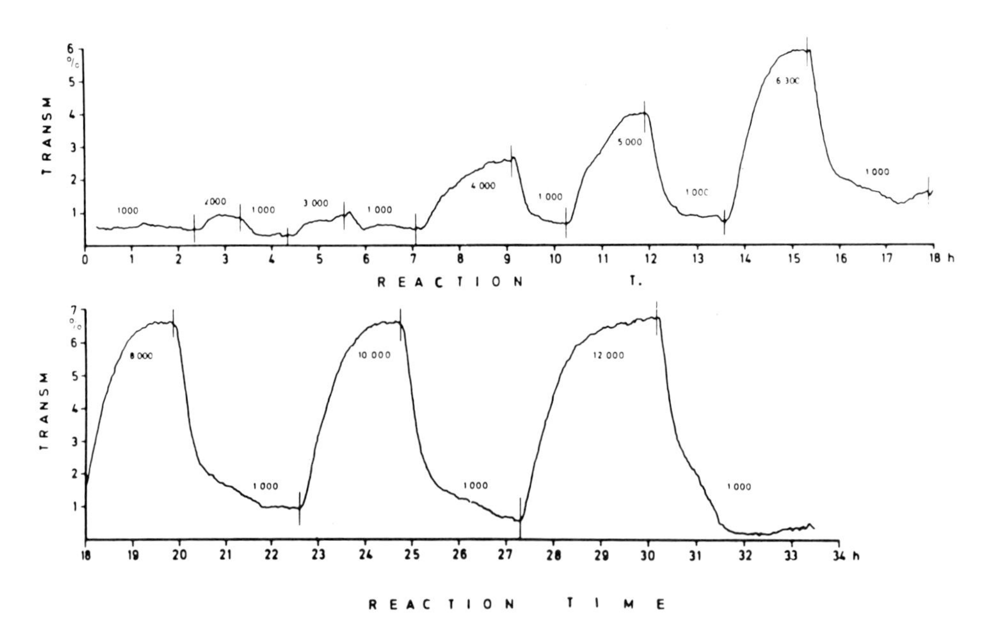

Figure 16. Typical plots of the high intensity movement and of the low intensity movement under different white light fluence rates (measured in lux units) of the stimulus light. Abscissa: reaction time. Ordinate: transmittance change.

The experimental set-up described in this experiment is a compromise with respect to stray light, sensitivity and stability. It has no movable parts, it is portable and it can be easily operated in the field, powered by a car battery. For ease of data processing, a data logger or a laptop computer with an A/D converter board could be connected to the system.

EXPERIMENT 27: PHOTOTAXIS AND MOTILITY OF THE GREEN ALGA *CHLAMYDOMONAS REINHARDTII*

OBJECTIVE

This laboratory exercise gives details on how to measure the phototaxis and motility of microorganisms. Using the green alga *Chlamydomonas reinhardtii* as an example, suggestions are also made on how to grow microorganisms under constant conditions.

GENERAL CONSIDERATIONS AND METHODS

Several methods of determining the phototactic response of unicellular organisms have been developed and allow quantitative results. A rather simple and inexpensive way is the population method in which the *phototactic movement of a whole population* of organisms is measured. The rate of phototactic movement, however, also depends on the motility, i.e., percentage of motile cells, and speed of their movement. Thus the cell motility should be measured simultaneously. Moreover, the phototactic signal obtained from a population measurement depends on the cell density. Therefore, provisions must be made to keep the cell number and growth culture conditions constant.

LIST OF MATERIALS

Chlamydomonas from the Algal Culture Collection at the University of Texas, Austin (or an alternative suitable source)
Manual or automated cell counting device
Nutrient medium for *Chlamydomonas* (see below)
White light fluorescent lamps (5 $W \cdot m^{-2}$)
Thermostated water bath
Erlenmeyer flasks (2 liter)
Air pump
Sterile air filters (0.25 - 0.4 μm pore diameter)
Glass tubes (about 3 mm outer and 1.5 mm inner diameter)
Silicon tubing
Optional: a complete culture controller/turbidostate system (Kniese)
Slide projector e.g., type "Prado" (Leitz)
Variable transformer (2 amps, Philips)
For precise measurements: either an AC stabilizer or, better, a DC power supply e.g., 6024A (Hewlett-Packard)
Recorder, general purpose, e.g., type SE110, sensitivity between 0.1 and 20 V at chart speeds of about 12 mm/min (BBC)
Photoresistors, or alternatively, photodiodes
Glass or pyrex cuvette with inlet and outlet tube
Red LED (light emitting diode), e.g., HLMP-8100 (Hewlett-Packard)
Current source (max 20 mA) for the LED
Glass compound filter, peak transmittance at 496 nm, e.g., SFK 9 (Schott)
Cut-off filter, e.g., RG 630 (Schott)
Set of neutral density glass filters
Black cardboard
Power supply, type 6227 B (Hewlett-Packard)

Photomultiplier with integrated power supply (PTI)
LED, red emitting, small angle, 12° e.g., type TLXR 5101 (AEG-Telefunken GmbH)
Photodiode BPW 20, (Siemens)
Amplifier, type 427 (Keithley)
Glass compound filter, peak transmittance at 670 nm e.g., SFK 16, (Schott)
Cutoff filter, RG 645 (Schott)
Pinhole diaphragm (Zeiss)
Recorder, general purpose or SE110 with speed settings between 200 and 600 mm/min (BBC)
Measuring cuvette
Microscope, normal or inverted as used to measure chromatophore movements

PREPARATIONS

Nutrient Medium for *Chlamydomonas reinhardtii*: The following are dissolved per liter of solution (solvent is doubly distilled water): 1011.1 mg of 10 mM KNO_3; 621 mg of 4.5 mM $NaH_2PO_4 \cdot H_2O$; 89 mg of 0.5 mM $Na_2HPO_4 \cdot 2H_2O$; 246.5 mg of 1 mM $MgSO_4 \cdot 7H_2O$; 14.7 mg of 0.1 mM $CaCl_2 \cdot 2H_2O$; 6.95 mg of 25 µM $FeSO_4$ as EDTA complex (see Nultsch et al., 1971) and 1 ml of a trace element solution. The following per liter of the trace element solution is dissolved: 61 mg of H_3BO_4; 287 mg of $ZnSO_4 \cdot 7H_2O$; 169 mg of $MnSO_4 \cdot H_2O$; 2.49 mg of $CuSO_4 \cdot 5H_2O$ and 12.35 mg of $(NH_4)_6Mo_7O_{24} \cdot 4H_2O$. See Nultsch (1979) for additional details on the preparation of the trace element solution.

Preparation of *Chlamydomonas* Culture: Inoculate *Chlamydomonas* cells into the Erlenmeyer flasks 2/3 filled with the *Chlamydomonas* nutrient medium. Close the flasks with a cotton stopper. Insert a sterilized glass tube of small diameter into the culture suspension and aerate the culture with filtered air by means of an air pump. Better growth is obtained if the air is mixed with 2% CO_2. Expose the cultures to either daylight of about 5 $W \cdot m^{-2}$, or, to avoid entrainment of circadian rhythms, to continuous fluorescent light. In order to keep the culture temperature constant, a water bath may be used. Good results will be obtained at temperatures of about 20°C. In order to make the results highly reproducible, a tubidostate with an automated cell density and dilution control (Pfau et al., 1971) is recommended. Such a system can be directly connected to the phototaxis and motility measuring system (for details see Pfau et al., 1983). The cultures should be checked for contamination each day. A culture density of 5×10^5 to 2×10^6 cells/ml is optimal.

Phototaxis Measurement System: Since instrumentation necessary to measure phototaxis is generally not commercially available, the construction details of such a system are provided here (see Fig. 12).

Stimulus Light: Mount a slide projector on a stage with 45° inclination. A projector equipped with a strong light source is recommended. Insert a glass compound filter with a peak transmittance of 496 nm (e.g., SFK 9) into the projector beam (this wavelength has proven to be most efficient for the phototaxis of *Chlamydomonas*). Power the projector from either an AC stabilizer and a variable transformer or, better, from a variable DC power supply (in the latter case the projector must be modified according to the rules given above for microscopes with built-in transformers). Cut an approximately 10 mm wide and 35 mm high slit into black cardboard and insert it into the projector slide plane to get a longitudinal projection field slightly wider than the cuvette itself. Set the lamp power to the highest desired output (e.g., 10 $W \cdot m^{-2}$) measured with a calibrated photodiode at the cuvette center. Insert appropriate neutral density glass filters for variation of the stimulus irradiation.

Measuring Light: This light source should be a low output (approximately 0.1 $W \cdot m^{-2}$) red light source which must be stabilized. A microscope lamp, powered from a suitable DC power supply and equipped with a filter of 670 nm peak transmittance (e.g., SFK 16) could, for instance, be used. A red LED with a 20 to 30 degree emitting light beam powered from a 0-20 mA DC current source is even more convenient.

Detection Set-up: The cuvette should be firmly mounted into a holder on top of a small plastic module box and held in place with spring clamps. A filter passing wavelengths beyond 630 nm (e.g., RG 630) is needed to block the stimulus light and to pass the red measuring light. For this purpose, an opening of exactly the filter size is cut into the lid of the box. Insert the cut-off filter and mount the two photoresistors underneath. (The photoresistors used to test this experiment were 2 mm in height, 10 mm wide and 30 mm long. Their resistance was determined to be 5 000 ohms at an illumination of 1000 lux). The bridge circuit must be completed by two fixed resistors (R_1 and R_2) and a variable resistor R_v for balancing. The resistors should be 10 kohms each. The bridge should be supplied by a stabilized voltage source providing, for example, 5 V. (Integrated voltage regulators of the μA7805-LP type are convenient). The capacitor smooths the output signal and should be at least 1 μF (bipolar). In order to make the system handy, we suggest the integration of the voltage source for the bridge as well as the current source for the measuring light LED into the plastic box along with the power supply for the operational amplifier.

Motility Measurement System: General instructions are provided for the construction of a system suitable for motility measurements. A typical block diagram of such a system is shown in Fig. 17.

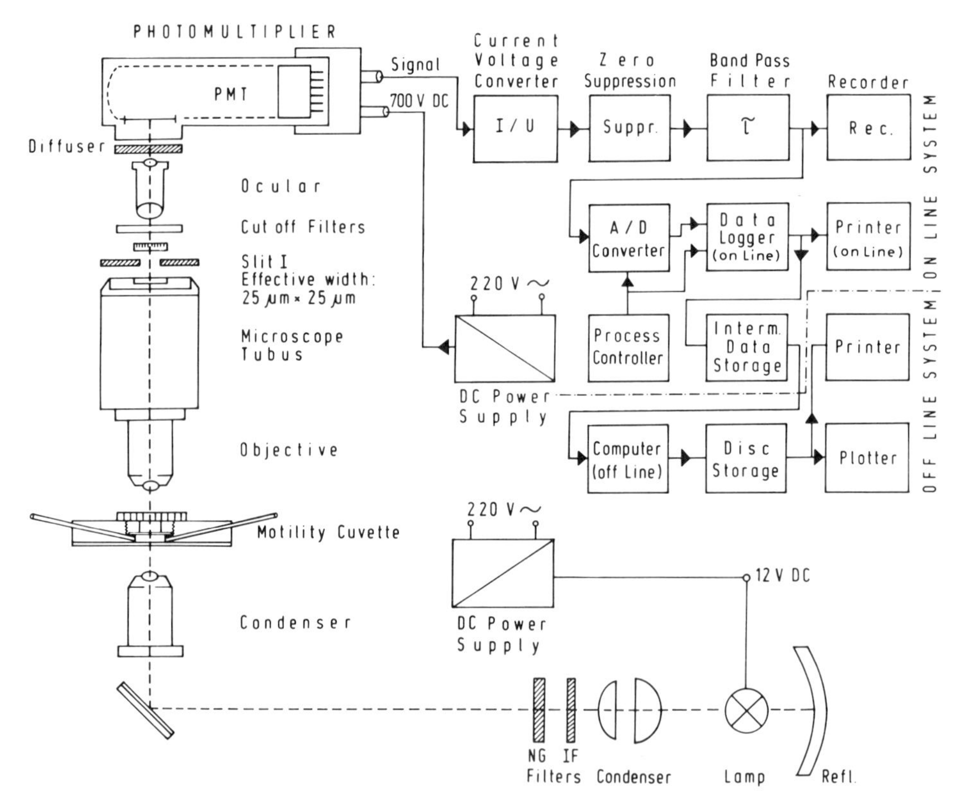

Figure 17. Block diagram of the motility assay. The motility noise signal in the photomultiplier is amplified, filtered and recorded. A data acquisition system digitizes the motility noise signal, calculates the spikes, integrates the signal and transfers the data to a disk for intermediate storage. Evaluation of those data is performed by an off-line computer system.

Modify either a monocular or a binocular microscope according to the design rules given above for the single beam microphotometer system, but remove the adapter box together with the detector and leave the eyepiece in the microscope. Put a microscope slide preparation on the stage and focus on it. The microscope is to be adjusted as follows: object plane in focus, field diaphragm sharply imaged in the object plane, condenser diaphragm half open. Now select one of the following two ways proposed to define the measuring area:

(1) Position a pinhole diaphragm (either circular or rectangular) as close as possible above the field diaphragm of the microscope and move the condenser unit until the diaphragm is in focus.

(2) Remove the eyepiece but leave the preparation on the object holder. Take a piece of transparent paper and find the "intermediate plane", near the end of the open microscope tube. In this plane you will find a sharp enlarged real image of the preparation. Get a pinhole diaphragm made that can be crimped into the tube of the eyepiece and move the diaphragm into the intermediate plane. Replace the eyepiece. The image of the preparation and the pinhole should then both be in focus.

The size of the real pinhole diaphragm depends on the magnification of either the condenser lens or the microscope objectives used. The effective size measured in the object plane should be on the order of 20 to 40 μm in diameter. We suggest a pinhole of about 0.8 mm used with a 16× objective or a pinhole of 0.4 mm with a 10× objective.

Place an SFK 16 compound (or equivalent) filter above the field diaphragm below the condenser and remount the adapter box with the detector onto the eyepiece. Insert a red cut-off filter (e.g., RG 645) in the adapter box to reduce stray light effects. Make the necessary connections to the high voltage power supply (in case a photomultiplier is used) and to the current amplifier. Set the time constant of the amplifier to 30 ms and switch the suppression current off. Connect the recorder and set it to 1V full scale and 300 mm/min.

The cuvette may be constructed as follows: a 20 mm bore, cut into an 80 × 25 × 5 mm plexiglass block, is sealed at the bottom with a microscope slide. At the top, a coverslip is pressed to a silicon-rubber seal ring by means of a plexiglass screw. Two bores with attached inlet/outlet tubes enable flow through and recycling of the culture suspension.

If it is intended to modify the inverted microscope of the dual beam system used for the chromatophore movement, the motility measuring field should be irradiated by a red LED with a 12° beam width. The LED should be positioned in place of the stimulus light source and focused on the object area. This diode must be supplied from a constant DC current source. Another elegant method of irradiation would be to replace the beam-mixer coverslip with a surface coated mirror. By this means, the measuring beam acts as field irradiation, and remounting efforts are reduced.

EXPERIMENTAL PROCEDURE

Phototaxis Measurements

1. Determine the cell density. Dilute the suspension to your "working densities" (between 5×10^5 and 2×10^6 cells/ml).

2. Fill the phototaxis cuvette either manually or automatically. For manual fill, set the stimulus light to 1 $W \cdot m^{-2}$ on the cuvette surface and close the stimulus beam shutter. Place the cuvette into the holder of the module box.

3. Balance the bridge to get a midscale deflection on the recorder.

4. Open the stimulus beam shutter in order to obtain a recording similar to that shown in Fig. 18. (For data processing purposes it might be sufficient to have the information of the phototaxis signal taken every 10 s).
5. After approximately 10 min the movement reaction should come to a steady state and the recording (data acquisition) can be terminated.
6. The phototactic activity may be derived from the electrical signal as the voltage value which is reached after 4 min. This value must be referenced to the first value at the start of the phototactic reaction (normally zero if the bridge was balanced). In an automated system the reference value is to be taken immediately after the stop of recycling the algal suspension.

Determination of the Signal Response in Relation to the Cell Concentration

7. Dilute the culture suspension to several defined cell densities (e.g., 1:1, 1:3, 1:10, 1:30, 1:100, 1:300 and 1:1000).
8. Place the diluted cultures under culture light conditions in order to keep the cells motile.
9. Repeat steps 1 to 6 with the different cell densities.
10. Plot the phototaxis amplitude against the log of the cell density.
11. Measure the fluence-rate response curve. Start with a stimulus irradiance of 10 $mW{\cdot}m^{-2}$. Proceed as described in steps 1 to 6.
12. Increase the fluence rate after each experiment up to 10 $W{\cdot}m^{-2}$. Refill the cuvette after each run.
13. Plot the phototaxis amplitude against the log of the irradiance.

Motility Noise Diagram Recording

14. Check/dilute a *Chlamydomonas* suspension to the "working density". Fill the cuvette (either manually or automatically), position it under the photomicroscope and focus onto a plane with freely moving cells.

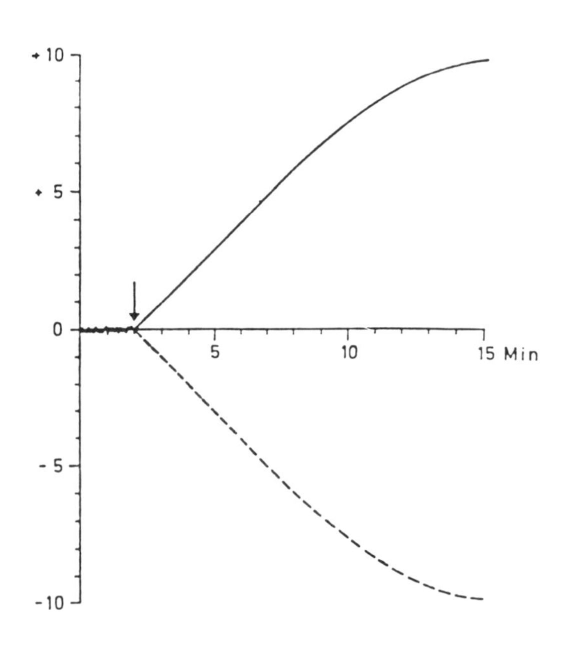

Figure 18. Recording of a positive (solid line) and a negative (broken line) phototactic reaction measured in relative units. Arrow indicates the beginning of the irradiation.

15. Switch the sensitivity of the amplifier to get a full range deflection on the recorder on the 1 V scale.

16. Switch the suppression current to compensate the recorder deflection, increase the recorder sensitivity approximately 5-fold and start the recording.

17. A recording, similar to the motility noise diagram in Fig. 19A, should be obtained. The spikes reflect the transmittance changes and thus give a measure of the number of cells crossing the irradiated field per unit time.

18. Count the number of positive and negative spikes in a predetermined time unit (e.g., 2 min). With an A/D converter available, and a bit of programming, a computer could be used to do the counting.

Control Recording

19. Fill the cuvette with plain water. The basic transmittance will increase slightly. Proceed as described in steps 14 to 16. A flat, essentially noiseless recording should result, such as shown in Fig. 19 B.

Measurement of the Time Dependence of the Motility

20. Fill the cuvette with a *Chlamydomonas* suspension of the "working density".

21. Perform recordings immediately, and 10 min, 30 min, 1h and 2h after filling. Due to unfavorable gas conditions in the closed cuvette, the motility may cease with time. Count the number of spikes and plot them against the elapsed time.

Measurement of the Dependence of the Motility on Cell Density

22. Dilute the culture suspension to several defined cell densities (e.g., 1:1, 1:3, 1:10, 1:30, 1:100, 1:300 and1:1000). Place the diluted cultures under culture light conditions in order to keep the cells motile.

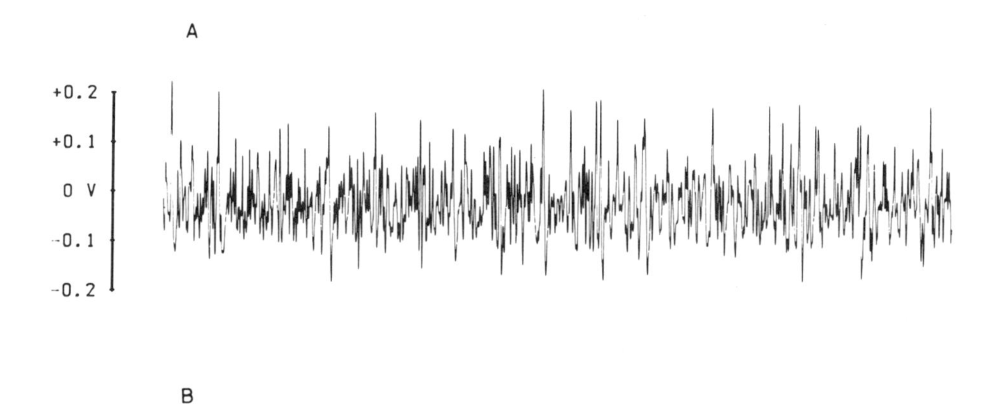

Figure 19. A: Typical motility noise diagram of a culture at a density of 2×10^6 cells·cm^{-3}. B: For comparison a diagram of the system noise (produced by an empty cuvette) is presented at the same scale. Abscissa: time, total = 2 min. Ordinate: electrical noise signal in volts.

23. Fill the cuvette with one of the suspensions and proceed as described in steps 14 to 16.

24. Count the numbers of spikes for each dilution and plot these against the log of the concentration.

DISCUSSION

While performing measurements with the phototaxis and motility systems, you may sometimes encounter unexpected results. This section lists possible sources of such errors.

With the phototaxis population system, one may obtain curves which are not as smooth as that shown in Fig. 18. One or several plateaus may be visible, as if the motion towards or away from the light is delayed, even though the measuring conditions are kept constant. If such phenomena occur repeatedly, the cuvette should be examined closely for clouds of different cell densities. This phenomenon is referred to as "pattern formation" and is believed to arise from thermodynamic effects.

The direction of movement is possibly dependent on the gas conditions in the suspension. Therefore, different results as to the sign of phototaxis may be obtained in a closed cuvette system and in an open Petri dish experiment.

In long-term experiments, the suspension used in the experiments should be examined daily for contamination by bacteria, even when an automatic culture apparatus is used.

Be aware of circadian rhythms in motility and phototaxis. These rhythms can be easily induced in homocontinuous cultures by imposing sudden changes in irradiation or temperature.

As previously mentioned, the "manual" measurement of motility by direct observation can be a very tedious task. Three main reasons why are expanded upon here. 1) Due to a thin water layer found between the microscope slide and the cover slip, the cells tend to stick to the glass surfaces with their flagella. Since the cells are not really immobile they tend to detach from the glass surface at random and start to swim again. Thus it is difficult to estimate the real number of immobile cells by a "manual" method. 2) The speed and direction of movement of single cells vary over a wide range, thus a large number of samples must be measured. 3) The swimming speed is frequently rather high (up to 200 $\mu m \cdot s^{-1}$ for the wild type of *Chlamydomonas*) which makes "manual" evaluation difficult.

The reproducibility of the results obtained from motility measurements is dependent on how accurately the microscope can be refocused to a certain plane in the cuvette. As some of the cells tend to settle at the bottom of the cuvette during the course of the motility measurements, it is best to select a focus plane near the top of the cuvette. When using an inverted microscope, however, stray light effects might cause signal deterioration. In this case a plane close to the bottom of the cuvette should be chosen. Thus the choice of the focus plane becomes a compromise.

The time constant (damping) of the current amplifier should be set to a low value (approx 30 ms) in order to obtain all the motility noise spikes. On the other hand, the low number of quanta per unit time transmitted through the pinhole needs a large amplification and gives rise to system noise as well (Fig. 19 B). Thus a computer program which extracts the number of spikes should have a software-threshold programmed such that only "significant" spikes are counted.

Low values in motility might not only be due to low frequency or amplitudes of flagellar beating. This is especially true when working with drugs. In such a case, the organisms should be checked under a microscope to make sure that they have not lost their flagella.

REVIEW QUESTIONS

1. In a three layer model system (*Dictyota*), both cortical layers are identical and each transmits 40%. The medullary cell layer transmits 60%. What is the total transmittance of the thallus?

2. You are working with three identical sets of neutral density glass filters with equal spacing in transmittance (90%, 80%, 70% ... 20%, 10%). What filter transmittances in stacked combination would you use if you need a total transmittance of 56, 24 or 17%. Outline a computer program for calculating which combination of filters (3 filters max) gives the closest match. (In some cases a perfect match is not possible with only three filters).

3. Verify the calculation implied in Fig. 8B, that out of 100 incident quanta, about 73 quanta are transmitted.

4. Why should (in the microphotometers described) the measuring wavelength coincide with an absorption band of the chloroplasts? What could happen if this were not the case.

5. In a single or dual beam photometer system, why is it that modulation of the measuring light can suppress the system sensitivity against incident stray light from the environment? (Think of a remote infrared TV controller).

6. Why should a *mechanical* chopper be chosen to modulate the measuring beam if a tungsten filament lamp is used as the measuring light in a double beam microphotometer system?

7. What will be the result in the recording, if during a motility measurement some of the organisms within the measurement area get 'lazy' and sink to the bottom of the cuvette?

ANSWERS TO REVIEW QUESTIONS

1.
$$T_{total} = T_{cortical} \times T_{medullary} \times T_{cortical}$$
$$T_{total} = 0.4 \times 0.6 \times 0.4$$
$$T_{total} = 0.096$$

2. Try the first set. If a fit can be performed with one filter, stop the program and display the filter value. If not, try the T = 90% filter of the second set successively combined with the filters of the first set. If no match is found, take the T = 80% filter of the second set and combine it with the first set. Repeat this procedure with the other filters of the second set. (Is it necessary to check the whole row?) If no match is found with two filters, take the filters from the third set and combine them with the filters of the first and second set. If no perfect match is possible, which is the closest match and how large is the error?

3. Sum up the transmittance of the single areas:

$$Q_{total(transmitted)} = \text{Sum of }(\text{Quanta}_{(in\ area\ -\ portion)} \times \text{Transmittance}_{(in\ area\ -\ portion)})$$
$$Q_{total} = (4/6) \times 100\text{Qnts} \times 100\% + (1/6) \times 100\text{Qnts} \times 25\% + (1/6) \times 100\text{Qnts} \times 12.5\%$$
$$Q_{total} = 72.92 \text{ Quanta}$$

4. As the calculation in question 3 shows, transmittance changes are inversely correlated to the absorption of the chromatophores. If the chromatophores don't absorb, no changes occur (ideally). In reality, however, light scattering effects due to the

interaction of the chromatophores with the plant tissue often produce results which are difficult to interpret.

5. Light from the environment is not modulated and therefore cannot pass the electrical filters incorporated in the detection system.
6. A tungsten lamp cannot be switched off and on rapidly enough to get a sufficient modulation depth.
7. A slow drift of the baseline in the motility signal recording will occur.

SUPPLEMENTARY READING

Song, P.S. and K.L. Poff (1989) Photomovement. In *The Science of Photobiology*, 2nd Ed. (Edited by K.C. Smith), pp. 305-346. Plenum Press, New York.

Häder, D.-P. and M. Tevini (1987) Light-dependent movement responses. In *General Photobiology*, pp. 247-278. Pergamon Press, Toronto.

REFERENCES

Britz, S.J. (1979) Chloroplast and nuclear migration. In *Physiology of Movements, Encyclop. Plant Physiol.*, (Edited by W. Haupt and M.E. Feinleib) Vol. 7, pp. 170-205. Springer Verlag, Heidelberg.

Dennison, D.S. (1979) Phototropism. In *Physiology of Movements, Encyclop. Plant Physiol.*, (Edited by W. Haupt and M.E. Feinleib) Vol. 7, pp. 506-567. Springer Verlag, Heidelberg.

Diehn, B., M.E. Feinleib, W. Haupt, E. Hildebrand, F. Lenci and W. Nultsch (1977) Terminology of behavioral responses of motile microorganisms. *Photochem. Photobiol.* **26**, 559-560.

Haupt, W. and G. Wagner (1984) Chloroplast movement. In *Membranes and Sensory Transduction*, (Edited by G. Colombetti and F. Lenci), pp. 331-375. Plenum Press, New York.

Nultsch, W. (1983) The photocontrol of movement of *Chlamydomonas*. In *The Biology of Photoreception*, (Edited by D.J. Cosens and D. Vince-Prue), pp. 521-539. Soc. Exp. Biol. Symp. XXXVI.

Nultsch, W., G. Throm and I. v. Rimscha (1971) Phototaktische Untersuchungen an *Chlamydomonas reinhardtii* Dangeard in homokontinuierlicher Kultur. *Arch. Microbiol.* **80**, 351-369.

Nultsch, W. and P.A. Benedetti (1978) Microphotometric measurements of the light induced chromatophore measurements in a single cell of the brown alga *Dictyota dichotoma*. *Z. Pflanzenphysiol.* **87**, 173-180.

Nultsch, W. and D.-P. Häder (1979) Photomovement of motile microorganisms. *Photochem. Photobiol.* **29**, 423-437.

Nultsch, W. and D.-P. Häder (1988) Photomovement of motile microorganisms - II. *Photochem. Photobiol.* **47**, 837-869.

Ojakian, G.K. and D. Katz (1973) A simple technique for the measurement of swimming speed of *Chlamydomonas*. *Exp. Cell Res.* **81**, 487-491.

Pfau, J., K. Werthmüller and H. Senger (1971) Permanent automatic synchronization of micro algae achieved by photoelectrically controlled dilution. *Arch. Microbiol.* **75**, 338-345.

Pfau, J., G. Throm and W. Nultsch (1974) Recording microphotometer for determination of light induced chromatophore movements in brown algae. *Z.Pflanzenphysiol.* **71**, 242-260.

Pfau, J., W. Nultsch and U. Rüffer (1983) A fully automated and computerized system for simultaneous measurements of motility and phototaxis in *Chlamydomonas*. *Arch. Microbiol.* **135**, 259-264.

Pfau, J., D. Hanelt and W. Nultsch (1988) A new dual beam microphotometer for determination of action spectra of light-induced phaeoplast movements in *Dictyota dichotoma*. *J. Plant Physiol.* **133**, 572-579.

Rüffer, U. and W. Nultsch (1985) High-speed cinematographic analysis of the movement of *Chlamydomonas*. *Cell Motil.* **5**, 251-263.

Rüffer, U. and W. Nultsch (1987) Comparison of the beating of cis- and trans-flagella of *Chlamydomonas* cells held on micropipettes. *Cell Motil. Cytoskeleton* **7**, 87-93.

Rikmenspoel, R. and R. van Herpen (1957) Photoelectric and cinematographic measurements of the motility of bull sperm cells. *Phys. Med. Biol.* **2**, 54-63.

Seitz, K. (1979) Cytoplasmic streaming and cyclosis of chloroplasts. In *Physiology of Movements, Encyclop. Plant Physiol.* Vol. 7, (Edited by W. Haupt and M.E. Feinleib), pp. 150-170. Springer Verlag, Heidelberg.

Sheingold, D.H., (Editor) (1980) *Transducer Interfacing Handbook.* Publ.: Analog Devices, Inc. Norwood, Mass., U.S.A.

Song, P.S. and K.L. Poff (1989) Photomovement. In *The Science of Photobiology*, 2nd Ed. (Edited by K.C. Smith), pp. 305-346. Plenum Press, New York.

Zurzycki, J. (1980) Blue light-induced intracellular movements. In *The Blue Light Syndrome*, (Edited by H. Senger), pp. 50-68. Springer Verlag, Berlin.

Effects of Light on Circadian Rhythms in Plants and Animals

Wilhelm NULTSCH, Gerhard HELDMAIER and Glenn WEAGLE*
Philipps-Universität Marburg
Germany
***The Royal Military College of Canada**
Canada

INTRODUCTION

Rhythmic oscillations of biochemical, physiological and behavioral characteristics occur in many different organisms: men, animals, plants and microorganisms with the exception of procaryotes which lack a true nucleus (bacteria, cyanobacteria). The period length can vary between minutes (*Desmodium* leaf movements) and a year. If we disregard rhythmic phenomena with short periods, we have to deal with daily (circadian) rhythms (24 h), lunar rhythms (28 days) and yearly rhythms.

It has to be emphasized, however, that circadian rhythms cannot exclusively be entrained by light-dark changes. Other factors, such as chemical stimuli and changes in temperature, can also influence the circadian clock, provided that they exceed a distinct threshold value. Thus, chronobiology is not simply a photobiological discipline. On the contrary, photochronobiology is rather a special, though probably the most important, discipline of chronobiology.

Circadian Rhythms in Plants

Daily rhythms in plants were detected as early as the middle of the 19th century (for reviews see Bünning, 1953, 1959, 1967; Feldman, 1989; Sweeney, 1977). Those higher plants which possess a so-called, pulvinus (leaf joint) are able to move their leaves. As mentioned in the photomovement chapter, they can respond directly to changes in fluence rates (photonasty). In addition, however, their leaves can occupy different positions during the day and night (Fig. 1). These movements are called nyktinastic movements (sleep movements). Under natural conditions they are synchronized by the day-night change, so that one could suppose that one position is occupied in light, the other in darkness. This, however, is not strictly true. If the nyktinastic movement is initiated by a light-dark regime, e.g., 12

WILHELM NULTSCH AND GERHARD HELDMAIER, Fachbereich Biologie der Philipps-Universität Marburg, Karl-von-Frisch-Strasse, D-3550 Marburg/Lahn, Germany
GLENN WEAGLE, Department of Chemistry and Chemical Engineering
The Royal Military College of Canada, Kingston, Ontario, K7K 5L0, Canada

Photobiological Techniques, Edited by D.P. Valenzeno *et al.*
Plenum Press, New York, 1991

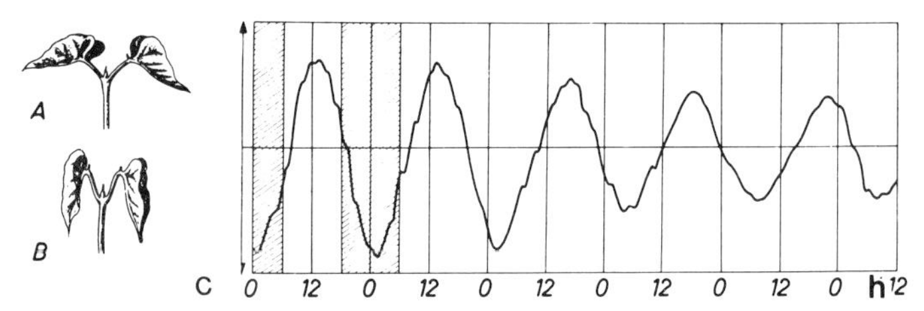

Figure 1. Nyktinastic leaf movements of the French bean (*Phaseolus coccineus*). A. position during the day, B. position during the night, C. diagram of circadian rhythms, initiated by a 12/12 h light:dark change (left), with a free-running period of 27 h in continuous dim light. The amplitudes damp out with time (after Nultsch, 1971, modified).

h light /12 h darkness, the movements follow exactly the daily rhythm, though the times at which the leaves start to change position need not exactly coincide with the light on / light off times. If the plants are transferred to continuous light (or in some cases to continuous darkness), the movements continue though with a different so-called free-running period, e.g., 27 h in French beans (*Phaseolus coccineus*). Therefore, they are called circadian (diurnal) rhythms and circadian (diurnal) movements (circa = approximately, dian = day). In this case, light acts only as the "Zeitgeber", i.e., it sets the circadian clock. If it then runs freely, it reveals a 27 h period. However, under natural conditions, it is set daily by sunrise and sunset, respectively, so that the period length is 24 h.

Another well known example of circadian rhythms is the opening and closing of plant leaves or petals which lack special joints. This movement occurs by a differential growth of both flanks, as shown, for example, by the inflorescence of *Leontodon hastilis* in Fig. 2. Some flowers open before sunrise, others at sunrise, others at several different times after sunrise and others during the night. The same is true for closing. Thus, if you are an expert in this field and you walk across a meadow, you are able to determine the time of day by looking at which plant species have opened and closed their flowers. This phenomenon is known as the flower clock. Again, if circadian flower leaf movement is induced, it continues in constant light and/or darkness (Fig. 3).

One can demonstrate circadian phenomena in phototaxis of *Chlamydomonas* and with the chromatophore movement in *Dictyota dichotoma* using the same equipment as described in detail in the photomovement chapter. If we transfer a homocontinuous culture of *Chlamydomonas reinhardtii* displaying almost constant reaction values for 72 h into darkness, the reaction values decrease immediately, reaching a minimum value (25-30% of the maximum reactivity) after about 2.5 h (Fig. 4). The values then increase again, i.e., a

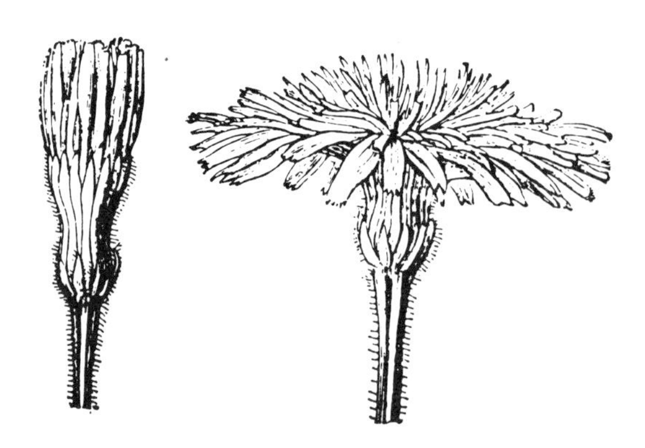

Figure 2. Inflorescences of *Leontodon hastilis* (dandelion). Left: in darkness, closed. Right: in light, opened.

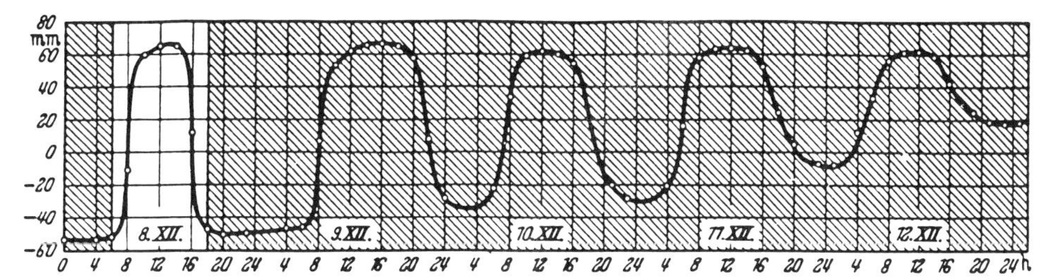

Figure 3. Continuation of petal movements in *Kalanchoe blossfeldiana* in constant darkness (shaded area). There is only slight damping. Ordinate: rising values = opening, falling values = closing of flowers. The absolute values refer to the recording system (after Bünsow from Bünning, 1967).

circadian rhythm with a period length of about 24 h is initiated which, however, damps out after a few days in darkness (Nultsch and Throm, 1975). As shown in Fig. 5, even one dark period of 8 h is sufficient to induce circadian rthythmic oscillations of phototactic reactivity, and as shown in the same figure, in continuous light the free running period is also 24 h, but the circadian rhythms damp out much later, e.g., after 7-14 days. As we have coupled the phototaxis assay with a mobility assay, we can be sure that the rhythms are due to changes in phototactic reactivity and not in mobility.

In addition to movement, there are many other processes governed by the circadian clock. In *Gonyaulax polyedra* both cell division (Fig. 6) and bioluminescence (Fig. 7) follow diurnal rhythms after they are initiated (Sweeney, 1977) Moreover, even circadian rhythms of metabolic rates (CO_2 output or O_2 consumption) can be continued under constant conditions. If *Bryophyllnum calycinum* plants are exposed to a light/dark regime, the CO_2 output of detached leaves follows circadian rhythms in continuous darkness (Fig. 8). Finally, the phycomycete *Pilobulus spaerosporus* shoots its sporangia after exposure to D/L = 12/12 or 15/15 in continuous darkness according to the entrained rhythm (Fig. 9).

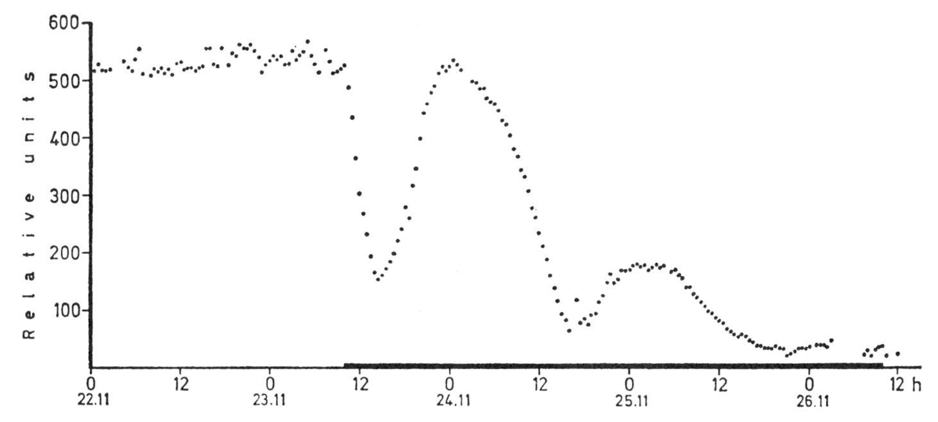

Figure 4. Initiation of circadian rhythms in phototaxis of *Chlamydomonas reinhardtii* by transfer to 72 h darkness. The rhythms damp out after a few days. Each point represents one experiement. Ordinate: magnitude of the signal (after Nultsch and Throm, 1975).

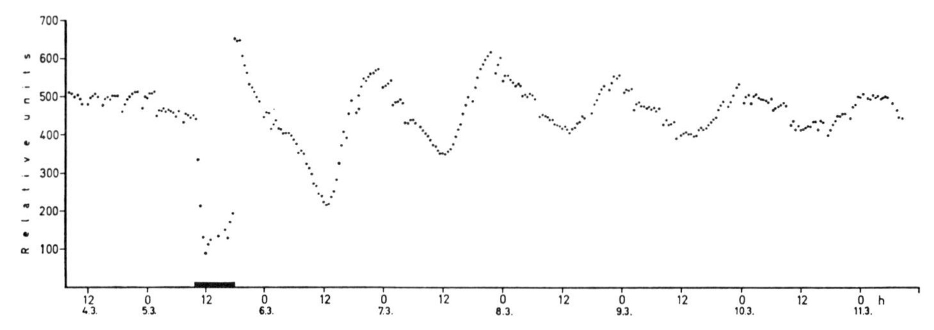

Figure 5. Initiation of circadian rhythms in phototaxis of *Chlamydomonas reihardtii* by 8 h of darkness (after Nultsch and Throm, 1975).

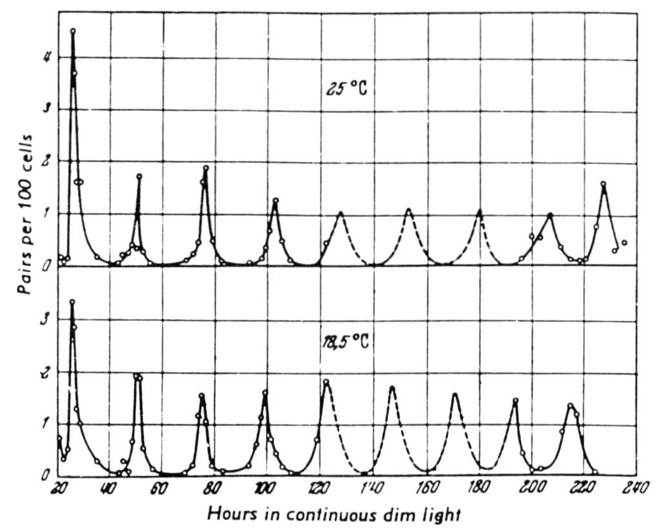

Figure 6. Diurnal rhythm of cell division of *Gonyaulax polyedra* at 18.5 and at 25°C, as measured by the percentage of cell-pairs present in the suspension in constant dim light and at different temperatures (after Hastings and Sweeney, from Bünning, 1967).

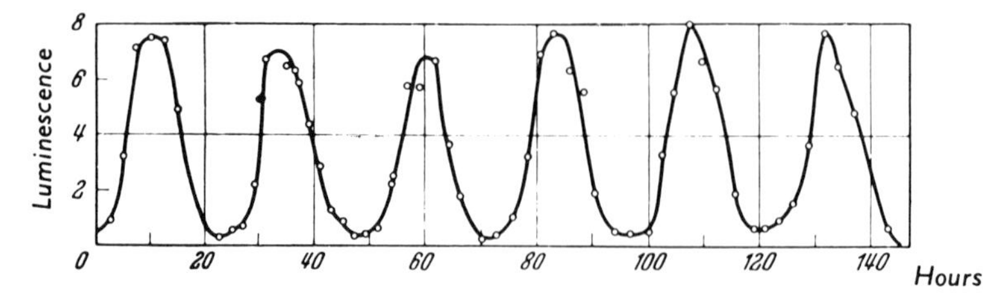

Figure 7. Bioluminescence rhythm of *Gonyaulax polyedra* cultures maintained in constant dim light (after Hastings and Sweeney, from Bünning, 1967).

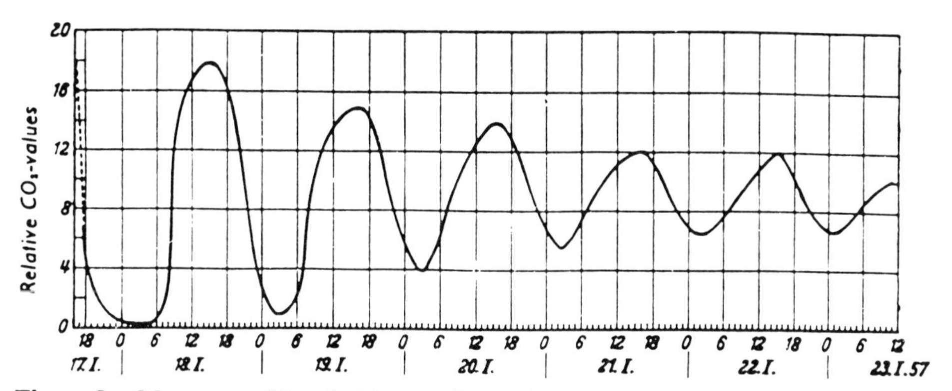

Figure 8. CO_2 output of detached leaves of *Bryophyllum calycinum* in continuous darkness (after Wolf, from Bünning, 1967).

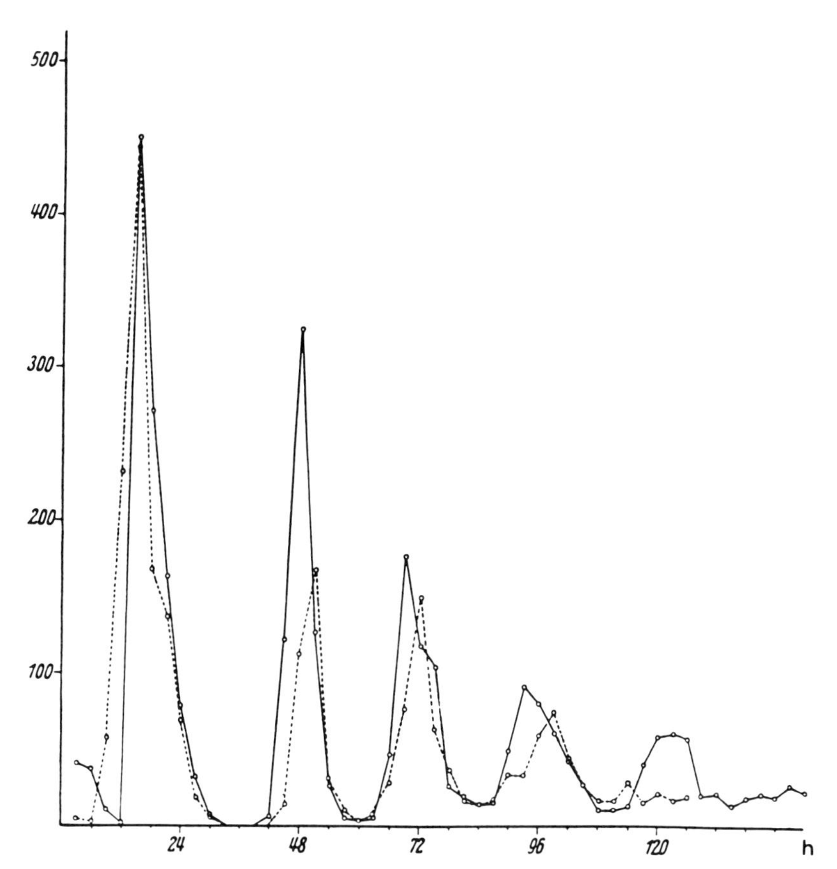

Figure 9. *Pilobolus spaerosporus*. Circadian shooting of sporangia after entrainment by 12/12 (solid line) and 15/15 h (dashed line) light/dark change (after Schmidle, from Bünning, 1963).

All these experiments confirm that a circadian clock is present in plants and eucaryotic microorganisms that is characterized by the following features: period length of about 24 h; oscillation rhythm is retained under constant conditions; the rhythm is endogenous but can be entrained by an exogenous rhythm; the rhythms are temperature compensated. This clock is comparable to a clock with a pendulum. If one pushes it, e.g. by a light/dark change, it measures the time and shows its own rhythm of 24 h ± a few hours, but a rhythm of 24 h when synchronized. This is also true for animals and humans.

Circadian Rhythms in Animals

In animals and humans, circadian rhythms have influence over all the behavioral and physiological responses of the organism. They can be observed as sleep-wake cycles, circadian changes in locomotor activity, body temperature, predation, feeding, urinary excretion, circulation, endocrine activity, etc., suggesting that circadian organization is an innate property of animal life (Aschoff, 1981; Rusak, 1981). An example of a circadian locomotor activity rhythm in a mouse is shown in Fig. 10. As for the majority of small mammals, mice are night active (nocturnal), whereas humans and most birds are day active (diurnal) (Fig. 11). A third pattern of circadian organization may also be observed in many small mammals where activity occurs around dawn and dusk, and accordingly their activity rhythm is called crepuscular. Figures 10 and 11 show that the circadian rhythm is precisely synchronized with the environmental light/dark cycle. In constant darkness the rhythm is maintained but runs with a frequency slightly different from 24 h ("circa" dian). This clearly shows that circadian rhythms are based on self sustained oscillations of an endogenous clock

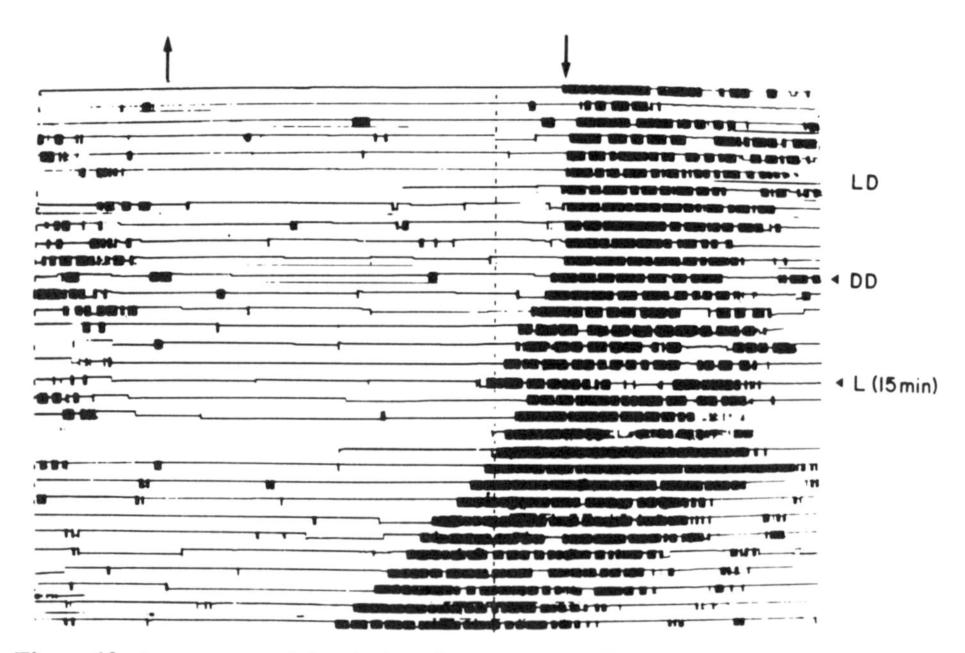

Figure 10. Locomotor activity rhythm of a mouse. The first few days the mouse was kept in a light dark cycle (LD). Following transfer to constant darkness (DD) the rhythm starts free running with a period length shorter than 24 h. In DD a light pulse of 15 min (L(15 min)) was applied causing a phase shift of the circadian rhythms on consecutive days (from Menaker, 1982).

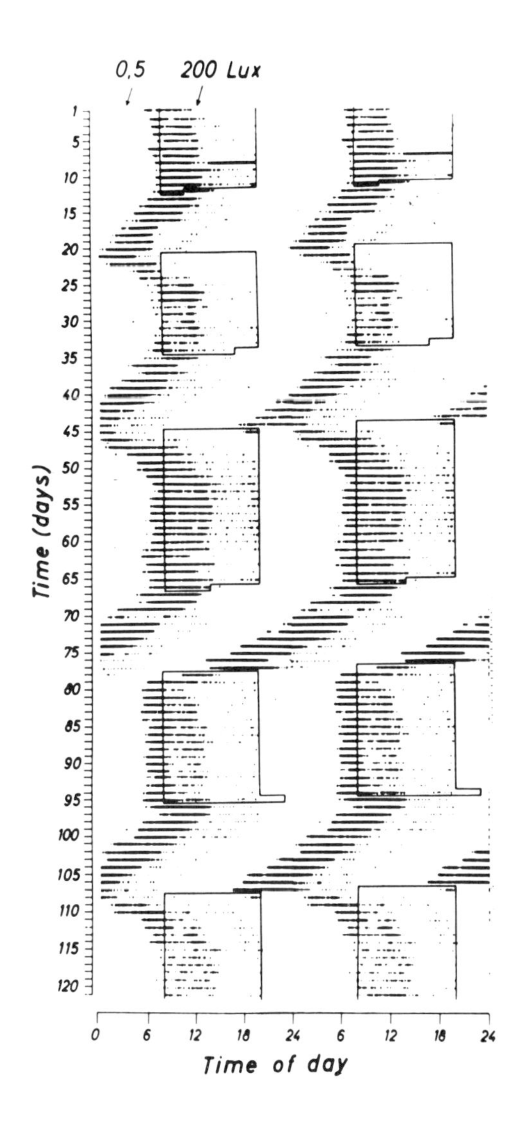

Figure 11. Locomotor activity rhythm of a chaffnich, *Fringilla coelebs*, alternately kept in a light/dark cycle and constant dim illumination (from Aschoff *et al.,* 1982).

and synchronization is required permanently to entrain them with environmental light/dark cycles (Aschoff *et al.*, 1982). In most species 24 hourly changes in light intensity are the most important "Zeitgeber" for circadian entrainment, but other signals may also act as a Zeitgeber, e.g., temperature in ectotherms or social synchronization in humans whose circadian rhythm is only weakly influenced by light.

The endogenous nature of the circadian clock is not only visible by free running rhythms in constant conditions but also during entrainment (Aschoff *et al.*, 1982; Pittendrigh, 1982). The onset of activity does not coincide with lights off (or lights on in a day active animal) but may precede or succeed this stimulus. However, entrainment causes a constant phase relation between the Zeitgeber and the circadian activity rhythm. Although this behavior reflects the physical laws of interactions between self sustained oscillators, it would not be observable when circadian activity changes occur as passive responses to changes in light intensity. These properties of entrainment can be further analyzed by phase response curves, when animals are kept in constant darkness to display their free running rhythms. At various times of their circadian clock, light is switched on and it can be observed that this causes a resetting of the circadian clock (Fig. 10). Depending on the time of day when this light stimulus is applied, the animals may show a phase delay or a phase advance of their activity rhythm in response to light.

Entrainment not only achieves a stable phase relationship between an animal and time structures in its environment but it also causes stable internal phase relationships between circadian rhythms of different body functions (activity, sleep-wake cycle, body temperature, excretion, etc.) which are always kept in synchronization. The physiological basis of this internal synchronization is unknown, but internal desynchronization and destructions of circadian organization may be observed following long term exposures to low light intensities or constant bright light. This internal desynchronization suggests that several circadian clocks may exist within an organism, which are phase locked by a circadian master clock.

The physiological nature and the site of the circadian master clock in animals is still not fully understood. In birds and lower vertebrates the pineal gland plays a major role during this process. Pinealectomized birds may loose their circadian behavior of activity and body temperature in constant conditions (Menaker and Binkley, 1981), although they can still be entrained by a light/dark cycle. Pineal glands kept in cell culture can maintain a circadian rhythm in the secretion of their major hormone, melatonin, which can be regarded as one of the key neuroendocrine transducers of circadian time in birds. In mammals this situation is slightly different. They also possess a pineal gland with a strong nocturnal pattern of melatonin secretion, but removal of the pineal gland does not disrupt free running activity periods. This, however, is observed when the suprachiasmatic nucleus (SCN) in the hypothalamus is destroyed, indicating that the mammalian circadian clock is in close connection with the SCN (Fig. 12, Rusak, 1982). In birds, destruction of the SCN also abolishes free running rhythms, also suggesting that the SCN is part of the circadian master clock in this group of vertebrates (Takahashi, 1982).

If the daily rhythm is changed, e.g., in humans by long distance travel, all these functions go out of phase, impairing efficiency and activity (jet lag). Depending on the individual constitution, resetting of the clock takes 2 days to 2 weeks. Advance phase shifts take longer to complete than delayed ones, but the reasons are unknown. Frequent phase shiftings, as in the case of shift workers, can cause serious sickness.

Cellular Mechanism of the Circadian Clock

Up to now, we do not know the chemical nature of the circadian clock. DNA, RNA, proteins, especially enzymes, cAMP and others have been suspected to be the molecular basis of the clock. Though circadian oscillations of the concentrations and/or activities of many of these substances have been observed, we are not able to decide whether their oscillations are the clock itself or are governed by the clock.

It seems highly probable, however, that there is a common molecular basis for the oscillator in the different organisms. This is supported by the discovery that in *Drosophila* mutants with altered circadian rhythms are mutated loci alleles of the so-called *per* locus which controls both rhythms of hatching and of spontaneous locomotor activity in adults (Konopka and Benzer, 1971). *per*° flies have no rhythm at all. Cloning and determination of the 4.5 k base sequence of the *per* gene by Jackson et al. (1986) led to the prediction that the gene product is a proteoglycan of 1218 amino acid residues. The putative *per* protein includes a potential site for phosphorylation by a cAMP-dependent protein kinase. The *per* DNA shows sequence homology to DNA fragments from many other organisms, e.g., the plastid genome of the green alga *Acetabularia mediterranea* and the nuclear genomes of some higher plants, such as rape (*Brassica napus*) and spinach (*Spinacea oleracea*). The exact function of the gene product, however, is not yet clear.

In general, protein synthesis seems to play an essential role in circadian timing, since inhibitors of cytosolic (80S) protein synthesis such as cycloheximide can reset, or phase shift, the clock. According to the coupled-translation-membrane hypothesis (Schweiger and Schweiger, 1977) the oscillations are due to a two step mechanism. In the first step, one (or a few?) specific polypeptide is synthesized, which in the second step is integrated into a membrane. By this integration the properties of the membrane are altered so that the synthesis of this protein ceases. The amount of the polypeptide in the membrane decreases by turnover. If it falls below a threshold value, the synthesis of protein is restored, and a new cycle begins. Hartwig *et al.* (1985) isolated a 230 kDa protein from *Acetabularia* chloroplasts whose rate of synthesis oscillates with a period of about 24 h. Its synthesis is inhibited by cycloheximide, and after this treatment a phase shift in its synthesis is observed. Thus it fulfills the requirements of the above mentioned model. How all the observed circadian phenomena are coupled to this process, however, is still unclear.

The site of the circadian clock inside the cell is also an open question. Since procaryotes do not show circadian rhythms, the nucleus has been suggested to play a key role. This, however, is in contrast to the finding that in the green alga *Acetabularia*, circadian rhythms continue even if the rhizoid which contains the nucleus is removed. Therefore, in accordance with the above mentioned model, several different biomembranes have been suggested as being the site of the clock, e.g., the endoplasmic reticulum, but this has also not yet been proven.

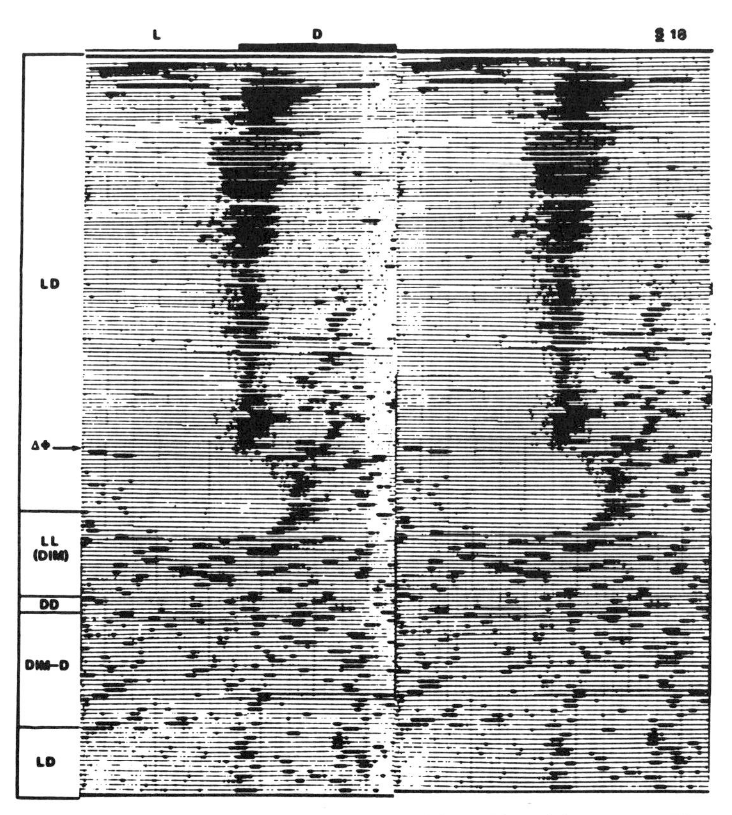

Figure 12. Locomotor activity of a SCN ablated hamster. He could be entrained to a light dark cycle but in constant illumination the circadian rhythm was lost.

Photoperiodism

Photoperiodic phenomena are well known in higher plants growing in the moderate latitudes. Seeds germinate and leaves appear in the spring, plants develop during the following months and flower either during spring or summer, and finally die or drop their leaves in the autumn. People who are not familiar with chronobiology often regard these phenomena as being coupled to the seasonal changes of the environmental conditions: rising temperature in spring, warm summer and decreasing temperature during autumn. However, temperature is an unreliable measure of the season, as sometimes the springs are warm and the summers are cold. Thus, how can the plants and animals recognize the seasons? The only reliable measure of the season is the daylength, and the recognition of season by measuring the length of the day is called photoperiodism. We have to distinguish between the following two types of plants:

(1) *Short Day Plants:* They flower when the length of the day (duration of illumination in the lab) does not exceed a distinct value, the so-called critical daylength which varies from species to species. Examples are *Kalanchoe blossfeldiana* and tropical plants in general.

(2) *Long Day Plants:* They flower when the length of the day exceeds the critical daylength. Examples are plants from moderate latitudes, e.g., *Hyoscyamus niger, Lactuca* (green salad) and our cereals.

The critical daylength of both short and long day plants can vary from species to species. In this connection, however, it must be mentioned that each daylength, with the exception of the shortest and longest, occur twice a year. Therefore, the plants must also be able to distinguish between short days in spring and short days in autumn. The first ones are, of course, short day plants. The second ones are the so-called long-short day plants. They must have long days (summer) first and then short days, i.e., they flower in the autumn.

The timing mechanism that monitors the annual changes in the length of the day is again the circadian clock. Obviously, it oscillates between two phases of different light sensitivity. Short day plants have a phase in which they must not see light, otherwise they do not flower. Long day plants have a phase in which they must see light, otherwise they do not flower. This is shown in Fig. 13 for the short day plant *Kalanchoe blossfeldiana.* It was exposed to 11.5 h of light and 60.5 h darkness. An additional light of one hour was given at different times after the first irradiation was finished. If it was applied at a time when the plants must not see light, no flower initiation was observed, but vegetative development was promoted. If it was applied in the photophilic phases, however, flower initiation was observed.

The contrary is true for long day plants, as shown in Fig. 14. In this experiment the long day plant *Hyoscyamus niger* was exposed to 7 h light and two additional hours at different times after the light was switched off. If this light was applied after the critical daylength, flowering was induced, but not between the 22nd and 33rd h, but again between the 36th and 47th h. This is an unambiguous proof that daylength is measured by the circadian clock. It must be mentioned, however, that autonomous development rhythms are also observed in tropical latitudes. In this case, the developmental cycles deviate from 12 months, as they are not synchronized by the daylength. Thus they follow an endogenous, circannual rhythm.

The daylight is measured by the green leaves and the stimulus is transferred to the apex of the stem. The photoreceptor pigment is phytochrome. In fungi, such as *Neurospora*, violet, blue and blue-green light is active. The action spectrum is indicative of a flavin. Light of shorter wavelengths, between 350 and 500 mm, without distinct peaks controls the eclosion rhythm of the insect *Drosophila.* In the alga *Gonyaulax*, short (470 mm) and long

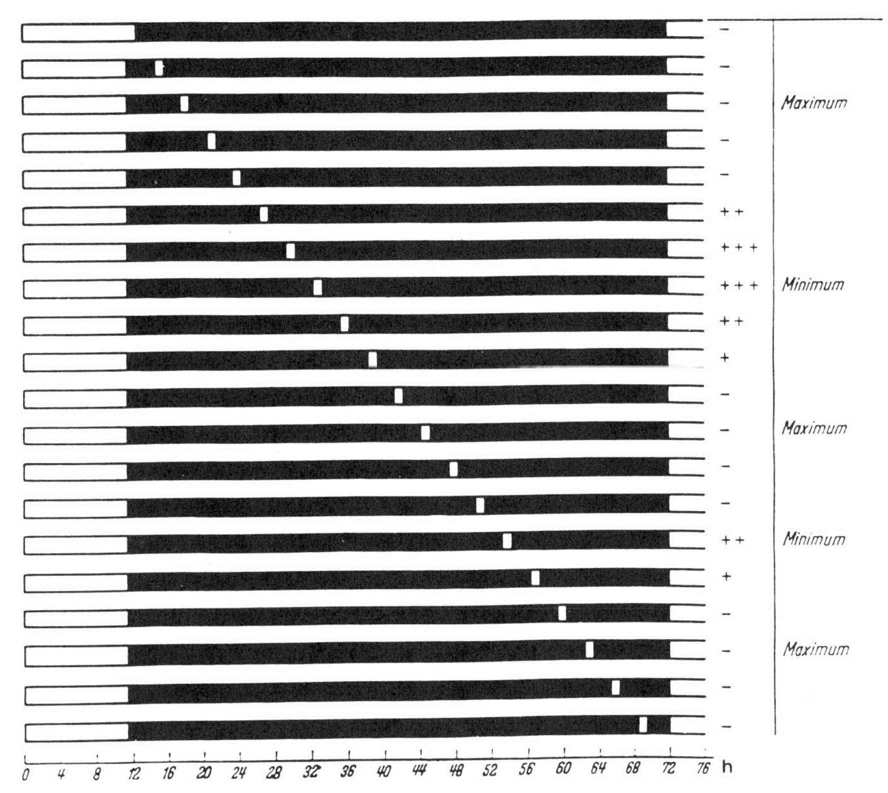

Figure 13. Flowering behavior of the short day plant *Kalanchoe blossfeldiana* after 11.5 h illumination with white light and 1 h light after different times (0-60 1/2 h). Light periods white, dark periods black. Symbols: +, ++, +++ flowering to increasing extent, – no flowering, indicating photophilic and skotophilic phases (after Carr, from Bünning, 1953).

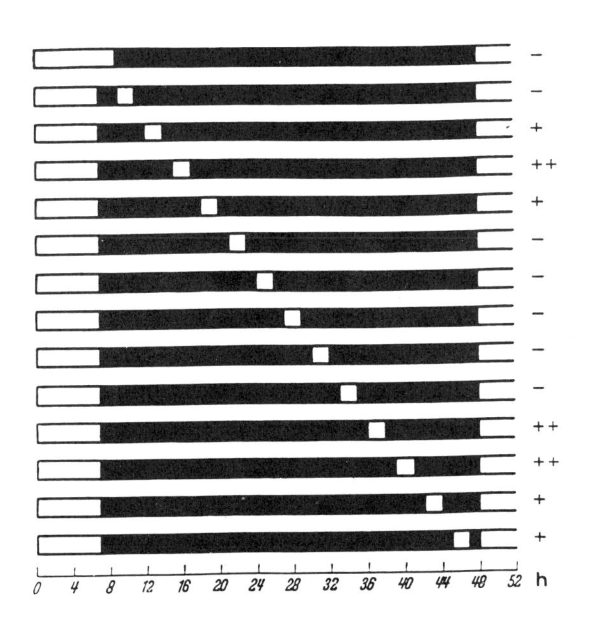

Figure 14. Flowering behavior of the long day plant *Hyoscyamus niger* (henbane) after 7 h illumination and 2 h light after different times (0-45 h). Light periods white, dark periods black. Symbols: ++ all plants, + a part of the plants, – no plants flower, (after Claes and Lang, from Bünning, 1953).

wavelength light (650 nm) can initiate circadian rhythms, but chlorophyll is apparently not the photoreceptor.

In mammals, photoperiod is the most important cue for entrainment of seasonal changes in behavior, e.g., reproduction, migration, hibernation, fur changes, etc. The photoperiod is received by the eye and then transmitted to the SCN by the retinohypothalamic tract. Light information is then further conveyed to the pineal gland where the hormone melatonin is synthesized during the hours of darkness. This is the most important neurochemical transducer for the photoperiod, since all seasonal changes in behavior such as reproduction, migratory behavior or thermoregulation are controlled by the duration of nighttime melatonin secretion (Gwinner, 1989; Heldmaier *et al.*, 1989). In birds and lower vertebrates, photoperiodic responses are also controlled by the pineal gland. However, in these species, specialized extraretinal photoreceptors in the brain or in the pineal gland are used to measure the lighting regime, independent of visual photoreceptors in the retina. In recent years evidence has accumulated that shows humans may be affected by the photoperiod. In short days man may show symptoms of depression, reduced physical performance, and hyperphagia ("carbohydrate craving"). This pattern of symptoms was named "seasonal affective disorder" and has been treated successfully by an extension of daylength with bright light.

EXPERIMENT 28: CIRCADIAN LEAF MOVEMENT IN *PHASEOLUS COCCINEUS*

OBJECTIVE

This experiment demonstrates circadian movements of primary leaves of *Phaseolus coccineus* (see Fig. 1) that were entrained by the natural day-night change (or by another or even inverse light/dark change entrained in a dark room with fluorescent light). The entrained rhythm is continued in continuous dim light.

The kymograph system described below is a low cost means of performing the experiment described. It is an effective and sensitive detection system for photomovement. Of course, movement can also be followed and measured with a modern video system if one is available.

LIST OF MATERIALS

Phaseolus coccineus (French bean), 21-26 day old plants grown in a greenhouse
Kymographs (number according to the number of experiments planned)
Kymograph paper
Apparatus for sooting the kymograph paper
Stands with clamps
Sticks
Scissors
Sewing thread
Sewing needle
Ethanol, 1 liter
Shellac, 200 g
Olive oil, 2 ml

PREPARATIONS

Plant Preparation: Ten days before the experiment, remove the plumula (terminal bud of the stem) from the plants.

Drum Fixing Solution: Prepare the fixing solution at least 2 days before fixing the soot on the kymograph drum. Mix 1 liter of ethanol with 200 g of shellac and 2 ml of olive oil. Stir the solution repeatedly until used.

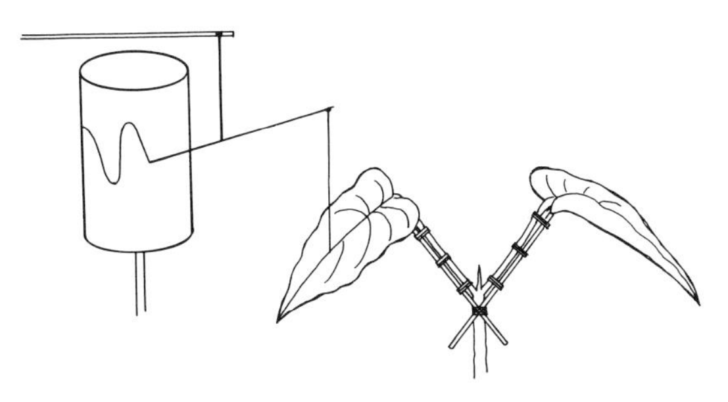

Figure 15. Set-up for measurements of circadian movements of bean leaves.

EXPERIMENTAL PROCEDURE

1. Cover the kymograph drum with paper. Soot the paper.
2. Fix the leaf stalks of each bean plant by binding them to a cross of sticks (Fig. 15) so that only the pulvinus (leaf joint) between the stalk and the leaf is movable while the primary joint between the stem and stalk is fixed.
3. Bind the middle rib of one leaf on each plant using a sewing needle and thread, about a third of the way from the tip of the leaf.
4. Tie the end of the string to one arm of a lever and suspend the lever from a gallow. A thread hanging from the other arm of the lever, weighted and equipped with a needle, transfers the movement on the kymograph and writes the curve by scraping the soot.
5. For the determination of period length, scrape time marks at regular intervals.
6. After several days or a week, remove the paper from the kymograph drum and fix the soot with the fixing solution. Dry the drum for several h.
7. For the determination of the period length, measure the distances between the maxima of the descending movements.

DISCUSSION

As long as the movements are measured during natural day/night changes or in a dark chamber using a suitable light/dark cycle (12/12, 8/16 h or others), the period length is exactly 24 h. If the plants are then kept under continuous dim light, the length of the free running period increases up to 27 h. If the plants are exposed to an inverse light/dark change, the rhythm is inverted. However, rhythms of a period length deviating from 24 h can also be entrained in many organisms, e.g., 15/15 h light/dark (cf. Fig. 9).

EXPERIMENT 29: CIRCADIAN ACTIVITY RHYTHMS IN SMALL MAMMALS

OBJECTIVE

The locomotor activity of mammals and birds is most frequently used to demonstrate circadian rhythms. In small mammals activity is frequently recorded by running wheels. This rather specialized behavior requires expensive running wheel cages and may not be suitable for the observation of overall activity. Recent developments in infrared sensing techniques, however, allow a simple way of measuring overall activity in small mammals in their standard laboratory cages or any other environment. It only requires a passive infrared detector which is mounted to the top of the cage (~$50 US). Such detectors are manufactured by a large number of companies and can be purchased in most electronic shops or through companies which deal with home security equipment (movement detectors). An inexpensive alternative sensor (~$5 US) would be a mechanical motion detector commonly used in security equipment.

LIST OF MATERIALS

Mouse or hamster kept in an individual standard cage
Passive infrared sensor (or mechanical motion detector, e.g., Radio Shack type 490/0521)
Power supply (6-12 VDC)
Strip chart recorder (alternatively event recorder, or counter, or computer with analog or digital interface card)

PREPARATIONS

Motion Detection: Maintain the mouse or hamster in a standard small animal cage with a top-feeding trough. For IR motion detection, place the sensor on the top of the cage so that movement during feeding periods will be detected. When animals are active the sensor will be triggered. For motion detection using a mechanical motion detector, place the sensor on top of the rodent's food. When animals are eating, the sensor will be triggered by the motion of the food.

EXPERIMENTAL PROCEDURE

1. After completion of the set-up, place the cage with the mouse or hamster in a room with a known light/dark cycle (or natural conditions), and record activity for several days (e.g., 1 week).
2. At the end of the chosen period, cut the paper strip from the recorder to separate each day. Then paste the consecutive days below each other (~1 cm apart from each other; see Fig. 10).
3. For visual evaluation of such records "double daily plots" are most convenient (Figs. 11 and 12). Therefore, make 2 photocopies of your pasted record (if possible reduce to 50% of original size) and paste the two copies alongside each other. Thereby, the right half of your copy should be moved up one day, in order to ensure that day 1 on the left side is continued by day 2 on the right half, etc.

DISCUSSION

This simple record already allows you to evaluate a few important features of the circadian system. The graph allows one to answer the following questions:

1. Is the activity pattern of the species studied day- or night-active or does it display crepuscular activity?
2. Does the onset of activity precede or follow the onset of darkness? How stable is this phase relationship on individual days (precision of circadian rhythms and entrainment)?
3. Is the amount of activity constant each day?
4. Calculate the hours of activity and the hours of rest each day. How does the amount of activity change with time of day (monophasic or polyphasic activity pattern)?

After 1 week of recording you should try to change the light/dark cycle and follow the animal's response for 1 or 2 weeks. This change could show a phase shift of the Zeitgeber for about six hours (advance or delay) or you could try to keep them in constant darkness. Constant darkness is sometimes not easy to obtain, since the room cannot be entered for inspection. It is therefore usually better to use constant dim light (below 10 lux) or a red dark room light. Under proper conditions, you will observe a free running circadian activity pattern and you can measure the period length of the animal's endogenous circadian clock (as an example, see Fig. 10).

REVIEW QUESTIONS

1. How can flowering plants recognize the time of year?
2. How can we prove whether daily rhythms are governed by an endogenous oscillator?
3. How can animals measure environmental light dark cycles as Zeitgeber for circadian entrainment?

ANSWERS TO REVIEW QUESTIONS

1. They measure the length of the day. Higher plants use phytochrome as photoreceptor.
2. In the absence of an external Zeitgeber (i.e., in constant light and constant temperature) the rhythm will be free-running with a period length which is different from 24h.
3. Light is sensed by retinal or extraretinal photoreceptors. Light/dark cycles will entrain e.g., circadian activity rhythms and adjust a constant phase-relationship between the Zeitgeber and the entrained activity rhythm.

SUPPLEMENTARY READING

Feldman, J.F. (1989) Circadian rhythms. In *The Science of Photobiology*, 2nd edition (Edited by K.C. Smith), pp. 193-213. Plenum Press, New York.

REFERENCES

Aschoff J., S. Daan and K.I. Honma (1982) Zeitgebers, entrainment, and masking: Some usettled questions. In *Vertebrate Circadian Systems.* (Edited by J. Aschoff, S. Daan and G.A. Groos), pp. 13-24. Springer Verlag, Berlin.

Aschoff, J. (1981) A survey on biological rhythms. In *Handbook of Behavioral Neurobiology, Vol 4, Biological Rhythms* (Edited by J. Aschoff), pp. 3-10. Plenum Press, New York.

Bünning, E. (1953) *Entwicklungs- und Bewegungsphysiologie der Pflanzen*, Springer Verlag, Berlin, Göttingen, Heidelberg.

Bünning, E. (1959) Tagesperiodische Bewegungen, In *Encyclopedia of Plant Physiology*, Vol. XVII/1 (Edited by W. Ruhland, E. Ashby, J. Bonner, M. Geiger-Huber, W.O. James, A. Lang, D. Müller and M.G. Stalfelt), pp. 579-657. Springer Verlag, Berlin, Göttingen, Heidelberg.

Bünning, E. (1963) *Die Physiologische Uhr*, 2nd Ed. Springer Verlag, Berlin, Göttingen, Heidelberg.

Bünning, E. (1967) *The Physiological Clock*, 2nd Ed. Springer-Verlag, New York.

Feldman, J.F. (1989) Circadian rhythms. In *The Science of Photobiology*, 2nd Ed. (Edited by K.C. Smith), pp. 193-214. Plenum Press, New York.

Gwinner, E. (1989) Photoperiod as a modifying and limiting factor in the expression of avian circannual rhythms. In *Biological Clocks and Environmental Time.* (Edited by S. Daan and E. Gwinner), pp. 125-138. The Guilford Press, New York.

Hartwig, R., M. Schweiger, R. Schweiger, H.G. Schweiger (1985) Identification of a high molecular polypeptide that is part of the circadian clockwork in *Acetabularia. Proc. Natl. Acad. Sci. USA* **82**, 6899-6902.

Hastings, J.W. and B.M. Sweeney (1959) The *Gonyaulax* clock. In *Photoperiodism and Related Phenomena in Plants and Animals* (Edited by Withrow), pp. 567-584. Washington, D.C.: Amer. Assoc. Adv. Sci.

Heldmaier, G., S. Steinlechner, T. Ruf, H. Wiesinger and M. Klingenspor (1989) Photoperiod and thermoregulation in vertebrates: Body temperature rhythms and thermogenic acclimation. In *Biological Clocks and Environmental Time.* (Edited by S. Daan and E. Gwinner), pp. 139-153. The Guilford Press, New York.

Jackson, F.R., T.A. Bargiello, S.H. Yun, M.W. Young (1985) Products of *per* locus of *Drosophila* shares homology with proteoglycans. *Nature* **320**, 185-199.

Konopka, R.J. and S. Benzer (1971) Clock mutants of *Drosophila melanogaster. Proc. Natl. Acad. Sci. USA* **68**, 2112-2126.

Menaker, M. (1982) The search for principles of physiological organization in vertebrate circadian systems. In *Vertebrate Circadian Systems.* (Edited by J. Aschoff, S. Daan and G.A. Groos), pp. 1-12. Springer Verlag, Berlin.

Menaker, M. and S. Binkley (1981) Neural and endocrine control of circadian rhythms in the vertebrates. In *Handbook of Behavioral Neurobiology, Vol. 4, Biological Rhythms* (Edited by J. Aschoff), pp. 243-255. Plenum Press, New York.

Nultsch, W. (1971) *General Botany*, Academic Press, New York.

Nultsch, W. and G. Throm (1975) Effect of external factors on phototaxis of *Chlamydomonas reinhardtii I. Light Arch. Microbiol.* **103**, 175-179.

Pittendrigh, C.S. (1981) Circadian systems: General perspective. In *Handbook of Behavioral Neurobiology, Vol 4, Biological Rhythms* (Edited by J. Aschoff), pp. 57-80. Plenum Press, New York.

Rusak, B. (1982) Physiological models of the rodent circadian system. In *Vertebrate Circadian Systems.* (Edited by J. Aschoff, S. Daan and G.A. Groos), pp. 62-74. Springer Verlag, Berlin.

Rusak, J. (1981) Vertebrate behavioral rhythms. In *Handbook of Behavioral Neurobiology, Vol 4, Biological Rhythms* (Edited by J. Aschoff), pp. 183-213. Plenum Press, New York.

Sweeney, B. (1977) Chronobiology (Circadian rhythms). In *The Science of Photobiology*, 1st Edition (Edited by K.C. Smith), pp. 209-226. Plenum Press, New York.

Takahashi, J.S. (1982) Neural mechanisms in avian circadian systems: Hypothalmic pacemaking systems. In *Vertebrate Circadian Systems.* (Edited by J. Aschoff, S. Daan and G.A. Groos), pp. 112-119. Springer Verlag, Berlin.

Bioluminescence: Biochemistry for Fun and Profit

John LEE
University of Georgia
U.S.A.

INTRODUCTION

Bioluminescence is often called Living Light because it is encountered as the emission of light by certain animals (Harvey, 1940). A more scientific definition is that bioluminescence is light emission of high efficiency that occurs from certain organisms and that serves a biologically significant purpose for that organism. This definition is made to include the fact that bioluminescence also occurs in plants and fungi, and in bacteria. The definition excludes the very weak light emission that is observed in many actively metabolizing cells. We call that *biological chemiluminescence* and the reader is referred elsewhere for a discussion of that subject (Slawinska and Slawinski, 1985).

The chapter as seen by its title, is divided into two parts. The first is the "fun" part, to do with the obvious beauty of the phenomenon and the purely intellectual questions that come out of it (Lee, 1989). Why do these organisms emit light? This is a biological question and the answer is convincing in the case of a few of the bioluminescent organisms. How do they do it and especially, how do they do it with such efficiency? The efficiency of conversion of chemical energy into light is almost unity in the case of the firefly.

Aside from the fact that bioluminescence research has provided a living for this author and his family for 30 years, the "profit" in bioluminescence is that the specificity of protein interactions combined with the extremely high sensitivity of light detection, have generated a great range of commercial applications of bioluminescence to "ultrasensitive" analysis. To illustrate this with the most up to date example, the chemical company Monsanto, has announced the ability to make transgenic corn, apparently not heretofore achieved. They used a "gene gun" to fire the gene for firefly luciferase into corn cells! The insertion of the gene into the corn cells was proven by the observation that the cells could generate bioluminescence on addition of the appropriate reagents.

Biology

Bioluminescence occurs most commonly in the ocean, particularly in the deep ocean, where the majority of species are luminescent. The terrestrial forms however, are the more

JOHN LEE, Department of Biochemistry, University of Georgia, Athens, GA 30602, U.S.A.

familiar, the firefly and the luminous fungi (foxfire), and there are a number of less well known examples. The "phosphorescence" that occurs on agitating ocean water is from the bioluminescence of protozoa or dinoflagellates, the latter is an alga that is responsible in part for red tide. The most common ocean species are the coelenterates, i.e., the soft corals (*Anthozoa*), jellyfish (*Hydrozoa*), and comb-jellies (*Ctenophora*). There are many fish that derive their bioluminescence by carrying around a culture of luminous bacteria in a special light organ, known also as a *photophore*. The flash-light fish for example, is displayed in some of the larger aquaria.

Because weak light emission, i.e., biological chemiluminescence mentioned above, is almost ubiquitous in aerobically metabolizing tissue, it has been suggested that bioluminescence evolved from biological chemiluminescence as an adaptation, enabling the species to compete more effectively in its biological niche (Seliger, 1975). The biological purpose of bioluminescence is to be seen so it could not have been until the development of primitive visual systems that survival would favor the diversion of valuable metabolic energy into the production of photons. Energy conservation would be an irresistible driving force for evolving the most efficient chemiluminescent system. The oxidation of hydrocarbons is by a free radical mechanism and is accompanied by weak chemiluminescence emission. It has been proposed that oxygenases were the ancestor bioluminescent proteins and that the accumulation of a suitably fluorescent metabolite would provide the means to amplify, or *sensitize* the chemiluminescence, at minimal energy cost.

The generally held view is that bioluminescence arose independently in many organisms. This is because the property is spread over such a diversity of organisms that it must have been a relatively recent event in evolution. For the number of bioluminescence proteins that have had their amino acid sequence determined, there is found to be no relationship between them nor with the other proteins in the available data banks. Also, the chemical reactants or substrates, for the bioluminescence reactions, usually have structures unique to that bioluminescence system.

A biologically significant function of bioluminescence that has been well established, is in the behavior of the firefly. The purpose is to enable sexual contact, and what can be more biologically significant (as well as more fun!) than the reproduction of the species! What is perceived as a flash by the human eye is actually a series of short flashes, a kind of Morse code that is characteristic of the species. It is of no biological value for a male of one species to find a female of a different one. The flying male flashes and is answered after a certain time interval, e.g., 2 s, by the female. An exceptional interspecies contact is in the case of a species of female that is insectivorous, feeding on the males of other species and luring its prey by imitating the female response of the prey species. Nourishment is also biologically significant!

Bioluminescence

The color difference among species of firefly is small compared to the spectral differences among bioluminescent systems in general. In the bioluminescent creatures now known, the color of the light ranges from red in the railroad worm found in Brazil, to the deep blue, characteristic of some marine creatures. The first question then, is about the identities of the light emitting molecules. The radiation is from the fluorescent state of these molecules and obviously for such large spectral differences, these emitters must have a different chemical structure depending on the organism. What are these molecules? Is the mechanism of these reactions to produce light similar, even though they differ in chemical structure? We shall perform some experiments later to look into these questions.

In more scientific terminology, the color of bioluminescence is the wavelength distribution of the spectral emission. Usually these spectra are broad and are characterized by the wavelength of the spectral maximum, and the half-width of the spectral distribution. It is relatively straightforward to obtain the *in vivo* bioluminescence spectrum. In most cases, just place the creature in front of the entrance slit of a fluorescence emission spectrometer and hope it glows for long enough to obtain a scan. Distortion of the spectrum can occur, however, if the light intensity is not constant during the scan or if there are pigments in the body of the organism that absorb part of the spectrum.

However, it is considerably more difficult to obtain extracts from organisms that *in vitro*, that is in solution, will react to produce the same bioluminescence. If the extraction is successful, then extensive effort is usually required to obtain the chemically pure components of the bioluminescence system. It is vital that this *in vitro* bioluminescence reaction have a high efficiency. This efficiency is measured as the quantum yield of bioluminescence, Q_B. The Q_B needs to be greater than 1% if there is to be a chance of identifying the chemical reaction responsible for light emission. Since a chemiluminescence reaction is one that proceeds with sufficient release of energy to raise some associated molecule to its electronic excited state, then bioluminescence can be viewed as a protein associated chemiluminescence. This protein is in many cases an enzyme, called *luciferase*, after the Greek, Lucifer – the bringer of light.

The first scientific investigation of bioluminescence, at least in the modern era, was by Robert Boyle in the 17th century. With his air pump he discovered that the bioluminescence from samples of luminous bacteria and fungi, was extinguished on removing the air. It was not until the end of the 19th century that an *in vitro* bioluminescence reaction was demonstrated. Dubois showed that two components could be extracted from firefly light organs, one with hot water and one with cold water. Mixing one with the other at room temperature, regenerated the bioluminescence. The cold water extract, which was heat labile, he named luciferase, and the hot water extract, luciferin. We now know that this hot water extract is not what we now call firefly luciferin, the organic molecule which becomes oxidized in the reaction, but adenosine triphosphate, ATP, an essential component in the firefly system. Firefly luciferin itself is also heat labile.

Dubois was also able to make hot and cold water extracts of the clam, *Pholas*, that produced bioluminescence when mixed, but which did not cross react with the firefly extracts. This was the first demonstration of the uniqueness of the reactants involved in bioluminescence systems. In this short chapter we will not be able to describe the diversity of chemical structures that are used by the multitude of bioluminescent types. We will restrict ourselves to two illustrative examples, the firefly and the bacteria. Most systems exhibit the classical enzyme substrate type of biochemistry, but many contain what is called a *photoprotein* which is a "precharged" system, specifically aequorin in the case of the jellyfish, *Aequorea*.

Chemiluminescence

The understanding of enzyme catalyzed reactions in many cases, has been achieved through study of the mechanisms of organic model reactions in homogeneous solutions. This is also true of bioluminescence reactions, since they are just chemiluminescence reactions that take place within the active site of a protein. The fact is often not recognized that some luciferases are the most catalytically active enzymes known, because they channel the reaction to produce an excited state 10^{10} times more efficiently than does the same uncatalyzed reaction.

There are two general questions that can be posed concerning the physical mechanism of chemiluminescence. These are about the production of the excited state and of the emission of the photon. In the gas phase at low pressure, the rate of energy release on chemical bond formation between atoms can be faster than the rate of "cooling" of the excess energy by collisions with the surroundings. What is called non-adiabatic crossing to the excited state of a product is a likely occurrence. In the liquid state however, the vibrational interaction with the medium is so predominant that it prevents any highly energetic or "hot" species being formed as a result of the rearrangement of atoms.

In solution only two mechanisms are regarded as feasible for excited state generation as the result of a chemical reaction. These two mechanisms are ones in which electrons jump: electron transfer between redox pairs or intramolecular electron rearrangement. The subject of *electrochemiluminescence* has therefore received much study. There are only a few studies of electrobioluminescence.

The chemiluminescence reactions of organic peroxides are good models for bioluminescence because the intermediacy of peroxides is inferred in many bioluminescence reactions (Gundermann and McCapra, 1987). Many cyclic peroxides called dioxetanes, have been synthesized and these break down often to produce chemiluminescence (Fig. 1). One of these compounds that yields a very high chemiluminescence quantum yield, $Q_C = 0.25$, is shown in Fig. 2 (Schaap et al., 1987). This compound is one now utilized in an analytical method for certain medically important enzymes.

A mechanism for the chemiluminescent breakdown of dioxetanes and other cyclic peroxides was proposed by both F. McCapra and by G. Schuster in 1976 (see Gundermann and McCapra, 1987). Experimental support for electron transfer was obtained in a study of the chemiluminescence of diphenoyl peroxide catalyzed by the presence of an easily oxidizable and fluorescent *activator* (Koo and Schuster, 1977). The reaction with rubrene as activator is shown in Fig. 3. This electron transfer mechanism was dubbed *Chemically Induced Electron Exchange Luminescence* (CIEEL).

Probably the chemiluminescent reaction most familiar to people, is the *Light Stick.* This is the reaction of oxalyl chloride by H_2O_2 in the presence of a dye, called the chemiluminescence sensitizer. Without the dye the Q_C is negligible; the dye amplifies the Q_C to about 0.34. The mechanism is quite obscure but the dye possibly acts as an activator for a CIEEL process.

Some generalities can be formulated at this point that are useful for understanding both chemiluminescence and bioluminescence reactions. It is important to measure the

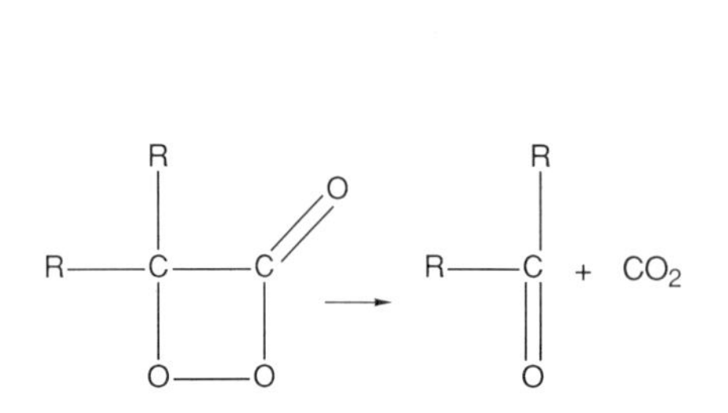

Figure 1. A dioxetanone (left) decomposes to a ketone and carbon dioxide.

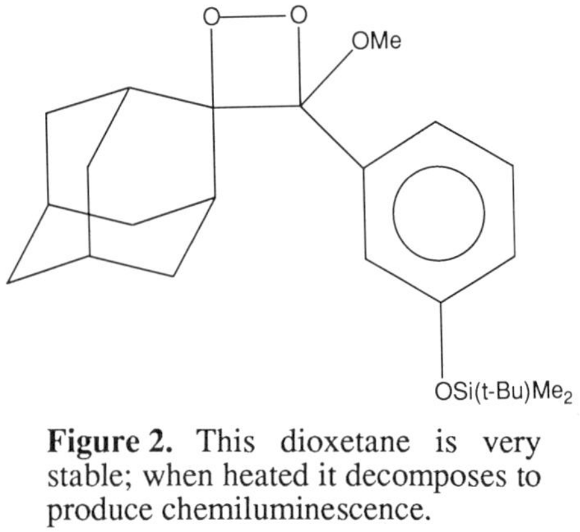

Figure 2. This dioxetane is very stable; when heated it decomposes to produce chemiluminescence.

chemiluminescence spectrum and if this corresponds precisely to the fluorescence of a product molecule, then it can be suggested that the reaction is a *direct* chemiluminescence:

$$A + B \rightarrow C^*$$

In many cases the chemiluminescence spectrum is the fluorescence of another molecule present in the system, not a product, and this is called an *indirect* or sensitized chemiluminescence:

$$A + B + D \rightarrow C + D^*$$

The mechanism of energy deposition onto the sensitizer may be by the weakly coupled long range electronic energy transfer mechanism, or by close range electron transfer, such as by the activator in the CIEEL mechanism. It is noted that the sensitizer is unchanged by the chemical reaction analogous to the function of a photochemical sensitizer.

The second essential measurement after the chemiluminescence spectrum, is the chemiluminescence quantum yield:

$$Q_C = \text{number of photons emitted / number of molecules of A consumed,}$$

where A refers to the reactant in the general equation above. This is the Q_C with respect to A, and it may also be measured with respect to the consumption of B, production of C, etc.

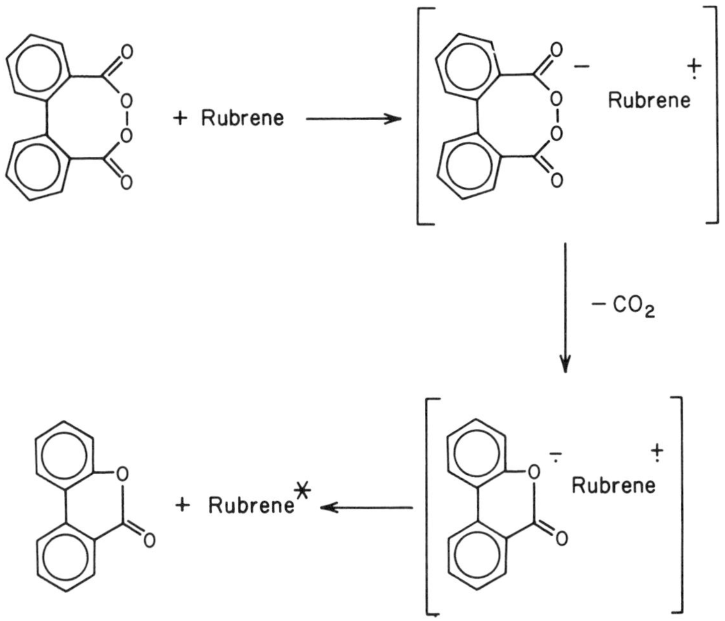

Figure 3. The rubrene sensitized "CIEEL" chemiluminescence mechanism.

The photon can be considered to be a reaction product, so it can be immediately written down that

$$Q_C \leq 1.0$$

The observed Q_C is an overall yield dependent on three separate efficiencies: Q_Y, the chemical yield along the reaction pathway; Q_E, the efficiency for populating the excited state over the ground state; and Q_F, the fluorescence quantum yield of the emitter:

$$Q_C = Q_Y \times Q_E \times Q_F$$

These yields are responsible for the observation that Q_C for most reactions is well below unity, in most cases orders of magnitude below.

Two rules that must be applied to any chemiluminescence investigation, come out of this relationship. The chemiluminescence reaction must go in 100% yield (i.e., $Q_Y = 1.0$) or at least comprise a major portion of the chemical process, if the chemistry is to be related to the light reaction. The second point is that not only must the emitter have the same fluorescence spectral distribution as the chemiluminescence, but the relationship must be satisfied that:

$$Q_F > Q_C .$$

In other words, once the emitter is in its excited state it should have no memory of how it was placed there. In the following sections and in the laboratory experiments, we will make application of these tests.

Luminol Chemiluminescence

The chemiluminescence reaction of luminol, 3-aminophthalhydrazide, was discovered in 1928 by Albrecht (Albrecht, 1928) and has since been the subject of many research studies. White and his colleagues (1964) established the mechanism of the reaction in basic dimethylsulfoxide, DMSO (Fig. 4). This achievement was greatly aided by the fact that the $Q_Y \approx 1.0$ and that the reaction is a direct chemiluminescence. In this case we can write:

$$Q_C = Q_E \times Q_F$$

and the observed

$$Q_C = 0.0124.$$

The chemiluminescence spectrum is the same as the fluorescence of the product, 3-aminophthalate dianion (Fig. 4), and $Q_F = 0.14$, which yields $Q_E = 0.09$. Note that the inequality above, $Q_F > Q_C$, is well satisfied.

The identity of spectra between the fluorescence of the product and the chemiluminescence is negated, however, under some solution conditions, for example if the DMSO contains a trace of water a new fluorescence band around 425 nm starts to appear before much change in the chemiluminescence spectrum is seen. Inorganic ions also have similar effects and explanations for these effects in terms of proton transfer rates, hydrogen bonding rearrangements and ion pairing, have been given. The differences are manifestations of effects on Q_E, in that the route to the final excited state of the product is affected differently by solution conditions.

Clean interpretations of the mechanism pertain to the reaction in dry DMSO and in aqueous solution. In water the Q_C depends on the pH, being maximum at pH = 11.6, Q_C = 0.0124, the same value as in DMSO. On the high pH side the Q_C decreases in exactly the same way as the Q_F, but at lower pHs, especially below 9, the Q_C decreases rapidly without any change in Q_F. This decrease is ascribed to competition from dark reactions.

The chemical mechanism of the aqueous luminol reaction is in Fig. 5. The oxidant is probably the same as in DMSO, the superoxide radical ion, O_2^-, which is formed rapidly from H_2O_2 in the presence of catalysts, such as a trace of hemoglobin, even better fresh blood. The product, 3-aminophthalate monoanion, has the same fluorescence spectrum as the chemiluminescence (Fig. 6) and the $Q_F = 0.30$ and $Q_C = 0.0124$, which makes $Q_E = 0.04$. The identity of Q_C for the DMSO reaction and aqueous one is tantalizing, but probably coincidental.

It was initially thought that the aqueous chemiluminescence reaction of luminol was an indirect one, sensitized by the reactant molecule, luminol. It was shown by Seliger (1961) using precise fluorescence spectral methods, that the fluorescence of luminol, although in the same blue region, could be distinguished from the chemiluminescence, and moreover, that the Q_F for luminol was quenched at the basic pHs that optimized the Q_C, in contrast to the Q_F behavior of the 3-aminophthalic acid anion.

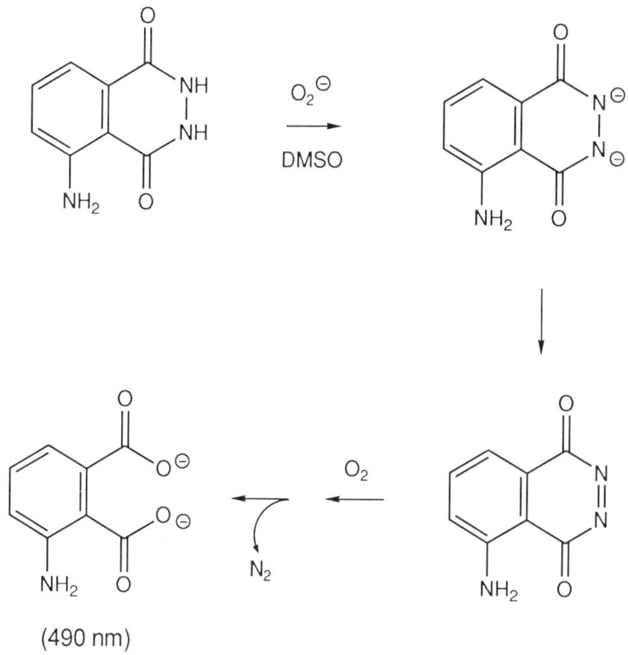

Figure 4. The chemiluminescence mechanism of luminol in dimethylsulfoxide.

An important application of the luminol reaction results from the fact that its Q_C has been measured accurately. This application is to the absolute calibration of photometers or as they are sometimes called, "luminometers", the instrument used to measure chemiluminescence and bioluminescence reactions. The determination of the response of such an instrument in terms of the photons coming from a sample, is a technically difficult task in itself. The standard of light intensity to which the calibration must be traceable is a *Standardized Tungsten Lamp* which is 10^6 times too intense for a luminometer to tolerate. An alternative method must be employed to calibrate these instruments. The luminol reaction provides a secondary standardization method and can be conveniently carried out in the instrument to be calibrated. The protocols for this procedure are detailed elsewhere and this will also comprise one of the planned experiments (O'Kane et al., 1986).

Firefly Bioluminescence

Firefly luciferase is a single subunit protein of size 60 kDa (i.e., the molecular weight is 60,000) and it contains no bound cofactors. Its primary structure, that is the amino acid sequence, is known and the recombinant form is commercially available in quantity. Firefly luciferase catalyzes two reactions, an acylation of firefly luciferin with the ubiquitous coenzyme adenosine triphosphate (ATP), followed by the bioluminescent oxidation by molecular oxygen (McElroy and DeLuca, 1985). The steps are as follows, where LH_2 is firefly luciferin and E stands for the luciferase:

$$LH_2 + ATP + Mg^{++} + E \rightarrow E\text{-}LH_2AMP + Mg^{++} + PP_i$$
$$E\text{-}LH_2AMP + O_2 \rightarrow\rightarrow E\text{-}LO + AMP + CO_2$$

where AMP is adenosine monophosphate, PP_i is pyrophosphate, Mg^{++} is needed because it is always involved in ATP reactions, and E–LO and CO_2 are the oxidized products of luciferin.

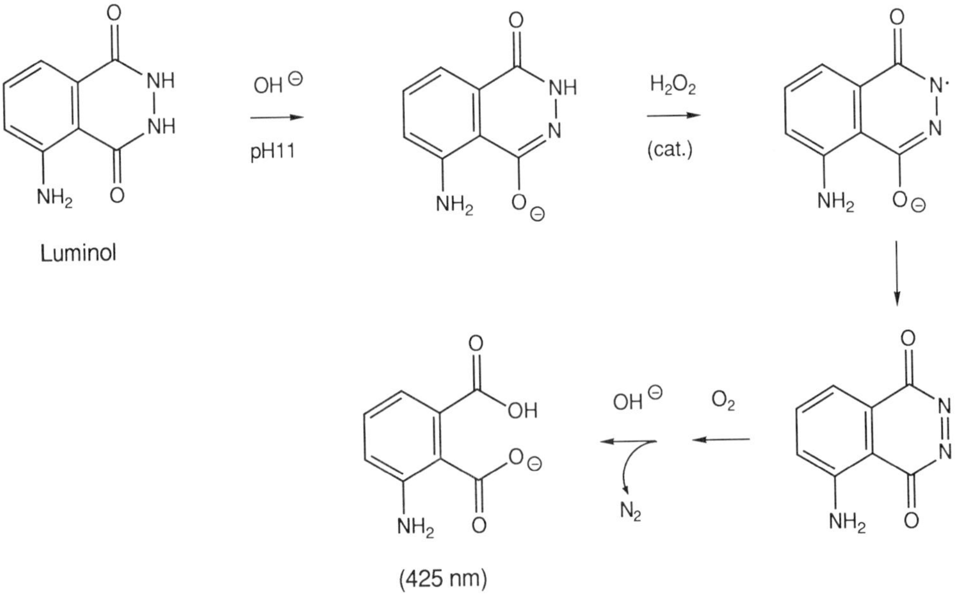

Luminol

$OH^{\ominus}$ pH11

H_2O_2 (cat.)

$OH^{\ominus}$ O_2 N_2

(425 nm)

Figure 5. The chemiluminescence mechanism of luminol in water.

Firefly luciferase has a high specificity for ATP, and this, together with the high sensitivity for the detection of photons, has established this bioluminescence reaction as the preferred method for the assay of ATP. We will practice the use of the firefly assay in the experimental section.

The chemical steps which occur in the active site of firefly luciferase, are shown in Fig. 7. Firefly luciferin, LH_2, was the first luciferin to have its chemical structure determined (White et al., 1961). The # mark is a chiral center and the luciferase is absolutely specific for the D-isomer. The key intermediate in the reaction is a high energy cyclic peroxide, the luciferin dioxetanone, shown in the square brackets because it has not yet been directly seen. Its intermediacy was implied from results of the bioluminescence reaction using the heavy isotope $^{18}O_2$, instead of ordinary $^{16}O_2$. One ^{18}O atom was found in the CO_2 product and the other in the oxidized luciferin, LO.

Firefly bioluminescence has a bioluminescence quantum yield, $Q_B = 0.88$, with respect to the luciferin. This Q_B is the highest found for any bioluminescence system; the Q_B with respect to O_2 consumption and CO_2 production are about the same value near unity. This would imply that both Q_Y and Q_E are near unity but unfortunately, the product E-LO is unstable and its fluorescence has not been accurately measured.

Several factors have been established that determine the emission spectrum of firefly bioluminescence. It has already been mentioned that a yellow to green color range is found with certain species. This range in terms of spectral maxima, is 552 - 575 nm. If the luciferase is purified from a species with *in vivo* maximum at 552 nm say, then the *in vitro* reaction with synthetic luciferin is also at this same wavelength. The spectrum is shifted by mutations of a few amino acid residues of the luciferase so that it is concluded that the range of observed emission maxima is an environmental perturbation of the LO excited state within the active site of the luciferase (Wood et al., 1989). This is analogous to the effect of solvent polarity on shifting fluorescence spectral maxima.

A more drastic spectral effect occurs in the *in vitro* bioluminescence on lowering the pH below 6, adding heavy metal ions, and also as the result of some mutations. The yellow emission changes to red, with a maximum at 615 nm. A chemiluminescence reaction in

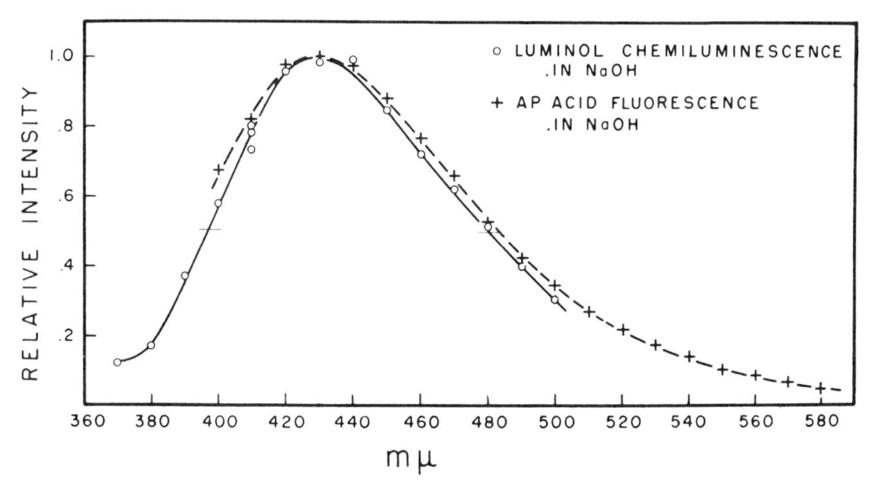

Figure 6. The chemiluminescence emission spectrum from luminol is the same as the fluorescence of the product, 3-aminophthalate.

solution can also be made using firefly luciferin. The reaction is with O_2 in a basic aprotic solvent, e.g., DMSO. The same red emission spectrum can be observed from this chemiluminescence reaction. However, if precautions are taken to keep the solution very dry and to make it sufficiently basic, a yellow-green spectrum is produced with a maximum at 562 nm, close to the bioluminescence. No chemiluminescence occurs at all in water solution.

These effects are explained by the protonation scheme shown in the lower right of Fig. 7. The yellow emission is from the dianion. It can undergo rapid protonation while still in the excited state to the monoanion, which has the red fluorescence. Evidently in the active site of firefly luciferase, the rate of protonation is slowed down and this allows the excited dianion to emit radiation before being protonated, and presumably quenched, by the surrounding water medium.

Bacterial Bioluminescence

The bioluminescent bacteria are mostly found from marine sources and are divided mainly into two genera, *Photobacterium* and *Vibrio*. The bioluminescence spectra are broad with the *Photobacterium* types having maxima in the range 472-495 nm, while the *Vibrio* maxima are in the 487-542 nm range. Bacterial bioluminescence is one of the most investigated bioluminescence reactions, one reason being that the luciferase from these bacteria is relatively easy to purify and the reactants are also readily available. The *in vitro* reaction requires reduced flavin mononucleotide, $FMNH_2$ (Fig. 8) a common coenzyme in all living cells, and a long chain aliphatic aldehyde (RCHO) of chain length longer than 8 carbons (Lee, 1985):

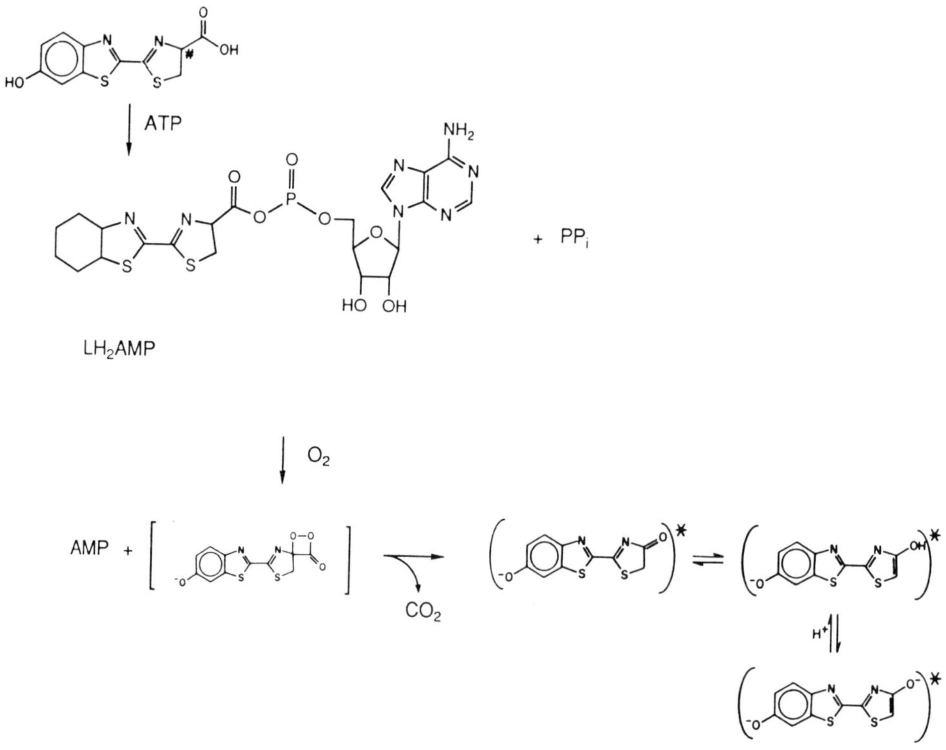

Figure 7. The mechanism of bioluminescence of firefly luciferin (top structure).

$$FMNH_2 + RCHO + O_2 \rightarrow\rightarrow FMN + RCOOH + H_2O$$

Tetradecanal gives the highest light intensity and this intensity is directly related to the chemical reaction rate. The evidence is good, but indirect, that these are the *in vivo* reactants, too. A fatty acid (RCOOH) reductase enzyme complex has been characterized from these bacteria and an NADH:FMN oxidoreductase is ubiquitous in procaryotes, providing the source of the $FMNH_2$. For the bacterial types with bioluminescence spectral maxima in the region of 487-505 nm, the *in vitro* spectra are the same or similar to the *in vivo* spectra.

The fluorescence of the final product, FMN, has a maximum at 535 nm, and does not resemble the bioluminescence, except for the type that emits with a maximum at 542 nm. Nor can the fluorescence of FMN be perturbed by solvents of low polarity, for example, to match the range of bioluminescence spectra. The question of the *in vitro* emitter was answered when it was found that, following mixing of the reactants together, a transient fluorescent species was formed that had a spectrum corresponding to the bioluminescence.

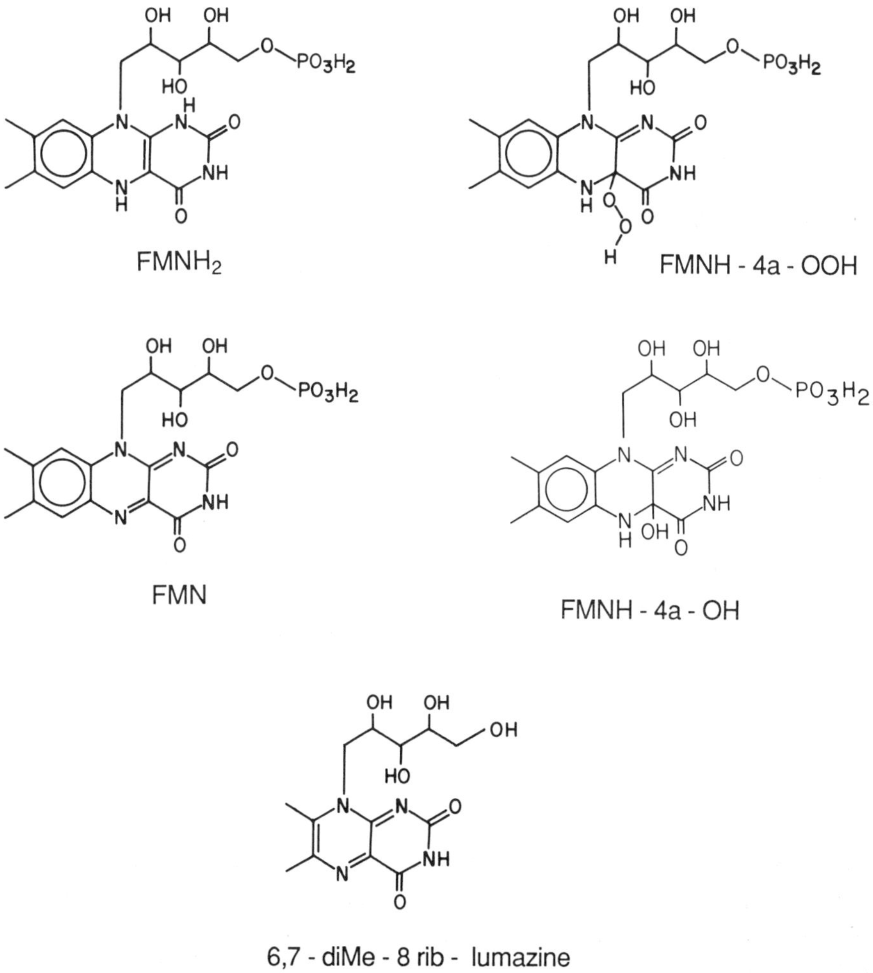

Figure 8. Structures of the flavin derivatives involved in bacterial bioluminescence.

The chemical mechanism of bacterial bioluminescence is not known and presents a fascinating challenge to investigators. One unusual property of luciferase is that, if the RCHO is omitted, an intermediate species is produced that can be stabilized for more than 12 h. It is one of the most stable enzyme intermediates known and, being available in millimolar concentration, the chemical structure of the bound flavin has been established by ^{13}C-NMR. The intermediate is 4a-peroxy$FMNH_2$ bound to luciferase (Fig. 8). The structure of the fluorescent transient is not known, but it has been suggested to be the FMNH-4a-OH also luciferase bound (Fig. 8).

The absence of CO_2 as a product makes it unlikely that a dioxetane is formed as the key energetic intermediate and no compelling model for the mechanism exists. The Q_B for bacterial bioluminescence is dependent on the type of luciferase but in every case it is < 0.1. Nevertheless, bacterial luciferase can be said to exert a tremendous catalytic effect because the reaction of $FMNH_2$, O_2, and tetradecanal alone has a $Q_C < 10^{-10}$.

For the *Photobacterium* types which generally have bioluminescence maxima around 475 nm, and for the *Vibrio* one at 542 nm, the bacterial luciferase reaction *in vitro* has a different bioluminescence with a maximum around 495 nm. This spectral difference was explained by the finding that in these types of cells, there is a second protein containing a bound fluorescent molecule that has the same fluorescence spectrum as the *in vivo* bioluminescence. Clearly these proteins are sensitizing the bioluminescence *in vivo*. The one in the *Photobacterium* cells is called *Lumazine protein* because its noncovalently bound group is 6,7-dimethyl-8-ribityllumazine (Fig. 8).

Lumazine protein has received detailed study. It has a mass of 21 kDa and it attaches one molecule of the natural lumazine derivative which can be replaced by a variety of artificial ligands containing carbohydrate tails. Lumazine protein's function as a bioluminescence sensitizer was proven by its effects on the bioluminescence reaction. If added to the reaction mixture, the spectrum blue shifts from a 495 nm maximum finally to 475 nm by the time the concentration exceeds about 10^{-5} M, and this bioluminescence spectrum has the same spectral distribution as the fluorescence of the lumazine protein itself. Additionally, the rate of reaction is speeded up and the quantum yield is increased. The maximum quantum yield of the *in vitro* bioluminescence that can be achieved by lumazine protein addition, is $Q_B = 0.25$. Since for lumazine protein $Q_F = 0.58$, and assuming $Q_Y = 1.0$, then $Q_E = 0.43$.

Lumazine protein itself has very attractive properties for a protein spectroscopic study. As well as being very fluorescent, the fluorescence decay lifetime is quite long, 14.7 ns, and this enables the rotational correlation time, ϕ, for the protein to be accurately recovered by analysis of the decay of the fluorescence anisotropy. The ϕ value is a measure of rotational diffusion and, therefore, of the size of the protein. Lumazine protein has $\phi = 22$ ns at 2°C in aqueous phosphate buffer. Coincidentally, the value of ϕ in ns, is almost the same as the mass in kDa (e.g., lumazine protein, 21 kDa).

When bacterial luciferase (E; 77 kDa) is mixed with lumazine protein, ϕ is increased up to a maximum value with luciferase concentration, of about 100 ns. This is due to the formation of a protein-protein complex:

$$\text{E} + \text{LumP} \rightleftarrows \text{E}\bullet\text{LumP}$$

where LumP stands for lumazine protein. Typical experimental data are shown in Fig. 9.

In order for lumazine protein to perform as it does in the bioluminescence reaction, i.e., shift the bioluminescence spectrum, increase the Q_B, etc., it must form a protein-protein complex with the luciferase such that the lumazine ligand is close to the active site on the luciferase. This was shown again by measurement of anisotropy decay. Energy transfer provides a route of anisotropy loss and, indeed, when the fluorescent transient is mixed with lumazine protein, and excited at 370 nm where lumazine protein does not have much absorbance, ϕ is strongly reduced to below 30 ns. From such experiments it has been concluded that the lumazine is nearby the luciferase active site within a complex of the two proteins (Fig. 10).

Bacterial luciferase consists of α and β subunits. It contains no metals or other cofactors. The molecular biology of bacterial luminescence is a very active research field at present and is becoming well defined (Meighen, 1988). The responsible proteins originate in the *lux* operon and about seven genes have now been distinguished. *lux*A and *lux*B correspond to the luciferase subunits, three genes are for proteins responsible for the production of aldehyde, and one other is for a flavoprotein of presently unknown function. The operon has been cloned and inserted into many types of DNA, including that of

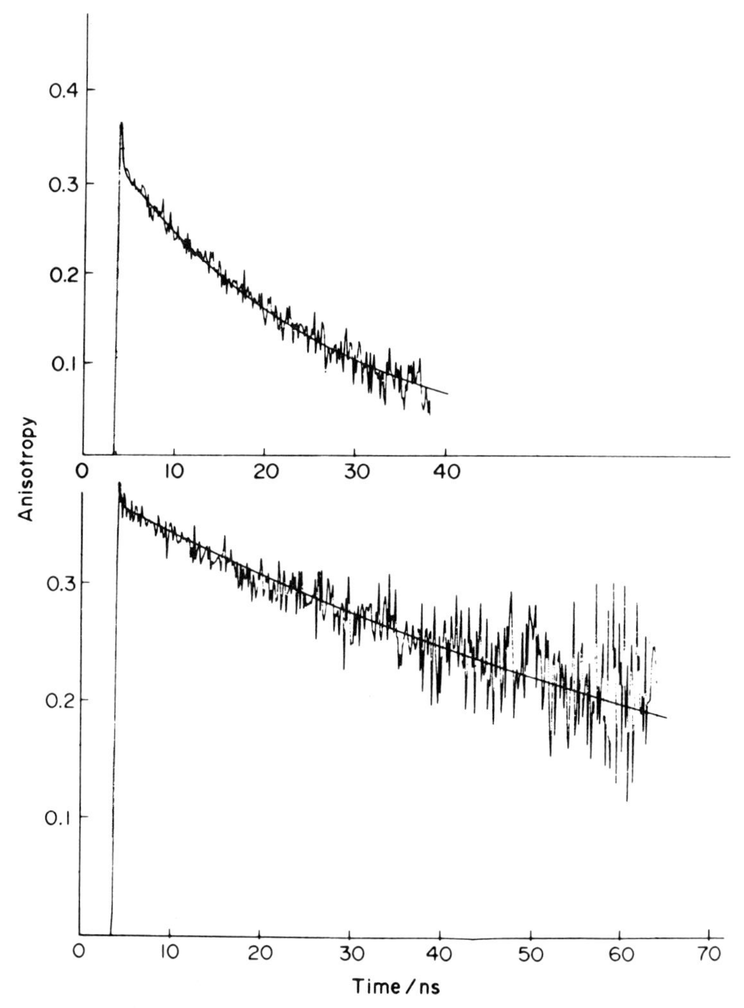

Figure 9. The decay of the emission anisotropy of lumazine protein (top) is much slower when it is mixed with bacterial luciferase (bottom) because it is in a much larger protein-protein complex.

eucaryotes. For the eucaryotes a fused *lux*AB gene has also been constructed artificially which has the advantage of being able to be controlled by a single promoter (O'Kane et al., 1988).

The lumazine protein gene has also been located and sequenced. It is not within the *lux* operon but instead has homology with the riboflavin synthetase gene. This comes as no surprise because the lumazine derivative is a known metabolite as the precursor to riboflavin in its biosynthesis. Both proteins use the same lumazine, one for its fluorescence as an antenna for the bioluminescence, and the other as a substrate for the synthesis of this vital coenzyme, riboflavin. The bioluminescent bacteria "borrowed" the protein to allow the lumazine to perform a different function.

Applications of Bioluminescence

Health care in the United States is a $500 billion/year industry. It is easy to understand then how successful use of bioluminescence methods in only a small slice of such an industry, could lead to great profits. Up to this time the most widely used application is for the assay of ATP using the firefly system. This is fast, specific, and more sensitive than any alternative method. Recombinant firefly luciferase is available commercially and is inexpensive. Only small amounts of luciferin are needed. The method itself is not the limiting feature in the assay of ATP, rather it is the instability of ATP outside the cell. Nevertheless, satisfactory protocols for the reliable extraction of ATP from tissue are available. The firefly bioluminescence method has been demonstrated to have sensitivity below 10^{-12} g of ATP, with linearity to 10^{-7} g.

If limiting quantities of ATP are added to a solution of firefly luciferase and luciferin in a buffered solution containing Mg^{++}, the bioluminescence intensity rises to a maximum, followed by a fast, then a very slow decay. The usual procedure is to wait a few seconds following the ATP addition, then integrate the light intensity over about 15 s. A straight line results when this integrated bioluminescence is plotted against the amount of ATP added, and this is the calibration line. For an unknown sample, the amount of ATP can be determined from its same integrated bioluminescence yield.

A novel method using the firefly bioluminescence has recently been developed, based on the use of a synthetic phosphate ester of firefly luciferin (Geiger et al., 1989). Assay of the enzyme alkaline phosphatase is involved in a number of clinical diagnostic techniques. It is also assayed joined, or conjugated, to an antibody as a marker for gene expression and in protein blotting. Alkaline phosphatase from tissue samples, or when it is used in conjugated form as a marker, is easily revealed by its ability to hydrolyze the luciferin phosphate to the substrate for the firefly luciferase. A detection limit of 10^{-14} g of alkaline phosphatase is claimed. Coupling to many other enzymes has also been described.

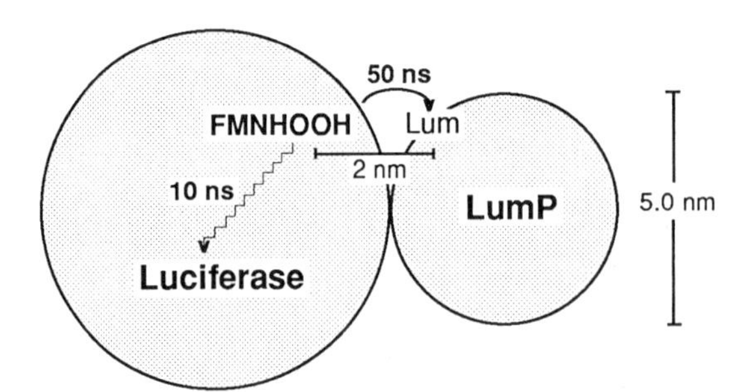

Figure 10. Scheme for energy transfer from the flavin site on the luciferase to the lumazine within the protein-protein complex.

An assay based on the same principle of measuring alkaline phosphatase but using chemiluminescence, is already in commercial development (Schaap et al., 1989). The key reactant is a dioxetane phosphate which evidently is quite stable until the phosphate group is cleaved off, e.g., by alkaline phosphatase. The detection limit in this chemiluminescence method has been demonstrated at less than 1000 molecules of alkaline phosphatase. In a real application, however, such a detection limit might not be achieved because of interfering substances in a sample.

In principle, the bacterial bioluminescence reaction *in vitro* should find very wide usefulness in analysis because the coenzymes NAD and NADP are widely distributed and are linked to many important enzymes. The bioluminescence emission from the luminous bacteria of the type *Photobacterium phosphoreum*, is quenched by many toxic substances such as might occur in drinking water supplies. A test kit called Microtox using these dried bacteria is available for water quality testing.

A more sophisticated use of the bacterial system is as a gene probe. An elegant method devised by S. Ulitzer and colleagues, is the insertion of the *lux* operon into phage specific for *Salmonella*. Adding the phage to a sample will produce infection specifically in viable *Salmonella* if present, not in other types of bacteria that may also be in a sample. From the bioluminescence that is produced by the transformed *Salmonella*, less than 10 bacteria can be detected within the incubation period of about 24 h.

The *lux* AB fused gene has also been inserted into the DNA of the nitrogen-fixing bacterium, *Bradyrhizobium japonicum* (O'Kane et al., 1988). Soybean roots infected with these bacteria formed nodules, and when the nitrogenase gene was expressed, so was the luciferase, detected by its bioluminescence on addition of decanal. The gene for firefly luciferase is also being actively employed as a gene marker, an example being mentioned in the Introduction to this chapter.

The reader is no doubt aware that many of the techniques employed in clinical analysis make use of antibodies, because of their specificity. Radioimmunoassay is one of the most important tools of the clinical biochemist, but the increasing concern with the disposal of radioactive waste, makes attractive the alternative and equally sensitive chemiluminescence and bioluminescence methods.

Experiments

In the following sections, a number of bioluminescence demonstrations will be described that can be carried out without access to elaborate laboratory facilities. There are also some experiments that require quantitative measurement, preferably with a luminometer, although any instrument that uses a photomultiplier can usually be adapted for this purpose. For example, a scintillation counter operated out-of-coincidence, a fluorometer or spectrophotometer with the lamp turned off will also do. For the visual demonstrations, the demonstrations are best seen with a 10-min dark adapted eye. All chemicals should be of the best grades available commercially.

EXPERIMENT 30: LUMINOL CHEMILUMINESCENCE

OBJECTIVE

Demonstrations of chemical light reactions can be a very effective teaching tool because of the range of colors that can be produced. Also, for a bioluminescence demonstration, students can be told that they can actually see an enzyme working! The book by Gundermann and McCapra contains a chapter listing chemiluminescence demonstrations and the luminol reaction immediately described here follows their recipe. In these times, many children are now familiar with the "light-stick" which uses the oxalyl chloride and hydrogen peroxide reaction, contained in a sturdy plastic tube.

LIST OF MATERIALS

Luminol, 50 mg
DMSO (dimethylsulfoxide), 100 ml, dry (new bottle)
Flask, 500 ml
Rubber stopper for flask
KOH pellets, about 25
Dark room

EXPERIMENTAL PROCEDURE

1. Dissolve about 50 mg of luminol in 100 ml of dry (new bottle) DMSO in a 500 ml flask.
2. Prepare 25 pellets of KOH so that they can be easily poured into the flask, but do not pour them in yet (CAUTION: very caustic!).
3. Take the flask and the KOH pellets into a dark room (or the darkest area available), and wait 10 min for your eye to dark adapt.
4. Pour the KOH pellets into the flask, and tightly stopper the flask.
5. Vigorously shake the mixture with the stopper in place.
6. Place the flask on a counter and observe it.
7. When the light emission is seen to be only from the surface of the liquid, shake the flask again and observe the change.

DISCUSSION

The liquid will commence to glow especially around the pellets. Fig. 4 indicates that the reacting species is the dianion of luminol. The second ionization of luminol only occurs under extremely basic conditions, at a pH even greater than 14! Such basic solutions can be made in dry DMSO with concentrated KOH. That is why the glow is seen mainly around the KOH pellets. As oxygen is consumed, the reaction ceases in the bulk of the solution, but continues at the surface. This is due to the slow diffusion of oxygen through the liquid surface. When the flask is shaken again, oxygen in the flask is mixed with the solution and the rest of the solution will light up.

EXPERIMENT 31: LUMINOMETER CALIBRATION

OBJECTIVE

The chemiluminescence reaction of luminol can be used to calibrate the absolute photon sensitivity of a luminometer. This is because, under standard conditions, the light yield from the luminol reaction is reproducible and the quantum yield is well established. An important feature is also that the position of the light emitting solution in respect to the detecting photocell, can be made to duplicate the bioluminescence reaction. There are extreme technical difficulties in calibrating a photometer directly by reference to a standard lamp, as used in other branches of photometry.

LIST OF MATERIALS

Luminol, 1 mg, reagent grade
K_2CO_3, 1 gm
Analytical balance (to 1 mg), and weighing accessories
Pipet, 5 µl
Pipet, 1 ml
Graduated cylinder, 10 ml
H_2O_2, 30%
Hemoglobin (or blood)
Hypodermic syringe and needle (or other device to measure 0.05 ml)
Spectrophotometer, scanning (250-600 nm)
Cuvettes, quartz (2) for spectrophotometer
Photometer vial
Bioluminescence luminometer or photometer (with lid or 1 sq. meter black cloth)

PREPARATIONS

Dilute H_2O_2: Add 1 ml of 30% H_2O_2 to 30 ml of distilled water.

K_2CO_3 Solution: Prepare a 0.1 M solution of K_2CO_3 by dissolving 0.7 gms of K_2CO_3 in 50 ml of distilled water.

Stock Luminol Solution: Prepare this solution by dissolving about 1 mg of reagent grade luminol in 10 ml of K_2CO_3 solution. Check its absorption spectrum by preparing a 10-fold dilution in K_2CO_3. The near UV maximum should be at 347 nm.

Hemoglobin Solution: Prepare a very dilute solution of hemoglobin in K_2CO_3 solution. Add only enough hemoglobin so that the solution just begins to show a brown color. (Alternately, very dilute fresh blood, also in K_2CO_3 solution, can be used.)

EXPERIMENTAL PROCEDURE

We will assume the standard volume for the bioluminescence assays is 1.0 ml.

1. Add 5 µl of the stock luminol to 1.0 ml of the K_2CO_3 solution contained in the reaction vial. The volume of luminol stock solution can be adjusted depending on the amount of light reaction you want, as long as the amount of luminol present is *known accurately*.
2. To this dilute luminol add 5 - 10 µl of 1% H_2O_2 freshly diluted from 30% stock.

3. Immediately (< 1 min) place the reaction vial in the photometer compartment and cover with the lid (or black cloth).

4. Initiate the chemiluminescence by adding only about 0.05 ml of the hemoglobin solution to the reaction vial. This is best done using a hypodermic syringe. (Fresh blood is an even better catalyst than hemoglobin and may be used, highly diluted.)

5. The light intensity will rise to a maximum in a few seconds then fall away, taking a few min to be exhausted. Another squirt of hemoglobin may be made to make sure all the light has been obtained.

6. If the kinetics are very fast or more than two squirts of hemoglobin are needed, then adjustments of the H_2O_2 and/or hemoglobin concentrations should be done. Be adventurous!

DISCUSSION

To calculate the sensitivity of the photometer you can apply the relation which can easily be derived from the extinction coefficient of luminol, 7640 $M^{-1} \cdot cm^{-1}$ at 347 nm, and the chemiluminescence quantum yield, 0.0125. For an aqueous solution of luminol having an absorbance of 1.0 at 347 nm, the chemiluminescence reaction yields a maximum of 9.75×10^{14} photons. To calculate the sensitivity for the bioluminescence reaction, which will probably have an emission to longer wavelength than luminol (max 425 nm), you will have to make an adjustment for the wavelength sensitivity of the photodetector. If this is a photodiode, then this sensitivity curve is fairly flat. Some photomultipliers, however, are a factor of two or more times less sensitive for photons having the firefly spectral distribution than for luminol. These sensitivity curves will be described in the manufacturer's literature.

The reaction in DMSO described in an earlier section can also be used for calibration. This yields a different chemiluminescence spectrum. It requires dry DMSO and potassium tertiary-butoxide dissolved in tertiary-butanol. The details of all these reactions are described in O'Kane et al. (1986).

EXPERIMENT 32: FIREFLY BIOLUMINESCENCE

OBJECTIVE

Firefly luciferase and luciferin will be extracted in a very crude form by grinding the light organs in buffer solution. The first reaction will show that addition of ATP generates the *in vitro* bioluminescence. The yellow light corresponds to the fluorescence of the dianion product in Fig. 7 (bottom right). By bubbling nitrogen through the solutions for long enough, it may be possible to show that no light is emitted unless oxygen is present. This is not always successful. By adding a small amount of acid, which may have to be adjusted, a red *in vitro* bioluminescence can be obtained, corresponding to the fluorescence of the monoanion product in Fig. 7.

LIST OF MATERIALS

Glycine, 1 g
NH_4HCO_3, 0.3 g
ATP, 0.15 g
$MgCl_2$, 0.06 g
HCl
NaOH
Fireflies, dried, about 10 (or 10 dried firefly tails)
Mortar and pestle
Clean sand, 1 g
Filter paper and funnel (or table top centrifuge and 1 ml tubes)
Nitrogen gas cylinder with regulator
Tubing for nitrogen
Acetic acid, 10% concentrated
Test tubes, 10 ml (about 6)
Large clear plastic bag
Pasteur pipet
Syringe, 1 ml, and needle (optional)
pH meter
Analytical balance and weighing accessories

PREPARATIONS

Firefly Buffer: Dissolve 0.8 g of glycine and 0.25 g of ammonium bicarbonate into 200 ml of distilled water. Adjust the pH to 7.6 using HCl and/or NaOH as needed.

ATP Solution: Dissolve 0.1 g of ATP and 0.05 g of $MgCl_2$ in 100 ml of firefly buffer.

EXPERIMENTAL PROCEDURE

All the preparation procedures with bioluminescence materials need to be done in the cold, say in a cold room, so that everything stays as cold as possible.

1. Take 10 dried fireflies, cut off the tails and use the mortar and pestle to grind them for about 10 min with about 1 g of clean sand in 3 ml of firefly buffer.
2. Filter the solution, or better centrifuge it, to get rid of all the particulate matter. The filtrate is the "cold-water" extract.

3. Divide the cold water extract into three portions in test tubes for experiments.

4. **All the light reactions should be carried out at room temperature, although the extract should be kept cold until that time.**

5. To one of the test tubes rapidly add 1 ml of ATP solution; a syringe is handy for these additions. A flash of yellow light will be observed.

6. Bubble nitrogen gas for about 10 min into one of the test tubes and another tube containing 1 ml of ATP solution. This is to try to get rid of the oxygen.

7. If the two tubes can be manipulated under a plastic bag containing flowing nitrogen, then mix them together. There will be no light seen until the mixture is shaken in air for a few seconds.

8. To 1 ml of the ATP solution use a Pasteur pipet to add 1 drop of 10% concentrated acetic acid. Mix this with the third test tube of firefly extract. A weak red glow will be seen.

DISCUSSION

Under the cold conditions used for extraction, the firefly luciferin and luciferase are both stable and are present in excess over ATP. Therefore, addition of ATP sets off the light reaction with the other reagents. The reaction requires very low concentrations of oxygen. Remember that air-saturated water contains about 0.26 mM oxygen. It is very difficult to reduce the oxygen concentration to a low enough level to show that its concentration also can control the light intensity.

The change of color on addition of the acid is explained by the scheme in Fig. 7. You will probably have to adjust the acid concentration a little for the best effect. This is best done if you have a pH meter.

EXPERIMENT 33: BACTERIAL BIOLUMINESCENCE

OBJECTIVE

It is relatively easy to culture bioluminescent bacteria from a natural source and these experiments give good training in microbiological techniques. The first test is to remove oxygen from above a culture to show again the requirement for oxygen for the bioluminescence emission. It is also relatively easy to extract luciferase from a liquid shake culture, but alternatively, a crude extract can also be purchased. The experiment will show the requirement for reduced flavin, which is generated via NADH, and for aldehyde. It is possible to distinguish a slightly green-blue color of this *in vitro* bioluminescence, corresponding to the fluorescence of the "fluorescent transient", whereas the bluer *in vivo* emission from *Photobacterium* is the fluorescence of lumazine protein.

LIST OF MATERIALS

NaCl, 3 g
Na_3PO_4, tribasic sodium phosphate, 0.6 g
Na_2HPO_4, dibasic sodium phosphate, 3 g
$(NH_4)_2HPO_4$, dibasic ammonium phosphate, 0.06 g
KH_2PO_4, monobasic potassium phosphate, 2.5 g
$MgSO_4$, 0.015 g
Bacto-peptone, 1.5 g (Difco)
Glycerol, 0.4 ml
Bacto-agar, 1.5 g
Distilled water
HCl
NaOH
NADH, 0.15 g
FMN, 0.05 g
Dodecanal, 2 drops
Steam sterilizer
Wire loop for microbiological work
Petri dishes (about 20)
Photobacterium, species (or your own culture, see preparations)
Nitrogen gas cylinder with regulator
Tubing for nitrogen
Bacterial luciferase: Bacterial luciferase can be purchased although an active extract can also be made by growing the bacteria in a liquid medium overnight, provided time is available and you have access to a steam sterilizer, flask shaker, and other bacteriological equipment.
Test tubes, 10 ml (3)
Dark room
Hot plate (or Bunsen burner and ring stand)
pH meter
Analytical balance and weighing accessories

PREPARATIONS

Microbiological Techniques: For a description of these the student should refer to any undergraduate laboratory manual, such as S.B. Primrose and A.C. Wardlaw (eds.) "Sourcebook of Experiments for the Teaching of Microbiology", Academic Press, London, 1982.

Bacterial Buffer: In 1 liter of distilled water dissolve 2.5 g of dibasic sodium phosphate and 2.1 g of monobasic potassium phosphate. Check that the pH is about 7.2. This is a standard Na/K phosphate buffer of pH around 7.2. The phosphate salts come in a variety of basicities and hydrations, so adjustments in weights must be made depending on which ones are used. The final phosphate molarity should be 50 mM.

Bacterial Nutrient Medium: Dissolve 2.5 g of sodium chloride, 0.5 g of tribasic sodium phosphate, 0.2 g of monobasic potassium phosphate, 0.05 g of dibasic ammonium phosphate, 0.01 g of magnesium sulfate, 1 g of Bacto-peptone (Difco), 0.3 ml of glycerol and 1 g of Bacto-agar in 100 ml of distilled water. Adjust the pH to 7.3 using HCl and/or NaOH as needed. Boil the mixture, and while still hot, pour about 5 ml into each of about 20 Petri dishes. Then sterilize all dishes.

"Wild" Bacteria: Allow a fish or squid to decay for several days; make sure you keep it in the fume hood! Luminous spots will appear on the surface of the fish from the contaminating bacteria and these can be sterilely transferred onto a plate. With several transfers a pure strain of these bacteria can be isolated. (See "Microbiological Techniques" above for details of transfer procedures.)

Bacterial Plates: Take the culture of bacteria you have purchased or grown and sterilely spread it on a plate that is at room temperature. The bacteria will take about three days to grow into a luminous "lawn".

NADH Solution: Dissolve 0.1 g of NADH (reduced nicotine adenine dinucleotide) in 10 ml of bacterial buffer.

FMN Solution: Dissolve 0.05 g of FMN (flavin mononucleotide) in 1 liter of water.

Dodecanal Suspension: Add 1 drop of pure dodecanal to 10 ml of bacterial buffer and shake vigorously for 1-2 min. You should obtain a cloudy suspension.

EXPERIMENTAL PROCEDURE

1. Take a plate containing the glowing bacteria and pass a stream of nitrogen into it using a piece of small tubing so that the lid can be kept almost closed. What happens to the intensity of luminescence during about 5 min of nitrogen superfusion?
2. Expose the bacteria to air again and observe the change in intensity of the luminescence. This is equivalent to the observation of Boyle using his air pump.
3. Dissolve 2 mg of bacterial luciferase in 1 ml of bacterial buffer in a test tube.
4. Add 1 drop of the dodecanal suspension, then 1 drop of FMN.
5. Take the test tube, the NADH solution, and a Pasteur pipet to a dark room (or the darkest area available) and wait about 10 min for your eyes to dark adapt.
6. In the darkened area, add 1 drop of the NADH solution to the tube prepared in steps 3 and 4.
7. Note the luminescence, both the initial and subsequent phases.

DISCUSSION

When the bacteria are depleted of oxygen the bioluminescence is extinguished. It shows that the reaction involves an oxygen oxidation. Note that as with the firefly reaction, it takes a long time to get the oxygen level down to show the effect. As soon as the plate is exposed to air the bioluminescence intensity will recover. It may even be brighter for a few seconds, since some of the other reactants have accumulated to a higher concentration.

When the NADH is added, these other reactants, notably $FMNH_2$, are increased to much higher concentration so that the light intensity will be brighter than before. It will glow constantly because the NADH is in such excess that it will supply the $FMNH_2$ at a constant level.

REVIEW QUESTIONS

1. Explain the observation that the red shift induced in the color of the firefly bioluminescence can be duplicated by adding small concentrations of Zn^{++}.
2. Suggest why oxygen is involved in all bioluminescence reactions. What implications does this have in the evolution of bioluminescence?
3. Design assays for glucose in microliter blood samples:
   i) based on a chemiluminescence reaction
   ii) using bacterial luciferase
   iii) using firefly luciferase.
4. Design a bioluminescence assay for alcohol dehydrogenase using firefly luciferase and using bacterial luciferase.
5. The quantum yield of the luminol reaction in DMSO is 0.0125 in air saturated solution. Calculate the total number of photons emitted from a sample of 1 ml of luminol at a concentration of 5 mM.

ANSWERS TO REVIEW QUESTIONS

1. The pH change affects the conformation of the luciferase. Evidently this can be duplicated by adding Zn^{++}, which is known to react with histidine and cysteine residues in proteins.
2. Oxidation reactions are very energetic and often go by radical pathways. To generate a photon of green light about 200 $kJ \cdot mol^{-1}$ of energy release is needed. Radicals also provide a ready source of free electrons for jumping, thereby producing excited states. These factors imply that oxygen would have to have been available in the atmosphere before bioluminescence evolved.
3. There are many possibilities for coupled reactions that can generate a chemiluminescent substrate and only one example will be given.
   i) Use glucose oxidase to form H_2O_2 from glucose. Then assay the H_2O_2 either by adjusting the pH to 11.6 and using the luminol-hemoglobin reaction, or at pH 7 using peroxidase-luminol.
   ii) a) hexokinase: G + ATP $\rightarrow$ G6P + ADP
      (glucose) (glucose-6-phosphate)

      b) G6P dehydrogenase:

      $$G6P + NADP^+ \rightarrow \text{D-6-phosphogluconolactone} + NADPH + H^+$$

      c) NADPH:FMN oxidoreductase:

      $$NADPH + FMN + H^+ \rightarrow FMNH_2 + NADP^+$$

d) The $FMNH_2$ is then assayed by bioluminescence reaction with bacterial luciferase.

iii) The hexokinase reaction above can be used and the change in ATP accurately determined with the firefly reaction.

4. Alcohol dehydrogenase:

ethanol + NAD^+ $\rightarrow$ acetaldehyde + NADH + H^+

Assay the NADH with the oxidoreductase-luciferase coupled reaction 3.ii c) d). The NADH could also be used to reverse the 3.ii b) a) reactions to produce ATP, which is then assayed with firefly luciferase.

5. One ml of this luminol contains 5 micromols = $5 \times 6.02 \times 10^{17}$ molecules. The total number of photons emitted is then this number of molecules multiplied by the chemiluminescence quantum yield:

$$\text{number} = 5 \times 6.02 \times 10^{17} \times 0.0125$$
$$= 3.76 \times 10^{16} \text{ photons}$$

ACKNOWLEDGEMENTS

This work was supported by a grant from the U. S. National Institutes of Health, GM-28139.

SUPPLEMENTARY READING

Lee, J. (1989) Bioluminescence. In *The Science of Photobiology*, 2nd edition (Edited by K.C. Smith), pp.391-417. Plenum Press, New York.

REFERENCES

Albrecht, H.O. (1928) Uber die Chemiluminescenz der Aminophthalhydrazid. *Z. Physik. Chem.* **136**, 321-330.

Geiger, R., R. Haubner and W. Miska (1989) New, bioluminescence enhanced detection systems for use in enzyme activity tests, enzyme immunoassays, protein blotting and nucleic acid hybridization. *Molec. Cellular Probes* **3**, 309-328.

Gundermann, K.-D. and F. McCapra (1987) *Chemiluminescence in Organic Chemistry*. Springer-Verlag, Berlin.

Harvey, E.N. (1940) *Living Light*. Princeton University Press, Princeton, NJ.

Koo, J-Y. and G.B. Schuster (1977) Chemically initiated electron exchange luminescence. A new chemiluminescent reaction pathway for organic peroxides. *J. Am. Chem. Soc.* **99**, 6107-6109.

Lee, J. (1989) Bioluminescence. In *The Science of Photobiology*, 2nd Edition (Edited by K.C. Smith), pp. 391-417. Plenum Press, New York.

Lee, J. (1985) The mechanism of bacterial bioluminescence. In *Chemi-and Bioluminescence* (Edited by J.G. Burr), pp. 401-437. Marcel Dekker, New York.

McElroy, W.D. and M. DeLuca (1985) Firefly bioluminescence. In *Chemi- and Bioluminescence* (Edited by J.G. Burr), pp. 387-399. Marcel Dekker, New York.

Meighen, E.A. (1988) Enzymes and genes from the lux operons of bioluminescent bacteria, *Ann. Rev. Microbiol.* **42**, 151-176.

O'Kane, D.J., W.L. Lingle, J.E. Wampler, M. Legocki and A.A. Szalay (1988) Visualization of bioluminescence as a marker of gene expression in rhizobium-infected soybean root nodules. *Plant Molec. Biol.* **10**, 387-399.

O'Kane, D.J., M. Ahmad, I.B.C. Matheson and J. Lee (1986) Purification of bacterial luciferase by high-performance liquid chromatography. *Meth. Enzymol.* **133**, 109-128 (1986).

Schaap, A.P., T-S. Chen, R.S. Handley, R. DeSilva and B.P. Giri (1987) Chemical and enzymatic triggering of 1,2-dioxetanes. 2: Fluoride-induced chemiluminescence from tert-butyldimethylsilyloxy-substituted dioxetanes. *Tetrahedron Lett.* **28**, 935-938.

Schaap, A.P., H. Akhavan and L.J. Romano (1989) Chemiluminescent substrates for alkaline phosphatase: application to ultrasensitive enzyme-linked immunoassays and DNA probes. *Clin. Chem.* **35**, 1863-1864.

Seliger, H.H. (1961) Some aspects of the luminol light reaction. In *Light and Life* (Edited by W.E. McElroy and B. Glass), pp. 200-205. The Johns Hopkins Press, Baltimore, MD.

Seliger, H.H. (1975) Origin of bioluminescence. *Photochem. Photobiol.* **21**, 335-361.

Slawinska, D. and J. Slawinski (1985) Low level luminescence from biological objects. In *Chemi- and Bioluminescence* (Edited by J.G. Burr), pp. 495-531. Marcel Dekker, New York.

White, E.H., F. McCapra, G.F. Field and W.D. McElroy (1961) The structure and synthesis of firefly luciferin. *J. Am. Chem. Soc.* **83**, 2402-2403.

White, E.H., O. Zafiriou, H.H. Kagi and J.H.M. Hill (1964) Chemiluminescence of luminol; the chemical reaction. *J. Am. Chem. Soc.* **86**, 940-941.

Wood, K.V., Y.A. Lam, H.H. Seliger and W.D. McElroy (1989) Complementary DNA coding click beetle luciferases can elicit bioluminescence of different colors. *Science* **244**, 700-702.

Photomedicine: Photodermatology

Brian E. JOHNSON
University of Dundee
Scotland, U.K.

INTRODUCTION

Photomedicine is concerned with any human reaction to sunlight, or UV or visible radiation from an artificial source. It, therefore, should include physiology and pathophysiology of vision, systemic effects not mediated through the skin, e.g., the treatment of winter depression, the supposed beneficial effects of sunlight on infectious disease and in vitamin D_3 synthesis, phototherapy for neonatal hyperbilirubinemia (jaundice of newborns), photodynamic therapy for tumors and possibly other diseases, and the major topic for this chapter, photodermatology.

The visual process is dealt with elsewhere in this volume. The pathological effects of UV radiation such as keratitis, due mostly to accidental exposure to short wavelength UV radiation from artificial sources such as germicidal mercury lamps, and UVA-induced cataract formation, are mentioned here in passing and further reference can be obtained from Epstein (1989). Of the other topics, only photodermatology will be considered here, and of this, only those parts which are easily amenable to laboratory demonstration will be in detail. Further information on the material outlined here can be found in the works cited in the Supplementary Reading at the end of the chapter. The most relevant material will be found in the Photomedicine chapter by Epstein (1989) in *The Science of Photobiology*.

Photodermatology – An Introduction

Photodermatology is the study of skin reactions to UV and visible radiation, their mechanisms, diagnosis, treatment and prevention. It encompasses the adverse, normal reactions such as sunburn (and suntan), sunlight induced premature aging and cancer of the skin, abnormal reactions of unknown etiology, the so-called idiopathic photodermatoses, specific genetic (DNA repair deficient xeroderma pigmentosum) and metabolic (photosensitive porphyrias) abnormalities which lead to increased photosensitivity and photosensitization by exogenous substances such as drugs, perfumes, plant materials, dyestuffs and polycyclic hydrocarbons. The beneficial effects of vitamin D photosynthesis and the use of UV and visible radiation alone or in photosensitized reactions for therapeutic purposes, especially for recalcitrant skin disease, are also included.

BRIAN E. JOHNSON, Department of Dermatology, Level 8, Polyclinic Area, Ninewells Hospital, University of Dundee, Dundee DD1 9SY, Scotland, U.K.

Photobiological Techniques, Edited by D.P. Valenzeno *et al.*
Plenum Press, New York, 1991

The major features of sunburn are a painful reddening (erythema) of the exposed skin, swelling, and even blistering which develop some hours after exposure. They are not produced if the exposure is through window glass, which acts as a filter for shorter wavelengths of light. This gives an approximate long wavelength limit of 320 nm for the skin response. Thus, although the longer wavelength UVA (320-400 nm) may have some complementary effect such as producing an early phase of erythema and, perhaps adding to or augmenting the shorter wavelength effects, the UVB (280-320 nm) between 295 and 315 nm is the essential causative wavelength region. Similarly, for most of the premature aging effects of chronic exposure, except for a UVA dependent skin sagging response demonstrated in hairless mice, the UVB of sunlight is held mainly responsible. Photocarcinogenesis is, again, mainly a UVB phenomenon although there is unequivocal evidence that sufficiently high exposure doses of UVA may also produce tumors in experimental animals. Photoaging and photocarcinogenesis can be studied in humans at the natural endpoint of the process, but experimental work is confined to animals. Studies of sunburn however, can be done on humans and consist mainly of examining the erythema response in terms of the threshold dose and dose response effects with various sources of broad spectrum and monochromatic UV radiation.

For suntanning, the UVB wavelengths are most effective. However, repeated less damaging exposures to UVA will produce a suntan but with little attendant protection against the damaging effects of UVB.

Dose-Response Study Techniques

The major investigative technique which provides the basis of any study in photodermatology is the determination of the skin reactions to the radiation concerned. Any investigation of abnormality must be preceded by the establishment of normal response levels. The laboratory studies presented in this chapter will be centered on this activity. Various sources of UV and visible radiation have been used in such studies. These include sunlight, carbon arcs, medium and high pressure mercury arcs, low pressure mercury arc lamps with different phosphor linings to the envelope to provide wavebands in the UVB, UVA and visible regions of the spectrum, and good solar simulation with a filtered xenon arc. Carbon arcs are too difficult to handle and sunlight is too difficult, in the main, to control, so that the mercury and xenon arcs are the lamps of choice. The wavelength characteristics of these lamps may be defined by the manufacturers or with emission spectroscopy. With solar simulators or broad spectrum sources, the spectrum used to irradiate skin may be varied using filters, or dispersion through a prism or grating monochromator system. The irradiance at the skin surface may be determined using a calibrated linear thermopile or other measuring device (Magnus, 1976; Johnson and Mackenzie, 1982; Holzle, *et al.*, 1987; Thomas and Amblard, 1988).

Broad band UVB may be obtained with a fluorescent tube such as the Philips TL12 or the old Westinghouse FS series with emission from approximately 280 nm to 335 nm, peaking at 315 nm. This source is useful for determining in a very simple way the normal skin sensitivity to sunburning. An irradiance at the skin surface of 3-4 $mW \cdot cm^{-2}$ allows a minimal erythema dose (MED) to be given to lightly pigmented skin in one to two minutes. Some variation in MED values with degree of skin melanin pigmentation might be expected and attempts to classify human skin according to its susceptibility to sunburn have been made (Table 1). A dose response study in terms of intensity of reaction is therefore feasible. Moreover, the lamps described here are commonly used for broadband UVB phototherapy of psoriasis and MED tests are required to determine the treatment exposure schedules.

Table 1

Classification of human skin types according to sunburn/tan sensitivity

Type I	Very sensitive, always burn easily.	Little or no tan.
Type II	Very sensitive, always burn.	Minimal tan.
Type III	Sensitive, moderate burn.	Gradual tan (light brown).
Type IV	Moderately sensitive, minimal burn.	Easy tan (moderate brown).
Type V	Variously pigmented Arabic and Asian skin with 3 main divisions of burning sensitivity, a, b and c.	
Type VI	Heavily pigmented Negroid and Australoid skin which is resistant to sunburning.	

Determinations of abnormal photosensitivity to the sunburning wavelengths of sunlight using this source are, however, restricted.

Dose-Response Studies of Abnormal Photosensitivity

For the idiopathic (unknown cause) photodermatoses, the MED for Polymorphic Light Eruption may differ little from the normal range and photoprovocation of the various types of abnormal reaction, principally itchy papules (spots), requires repeated exposures and perhaps the use of a high power UVA source. Phototesting for Hydroa vaccineforme and Actinic Prurigo with these UVB lamps has the same disadvantages. However, subjects with Photosensitivity Dermatitis/Actinic Reticuloid Syndrome (PD/AR), also known as Chronic Actinic Dermatitis (CAD) exhibit exquisite sensitivity to the UVB as well as increased UVA and even visible sensitivity so that a decrease in MED with the fluorescent tube sources is easily obtained. Moreover, the abnormal morphological response, resembling the characteristic cell mediated delayed hypersensitivity reaction of contact allergic dermatitis may be elicited with exposures little higher than the MED.

The wheal (a sudden elevation of the skin surface) and flare (reddening, spreading from the wheal) reactions of Solar Urticaria are mainly induced by the UVA and visible wavelengths so that irradiation with UVB fluorescent tubes rarely demonstrates an abnormality for this photodermatosis. However, the reactions are readily reproduced with an appropriate source such as a simple UVA producing lamp or, where the visible is effective, a slide projector. In many cases, simply for diagnostic confirmation, a controlled exposure to bright daylight proves sufficient.

In Xeroderma Pigmentosum, responses to UVB determined at 24 h may not differ from those of normal skin, but the abnormality of the response is manifest at later times after exposure, the reaction intensifying rather than fading with time and erythema developing up to 96 h after the exposure in test sites which were negative at 24 h. The UVB lamps are therefore obviously useful for phototests in Xeroderma Pigmentosum.

The cutaneous reactions in the photosensitive porphyrias are elicited by wavelengths around 400 nm in the main and responses to the UVB wavelengths are normal. A high power UVA or blue light source allows successful phototesting for the porphyrias.

Studies of the acute skin responses to UVB in mice may provide information about human sunburn, just as the results of UV-carcinogenesis experiments are extrapolated to human skin cancer. Knowledge of the nature and dose dependence of these reactions to single exposures is essential both for the comparison with human sunburn and for planning UV-carcinogenesis studies. The UVB fluorescent tubes are very suited to such work and the acute reactions may be observed as edema by the use of intravenous blue dye (erythema in mice is generally a high exposure dose reaction) or by measuring changes in ear thickness or dorsal skin fold thickness.

Phototoxicity and Photoallergy

Cutaneous photosensitization is commonly considered in terms of normally harmless doses of radiation, particularly UVA and visible, being absorbed by a "foreign" molecular species, the photosensitizer, present in the skin at non-damaging concentrations (Johnson, 1984). Skin reactions result from the involvement of cell and tissue constituents in the dissipation of the absorbed energy as in photosensitization generally. Rarely, the photosensitized reaction may be so efficient that UVB wavelengths will produce reactions with exposure doses lower than those required for sunburn. Alternatively, the reaction produced may differ in morphology and timing so that a UVB photosensitization can be detected.

Photosensitization reactions in the skin can be divided into 2 broad classifications. The acute, immediate or delayed inflammatory reactions of varying severity resulting directly from cellular damage are known as *phototoxicity*. Where the immune system is activated, the term *photoallergy* is used.

Phototoxic reactions should be produced in any subject who has sufficient photosensitizer present in the skin, and who is then exposed to high enough doses of the appropriate radiation. If a photosensitizer is distributed evenly throughout the skin, the reaction may take place at any level depending on the penetration of the active wavelengths. Photosensitizer confined to the blood produces damage restricted to the vascular endothelium. Where the photosensitizer is free within the dermis, the histamine containing mast cells are particularly sensitive. Photosensitizer applied to the skin surface produces damage to the epidermis. Immediate reactions derive from photosensitization in the dermis while the delayed, sunburn type of effects arise from damage to the epidermis alone. Complex patterns of phototoxic reaction should therefore, be expected. Nonetheless, it is possible to classify the major reaction patterns into 4 types.

Type A With coal tar, pitch and their constituent polycyclic hydrocarbons, contact with the skin and exposure to sunlight or UVA leads to a prickling or burning sensation but, as in Erythropoietic Protoporphyria, there may be no physical sign of photosensitization. If the exposure is extended, the intensity of burning increases and erythema develops in the exposed skin. With higher doses still, a wheal is produced which subsides after an hour or so with a flare reaction also fading to leave erythema restricted to the irradiated area. With the higher doses, the early phase erythema may also fade, but then becomes intense again reaching a maximum between 24 and 48 h. Hyperpigmentation may follow.

This pattern of reaction, which appears to be the major manifestation of photodynamic action in the skin, is also obtained with a number of dyestuffs and drugs (Table 2). When methylenc bluc, cosin, Rose Bengal, the polyacetylenes or alpha-terthienyl from plants of the *Asteraceae* are placed on the skin, no photosensitized reactions are obtained. However, a typical coal tar phototoxicity is obtained if they are injected intradermally or applied to scarified skin.

Type B — An exaggerated sunburn response, typical sunburn with normally harmless exposures to sunlight, or severe reactions with normally only mildly damaging exposures is produced by various drugs, but particularly moderate doses of demethylchlortetracycline and high dose chlorpromazine.

Type C — A reaction associated specifically with psoralen-induced photosensitization is seen in Phytophotodermatitis. Typically, the reactions are initiated by contact with sap from a psoralen containing plant and exposure to sunlight on a hot, humid day. The response is first seen some 24 h later as erythema, possibly painful, distributed in a pattern clearly related to contact with the plant. Blisters develop during the next 24 h which may coalesce to produce a bizarre pattern but subside relatively quickly. Where skin damage is not severe, the reaction develops as intense pigmentation which may persist for months. The intensity of erythema depends on both the amount of photosensitizer in the skin and the exposure dose. When these are minimal, only a faint erythema may develop with a latent period of 72 h or more, followed by hyperpigmentation. Repeated, sub-minimal exposures may result in hyperpigmentation only, often the case in Berlock Dermatitis due to the 5-methoxypsoralen in perfumes.

Type D — Although acute photosensitivity reactions may be obtained in Porphyria Cutanea Tarda, a major feature of this condition is an increased fragility of exposed skin leading to blistering after minor trauma. Photosensitization here may be chronic rather than acute and is also obtained with the drugs frusemide, nalidixic acid, high dose tetracycline, naproxen and amiodarone.

Phototoxic reactions may be elicited in human skin by applying the photosensitizer, preferably under occlusion (covered), to assist with its penetration through the stratum corneum, and then exposing the skin to the appropriate radiation (Frain-Bell, 1985). The reactions obtained may be recorded visually or by reflectometry and in the case of Type A (Table 2) phototoxicity, threshold pain levels may be determined. Reactions may also be obtained after administration of photosensitizing substances, usually drugs, by mouth or injection (Kligman and Kaidbey, 1982). Minimal phototoxic dose (MPD) determinations are valuable in determining the photosensitizing potential of various substances and are an important part of the European approach to the use of psoralens in photochemotherapy for psoriasis and other skin diseases, the treatment exposure schedules being based on a predetermined MPD (Wolff *et al.,* 1977).

Various animal models exist for experimental phototoxicity (Morikawa *et al.*, 1974) and quantification of the reactions may be obtained by measuring weight gain in mouse tails for instance (Ljunggren, 1984), or by measuring ear thickness as for the UVB reactions.

Photoallergy is also detected in human skin by photopatch testing, a positive reaction requiring that the subject has been immunologically sensitized earlier. This technique is

Table 2

Major patterns of cutaneous phototoxicity

Type	Skin Reactions	Photosensitizers or Diseases
A	Prickling or burning during exposure Immediate erythema; edema/urticaria With higher doses; sometimes delayed Erythema/hyperpigmentation	Coal tar; pitch; anthraquinone based dyestuffs; benoxaprofen; amiodarone; chlorpromazine; Erythropoietic protoporphyria
B	Exaggerated sunburn	Drugs such as chlorpromazine; chlorthiazides; quinine; demethylchlortetracycline
C	Late onset erythema; blisters with slightly higher doses. Low exposures give hyperpigmentation only.	Psoralens; Phytophotodermatitis; Berlock Dermatitis
D	Increased skin fragility giving blisters with trauma	Nalidixic acid; frusemide; tetracycline; naproxen; amiodarone; Porphyria Cutanea Tarda

difficult but reactions may occur in sensitized subjects with very low concentrations of photosensitizer and normal subjects produce no reactions at all. Although guinea pigs are favored in immunological studies, mice have proven useful. The photosensitization is carried out on abdominal skin, the challenge being on the ears where increase in thickness again provides a quantitative assay for the procedure.

Other Techniques in Photodermatology

The various procedures described here provide the relatively simple photobiological experimentation which is required to initiate a study. Generally, further study may consist of a direct extension of the procedures used, e.g., repeated exposures to the same radiation, for instance in the chronic exposure studies of photoaging and photocarcinogenesis. Alternatively, routine techniques of molecular biology and biochemistry may be used to study the changes in irradiated skin, as in the determination of photoproducts in cutaneous DNA, collagen or 7-dehydrocholesterol. Levels of 1,25 dihydroxycholecalciferol and 25-hydroxycholecalciferol, the active forms of vitamin D_3 may be determined in circulating blood.

Analytical pharmacological techniques may be applied to the perfusate or suction blister fluid (Black *et al.*, 1978) of irradiated skin to obtain evidence of stimulation of the arachidonic acid cascade to produce the prostaglandins or, more likely, related vasoactive agents and cytokines involved in the various inflammation processes known as sunburn and photosensitization. Histological, histochemical and histo-immunochemical studies at light and electron microscopic levels may be applied to irradiated skin specimens, just as they are used to elucidate the mechanisms in other cutaneous disease states (see for instance, Norris *et al.*, 1990). More complex animal studies involving immunologic sensitization and challenge techniques along with lymphocyte transfer have proven useful in determining the allergic nature of certain types of cutaneous photosensitization (Maguire and Kaidbey, 1983) and even more complex experimentation has revealed the importance of both cell mediated and humoral effects in photocarcinogenesis (Kripke, 1980; Roberts *et al.*, 1986).

Although complex and sophisticated experimentation is required to solve the problems of cutaneous photobiology, not one of these procedures is of use unless the results obtained can be related to what may be seen to have happened to the irradiated skin. Demonstrating that changes in the chemical constitution of dermal collagen are obtained with 20 $kJ \cdot cm^{-2}$ of UVB has no meaning if it is well known that the MED for UVB is 20 $mJ \cdot cm^{-2}$ and that a very severe sunburn reaction is obtained with only 200 $mJ \cdot cm^{-2}$. Experimental procedures designed to define the irradiation dose and wavelength dependence of the observed skin reactions are therefore not only useful in the diagnosis of the photodermatoses and in the use of UVB phototherapy and photochemotherapy for skin disease, but also remain the backbone of any investigation into photodermatology. By extension, similar studies should be available in other branches of photomedicine.

Two procedures, which differ markedly from the simple basic techniques outlined so far, have a particular role in photodermatology.

First, various models for studies of phototoxic compounds have been developed (Johnson *et al.*, 1987) and some are described elsewhere in this volume by Valenzeno (Chapter 8) and Arnason (Chapter 13). Perhaps the simplest and most significant for photodermatology, is the use of a yeast, *Candida albicans*, as first described by Daniels (1965). This model is useful as a qualitative screening test for furocoumarin type photosensitizers in plant and perfume materials and also provides a quasi-quantitative method for determining the photosensitizer content. The basic test has been ingeniously adapted for action spectrum studies by Gibbs (1987) and this adaptation has in turn been used to analyze the contribution of monoadducts and cross-links to the 8-methoxypsoralen photosensitized killing of *Candida albicans* (Baydoun and Young, 1987; Baydoun *et al.*, 1989).

Second, the association between the UVB wavelengths of solar radiation and the development of skin cancer, more than any other factor, has encouraged the study of environmental levels of these wavelengths. For epidemiologic and industrial medicine studies, a simple cheap method is required. This has taken the form of a polysulphone film badge, produced by Davis and his colleagues in the U.K. to provide a world-wide map of UVB which damages organic polymer packaging materials (Davis and Gardiner, 1982). Levels of UVB are determined from measurements of the change in optical density of the film. These film badges have been adapted for investigations of anatomical, personal and personnel variations in UVB exposure in therapeutic, industrial and vacation settings as well as in studies of UVB levels in relation to cutaneous vitamin D photosynthesis and photocarcinogenesis (Diffey, 1989).

Although the experiments described here would appear to be time consuming, if carefully set up and managed, they may all be initiated in one 3 h session. When required, the results may be observed over the following days.

EXPERIMENT 34: NORMAL HUMAN SKIN REACTION TO UVB 280-320 nm

All experiments described in this chapter involving human volunteers must be carried out under the supervision of a qualified dermatologist knowledgeable in photomedicine. Painful skin reactions may result! Local medical, legal and ethical codes must be strictly followed. Volunteer consent forms must be used (see Appendix 1).

OBJECTIVE

The reaction of normal human skin to UVB (280-320 nm) will be determined by simple phototesting. Responses will be examined visually to determine various threshold response doses and to provide subjective dose response relationships. Reflectometry will provide objective dose-response results.

LIST OF MATERIALS

Volunteers with advised consent (Appendix 1). If possible, various skin types (I, II, III, IV). (Types V and VI would not be expected to respond significantly.)
Flexible plastic sheet that is opaque to UV and visible light (~20 cm × 8 cm)
Shielding for rest of exposed skin
Protective eye glasses
Philips TL12 fluorescent tubes (emission approximately 280-335 nm, peak 315 nm) or equivalent: FL20SE Sunlamp
Calibrated linear thermopile or other method for determining irradiance. (Unless the lamps are very old, the irradiance at 20 cm should be approximately 3-4 $mW \cdot cm^{-2}$)
Reflectance erythema meter. DIA-STRON Limited, Unit 9, Focus 303, Business Centre, South Way, Andover, Hampshire SP10 5NY, U.K.; Minolta Chroma Meter or CR-200, Minolta, Osaka, Japan.

PREPARATIONS

Shutter Device: Construct the shutter device from the opaque, flexible plastic sheet. If necessary, multiple sheets may be used to ensure opacity. Cut 8 flaps (2 cm × 2 cm), leaving them attached. These are the illumination ports.

EXPERIMENTAL PROCEDURE

1. Select a test site. If possible, use the suprascapular area of the back (upper back on either side, but not including, the spinal column). If it is sunburnt or suntanned, the buttock area may be more useful. Shield all other regions of skin from accidental exposure.
2. Place the shutter device on the selected test site with all exposure ports open.
3. Close the ports sequentially to give the following exposure dose series: 14, 20, 28, 40, 56, 80, 112 and 160 s. (This exposure series is based on an irradiance of 4 $mW \cdot cm^{-2}$; if the actual irradiance is less, adjust accordingly: i.e., at 2 $mW \cdot cm^{-2}$ expose for 28, 40, 56, 80, 112, 160, 224 and 320 s).

4. Observe the test sites and record the reactions during the first hour, and ideally at 2, 4, 6, 12, 24, 48, 72 and 96 h. In practice, you may elect to record the reaction only at 24 h. Grade the reactions according to the following scale and/or using a meter.

   a. Visual Scale:

–	negative
±	doubtful
+	definite redness
++	redness with sharp margins
+++	intense redness
++++	redness with swelling

   b. Erythema meter measurements are carried out using the simple hand held reflectometer as described by Diffey *et al.* (1984). For the purposes of this experiment, the theory behind the instrument is not required. A reading of reflectance at the hemoglobin peak is obtained and may be expressed as the degree of redness or erythema index through subtraction of the reading obtained in adjacent, non-irradiated skin. Alternatively, the Minolta Chroma Meter may be used (Seitz and Whitmore, 1988).

5. From your results determine the threshold dose, i.e., the minimal erythema dose (MED). If you recorded both visual observations and meter readings, plot the visual observations versus the meter readings. How well do the two measurements correlate?

DISCUSSION

The exposure dose series used gives increments of approximately 40% and, although this is relatively large, variations in skin sensitivity are such that lower increments are probably not useful.

The results of phototesting expressed as MEDs in different skin types should reflect the variation in sensitivity to sunburning, but although there may be a general correlation from Type I to Type IV, it is usually difficult to differentiate neighboring skin types by MED determinations. Nonetheless, the results should be examined to see whether such a correlation has been obtained. If skin Types V and VI are tested, reactions, if obtained at all, will be difficult to visualize and will require high exposure doses.

Although the eye is very effective in determining a threshold response, visual grading of responses is not obviously efficient and may give rise to much variation in reports of different investigators and different laboratories. An instrumental recording of erythema may allow standardized recordings. The plot of the visual gradings against the reflectometer readings should illustrate this point. The slopes of the dose-response curves obtained with the reflectometer or with visual grading may correlate better with skin type than does the MED.

Another point about threshold dose determinations is the nature of the end point chosen. In many European centers, the MED is that exposure dose which produces a just perceptible erythema. In others and in the USA, it appears that the reaction chosen is erythema with clearly defined boundaries matching the irradiation field. The dose required for this reaction is generally 2 times that for the just perceptible erythema.

EXPERIMENT 35: NORMAL HUMAN SKIN PHOTOTOXIC REACTION (1)

All experiments described in this chapter involving human volunteers must be carried out under the supervision of a qualified dermatologist knowledgeable in photomedicine. Painful skin reactions may result! Local medical, legal and ethical codes must be strictly followed. Volunteer consent forms must be used (see Appendix 1).

OBJECTIVE

This experiment will use simple photopatch testing to study the phototoxic reaction. Photopatch testing involves application of a photosensitizer to a small patch of skin followed by illumination. An example of abnormal cutaneous photosensitivity Type A will be provided by photosensitization with two anthraquinone dyestuffs, benzanthrone and Disperse Blue 35. The reactions to be observed include prickling or burning during exposure, immediate erythema, edema/urticaria (i.e., swelling/hives) with higher doses, and sometimes delayed erythema/hyperpigmentation. Responses will be examined visually to determine various threshold response doses and to provide subjective dose-response relationships.

LIST OF MATERIALS

Volunteers with advised consent (Appendix 1)
Finn chambers for patch tests: Aluminium chambers (Epitest Oy, Hyryla, Finland) mounted on Scanpor tape (Norges-plaster A/S, Venesla, Norway). Alternately, small sticky plaster dressings may be used.
Filter paper disks
Disperse Blue 35 (ICI dyestuffs)
Benzanthrone (ICI dysetuffs). Alternatively anthracene may be used
Methanol
Blue ink
Slide projector with 250-450 W bulb
Schott OG551 cut-off filter
UVA fluorescent tubes (emission approximately 315-400 nm, peak 355 nm) Philips TL09, Sylvania TR74/09 PUVA or equivalent: General Electric F20T12 BLB or F20T9 BL)

PREPARATIONS

Disperse Blue Solution: Prepare a saturated solution of Disperse Blue 35 in methanol.

Benzanthrone Solution: Prepare a saturated solution of benzanthrone in methanol.

EXPERIMENTAL PROCEDURE

1. Apply approximately 5 μl of dyestuffs (2 sets of 7, one set each for Disperse Blue 35 and benzanthrone) and also of blue ink (1 set of 4) to separate filter paper disks in Finn chambers and stick them firmly to the skin of the forearm.

2. Leave for 1 h.

3. Remove the patches and carefully clean off excess dyestuffs and ink by wiping gently with a dry tissue.

4. Expose 4 sites of each material to projector radiation at 15 cm:

   Disperse Blue 35: 0, 1, 2, 3 min
   Benzanthrone: 0, 2, 4, 6 min
   Ink: 0, 2, 4, 6 min.

5. Expose the remaining 3 sites of Disperse Blue 35 and benzanthrone to OG551-filtered projector radiation at 15 cm:

   Disperse Blue 35: 1, 2, 3 min
   Benzanthrone: 2, 4, 6 min.

6. Record any reactions during exposure*: 1, 2, 5, 10, 20, 30, 45 min 1, 6 and 24 h after exposure.

   Visual recording to scale:

–	negative
±	doubtful
+	definite redness
++	redness with sharp margins
+++	intense redness
++++	redness with swelling

* A mild burning sensation may be experienced. If this becomes uncomfortable, terminate the exposure and note the duration.

7. Repeat the benzanthrone and Disperse Blue 35 tests using 3 patches for each and UVA lamps with an irradiance of 4-5 $mW \cdot cm^{-2}$ and exposure times of 4, 6 and 8 min.

8. Determine the natural history of the reactions obtained by following their development. Establish the lack of reaction with ink.

9. Establish threshold doses for various reactions observed. Establish waveband, rather than wavelength dependence for the reactions.

DISCUSSION

The reactions obtained here are typical of the Type A phototoxicity and are demonstrated easily with simple apparatus and methodology. In the first series, with light from the slide projector, reactions of varying intensity are obtained with Disperse Blue 35 depending on the exposure dose. There should be little difference in these reactions when the OG551 filter is placed in the light path. With benzanthrone, similar reactions may be obtained with higher exposure doses but the intensity should not be as great as with Disperse Blue 35. When the filter is in place, no reactions should be obtained with benzanthrone.

No reactions should be obtained with the ink.

When the UVA fluorescent tubes are used, reactions of varying intensity depending on the exposure dose should be obtained with benzanthrone, but no reactions should occur in the Disperse Blue 35 sites.

This experiment demonstrates in a very simple way the natural history of a Type A phototoxicity in skin. It also shows that when using projector light, the reactions obtained are not due to localized heating by light absorption since the test with ink is negative. The wavelength dependence for the phototoxicity with Disperse Blue 35 is shown to be in the visible above 400 nm and possibly above 500 nm, while that for benzanthrone is clearly limited to the UVA wavelengths.

The determination of an action spectrum for cutaneous phototoxicity is best done using an irradiation monochromator and is a painstaking and time consuming exercise. The use of specific waveband lamps is obviously helpful in establishing a general wavelength dependence for phototoxicity. When visible radiation is causative, a projector is a good source and wavelength dependence can be established with cut-off and/or interference filters. Where UV radiation is required, a xenon arc solar simulator with successive cut-off filtering or interference filters would provide a more defined wavelength dependence.

Further simple studies using this kind of test system could demonstrate the effect of: 1) photosensitizer concentration; 2) time between application of photosensitizer and irradiation and; 3) vehicle in which the photosensitizer is dissolved or suspended.

EXPERIMENT 36: NORMAL HUMAN SKIN PHOTOTOXIC REACTION (2)

All experiments described in this chapter involving human volunteers <u>must</u> be carried out under the supervision of a qualified dermatologist knowledgeable in photomedicine. Painful skin reactions may result! Local medical, legal and ethical codes must be strictly followed. Volunteer consent forms must be used (see Appendix 1).

OBJECTIVE

This experiment will use simple photopatch testing to study the phototoxic reaction. Photopatch testing involves application of a photosensitizer to a small patch of skin followed by illumination. An example of abnormal cutaneous photosensitivity Type C will be provided by photosensitization with 8-methoxypsoralen (8-MOP). The reactions to be observed include late onset erythema and blisters with slightly higher doses. Lower intensity illumination gives delayed hyperpigmentation only. Responses will be examined visually to determine the effect of varying concentration and to provide subjective dose-response relationships.

LIST OF MATERIALS

Volunteers with advised consent (Appendix 1)
Finn chambers for patch tests: Aluminum chambers (Epitest Oy, Hyryla, Finland) mounted on Scanpor tape (Norges-plaster A/S, Venesla, Norway). Alternately, small sticky plaster dressings may be used.
Pipetors capable of accurately measuring 2 μl, 0.5 ml, and 1 ml
Small capped containers, plastic or glass ~2 ml (8)
Filter paper disks
8-methoxypsoralen (8-MOP), 50 mg
Chloroform, 10 ml
Methanol
UVA fluorescent tubes (emission approximately 315-400 nm, peak 355 nm): Philips TL09, Sylvania TR74/09 PUVA or equivalent, such as General Electric F20T12 BLB or F20T9 BL
Calibrated linear thermopile or other method for determining irradiance
Analytical balance

PREPARATIONS

8-MOP Test Solutions: In a well ventilated fume hood, dissolve 50 mg of 8-MOP in 1 ml of chloroform. Be careful to keep these solutions tightly capped as much as possible. Prepare serial dilutions as follows:

	5.0		1 ml of 50		9.0 ml of methanol
To prepare	2.5		1 ml of 5		1.0 ml of methanol
8-MOP test	1.0	mg·ml^{-1} combine	1 ml of 2.5	mg·ml^{-1} 8-MOP with	1.5 ml of methanol
solution at	0.5		1 ml of 1		1.0 ml of methanol
	0.25		1 ml of 0.5		1.0 ml of methanol
	0.1		1 ml of 0.25		1.5 ml of methanol.

EXPERIMENTAL PROCEDURE

1. Further dilute each 8-MOP test solution by mixing 2 μl of each with 200 μl of methanol to prepare the series of solutions to be used in step 2.
2. Add 2 μl of each solution prepared in step 1 to duplicate series of filter paper disks (12 total) in Finn chambers, and allow the excess methanol to evaporate completely. After drying add 2 μl of distilled water to each disk.
3. Place the Finn chambers on the skin of the inside of the lower arm.
4. Leave occluded for 30 min.
5. Expose one set (6 concentrations) for 20 min at approximately 2 $mW{\cdot}cm^{-2}$ (or equivalent).
6. Record reactions during exposure*, 1, 2, 5, 10, 20, 30, 45 min, 1, 6, 24 and 48 h afterwards and at 72 and 96 h for real value of test.

   Visual recording to scale:

–	negative
+	doubtful
+	definite redness
++	redness with sharp margins
+++	intense redness
++++	redness with swelling

   * A mild burning sensation may be experienced. If this becomes uncomfortable, terminate the exposure and note the duration.

7. Determine the natural history of the reaction by following its development.
8. Determine the concentration of 8-MOP required to produce various grades of reaction.

DISCUSSION

The final amounts of 8-MOP applied to the skin in this experimental procedure are given in Table 3.

Table 3

8-MOP for Photopatch Tests

Stock dilution	8-MOP in 2μl pre-dilution	8-MOP in 2 μl post-dilution
5.00 mg/ml	0.100 μg	0.3500 $\mu g{\cdot}cm^{-2}$
2.50 mg/ml	0.050 μg	0.1750 $\mu g{\cdot}cm^{-2}$
1.00 mg/ml	0.020 μg	0.0700 $\mu g{\cdot}cm^{-2}$
0.50 mg/ml	0.010 μg	0.0350 $\mu g{\cdot}cm^{-2}$
0.25 mg/ml	0.005 μg	0.0175 $\mu g{\cdot}cm^{-2}$
0.10 mg/ml	0.002 μg	0.0070 $\mu g{\cdot}cm^{-2}$

The reactions obtained in this test are typical of phytophotodermatitis and, in the less severe type, of Berlock Dermatitis obtained with perfumes containing Bergamot Oil. The photosensitizers involved are mainly the psoralens, since it is difficult to obtain reactions with such low concentrations of the angular furocoumarin, angelicins. Moreover, the majority of the skin reactions seen are due to psoralen itself, 8-MOP (xanthatoxin) or 5-methoxypsoralen (bergapten). Examination of the concentration of 8-MOP required to produce a reaction and the knowledge that as much as 100 micrograms of 8-MOP may be extracted from only one gram of the stem of a plant like *Heracleum mantegazzianum* gives an indication of the potential for phytophotodermatitis with this plant. Similar relationships for perfumes might be established by photopatch testing with bergapten.

It is obvious from the results of these tests for phototoxicity with 8-MOP that the determination of a minimal phototoxicity dose (MPD) should be a preliminary step in the design of a dose schedule for photochemotherapy. In fact, clinical practice has allowed schedules to be derived from skin typing with a low incidence of burning accidents. Nonetheless, many groups still prefer to perform the MPD test to establish reaction levels in the skin to be treated.

EXPERIMENT 37: NORMAL MOUSE SKIN REACTION TO UVB

OBJECTIVE

Groups of mice will be exposed to different doses of UVB. The reaction of normal mouse skin will be quantified by measuring the increase in ear thickness and dorsal skin fold thickness.

LIST OF MATERIALS

License or approval from National and/or State, Local and Institutional Authority for animal experimentation
Albino mice, preferably female since they fight less (25)
Philips TL12 fluorescent tubes (emission approximately 280-335 nm, peak 315 nm) or equivalent
Calibrated linear thermopile or other method for determining irradiance
Pocket thickness gage 7309; Mitutoyo Corporation, 31-19 Shiba5-chome, Minato-ku, Tokyo 108, Japan. Alternately, any metal micrometer may be used.
Anesthetic
Waterproof markers

PREPARATIONS

Mice: Approval for animal experimentation must be obtained from the appropriate National, State, Local and/or Institutional Authority before beginning any experiments involving animals. Six days before the experiment, manually pluck the hair from the dorsal skin of the mice under light anesthesia. This will provide a large area of synchronous growth.

Irradiation: Lamp to target distance should give an irradiance at the skin surface of 1.5 $mW \cdot cm^{-2}$. This should be done with the calibrated linear thermopile or UVB meter according to the manufacturers instructions.

Mouse Identification: Divide the mice into 5 groups of 5. Use waterproof markers to mark the tail of each mouse. Use a system whereby you can easily tell which group each mouse belongs to and which individual he is within the group.

Standard Veterinary Intraperitoneal Injection: The requirements for this technique may vary from country to country. These experiments should be performed in registered premises, under supervision of qualified animal handlers. It is the experimenter's responsibility to ensure that all appropriate local and national regulations on animal use are observed.

EXPERIMENTAL PROCEDURE

1. For convenience, anesthetize each mouse with standard veterinary intraperitoneal injection before measuring ear and skin thicknesses. (**IMPORTANT:** See injection technique above.)
2. Measure and record the ear thickness of each animal.
3. Measure and record the dorsal skin fold thickness of each animal.
4. When all mice have recovered from the anesthetic, expose them to UVB according to the time and dose schedule given below. The mice can be in any standard housing

during exposure, as long as there is only a wide wire mesh between the animals and the light source.

Exposure times (s)	0,	45,	60,	90,	120
Exposure doses ($mJ \cdot cm^{-2}$)	0,	67.5,	90,	135,	189.

5. Record the reactions at 24 h by visual observation. Also measure and record the ear thickness and dorsal skin fold thickness of each animal.
6. If possible, determine the threshold response dose. Plot the thickness measurements against exposure dose.

DISCUSSION

The vasculature of mouse skin is less developed than that of human skin and an erythematous response is generally more difficult to obtain or visualize.

The edematous response which requires a blue dye for visualization can be detected through measurement of skin thickness, however, and this simple exercise allows some quantification of the reactions to UVB. The ear swelling response here may not be so well defined as that obtained in the dorsal skin. In many instances, hairless mice provide a good model for this exercise. However, by using haired mice an extra dimension is added to the model. When the hair is plucked from the flank or the dorsal skin, a new cycle of hair growth is initiated throughout the plucked area and this growth is synchronous. Damage to the DNA of the rapidly dividing hair matrix cells results in inhibition of growth which may be easily detected as soon as the non-affected hair appears above the skin surface. This reaction cannot be seen in the short term laboratory exercise but is useful for examining UV radiation effects alone and the results of photosensitization.

It is seen that mouse skin reactions to UVB from fluorescent tubes are very simply obtained and quantified to produce results similar to those for human skin erythema. Just as with human skin studies, the model here can be used to study wavelength dependency with monochromator irradiation or responses to solar simulator radiation.

EXPERIMENT 38: PHOTOTOXIC REACTIONS IN MOUSE SKIN

OBJECTIVE

Groups of mice will have photosensitizer solution applied to their ears and will then be exposed to appropriate radiation. The phototoxic reactions of mouse skin will be quantified by measuring the change in ear thickness.

LIST OF MATERIALS

License or approval from National and/or State, Local and Institutional Authority
Albino mice, preferably female since they fight less (25)
8-methoxypsoralen (8-MOP), 40 mg
Chloroform, 2 ml
Methanol, 10 ml
Benzanthrone (ICI dysetuffs)
Pipetors capable of accurately measuring 5 µl, 0.25 ml, 1 ml, and 1.75 ml.
UVA fluorescent tubes (emission approximately 315-400 nm, peak 355 nm) Philips TL09, Sylvania TR74/09 PUVA or equivalent General Electric F20T12 BLB or F20T9 BL
Small glass containers, with caps, ~ 2 ml, (3)
Calibrated linear thermopile or other method for determining irradiance
Pocket thickness gauge 7309; Mitutoyo Corporation, 31-19 Shiba5-chome, Minato-ku, Tokyo 108, Japan. Alternately, any metal micrometer may be used.
Waterproof markers
Analytical balance

PREPARATIONS

8-MOP Solutions: In a well ventilated fume hood dissolve 40 mg of 8-MOP in 1 ml of chloroform. This is the 8-MOP stock solution. Prepare the 8-MOP test solution by adding 0.25 ml of the 8-MOP stock solution to 1.75 ml of methanol.

Benzanthrone Solution: Prepare a saturated solution of benzanthrone in methanol.

Irradiation: Lamp to target distance should give an irradiance at the skin surface of 2 $mW \cdot cm^{-2}$. This should be done with the calibrated linear thermopile or UVB meter according to the manufacturers instructions.

Mouse Identification: Approval for animal experimentation must be obtained from the appropriate National, State, Local and/or Institutional Authority before beginning any experiments involving animals. Divide the mice into 5 groups of 5. Use waterproof markers to mark the tail of each mouse. Use a system whereby you can easily tell which group each mouse belongs to and which individual he is within the group.

EXPERIMENTAL PROCEDURE

1. Measure and record the ear thickness of each animal.
2. Apply 5 µl of the 8-MOP test solution (Be sure you have the test solution and not the more concentrated stock solution.) to one ear and 5 µl of benzanthrone to the other. (Ideally, plan the applications so that measurements are done without knowledge of which compound has been applied. That is, have one part of the group apply the

solutions and note which ear got which solution, and have the other part of the group make the measurements.)

3. Expose all mice in each group to UVA irradiance at 2 $mW \cdot cm^{-2}$ for: 0 min–group 1, 6 min–group 2, 8 min–group 3, 12 min–group 4 and 16 min min–group 5. This will result in a total exposure energy of 0, 0.72, 0.96, 1.44 and 1.96 $J \cdot cm^{-2}$ for each of the 5 groups. The mice can be in any standard housing during exposure, as long as there is only a wide wire mesh between the animals and the light source.

4. Measure and record the ear thickness at 5 - 10 min, 30 min, 1 h and 24 h after exposure, for each animal.

5. Plot ear thickness against radiation dose at different times after exposure to determine the natural history of the two photosensitized reactions. Determine the threshold exposure dose for each. Can you tell which ear had which photosensitizer from the reactions?

DISCUSSION

The difference between the phototoxic reactions due to 8-MOP and benzanthrone are clearly evident in human skin studies. Results with the mouse ear swelling model may not be as well defined, but the test itself is so important for studies of contact allergic reactions, and by extension, photocontact allergic reactions, that it is important to use it in the photodermatology laboratory. In this study, any reactions obtained can only be phototoxicity since the animals have not had the prior sensitization required for a photoallergic reaction to develop. Good ear swelling responses could have been induced with very low concentrations of tetrachlorosalicylanilide and UVA exposure if the mice had been sensitized 1 week earlier on the abdominal skin.

The results obtained in this study should indicate the relative usefulness of the model for phototoxicity in relation to photoallergy and also to the well established mouse tail test of Ljunggren (1984).

EXPERIMENT 39: MEASUREMENT OF ENVIRONMENTAL SUNBURNING UV RADIATION

OBJECTIVE

Environmental sunburning UVB will be monitored using polysulphone film badges. Changes in the optical density of the badges give a quantitative estimate of the available sunburning radiation and when badges are worn by students, an estimate may be made of the exposure actually encountered.

LIST OF MATERIALS

Polysulphone film badges
UV spectrophotometer
Volunteers

EXPERIMENTAL PROCEDURE

1. Record the absorbance at 330 nm of each film badge before use.
2. Have volunteers wear a badge during their daily activities (usually on the lapel area but in summer, the ventral belt area might be better).
3. If the subject sunbathes, the badge should be changed at the end of the exposure.
4. If there is no continuous exposure as such, the badge can be worn for an entire day, the absorbance at 330 nm recorded, and the badge reworn.
5. Expose a series of badges from 8:30 A.M. on a flat, unshaded surface. Change the badges every 2 h, and then during the midday, at 1 h intervals until 14:30 P.M. and then again at 16:30 P.M. each day.
6. At the end of each exposure period, measure and record the badge absorbance at 330 nm again.
7. Obtain the equivalent 305 nm exposure dose from the calibration table on the next page (Table 4) and from the formula below.

DISCUSSION

Monochromator studies have shown that the average minimal erythema dose (MED) at 305 ± 5 nm is 220 $J \cdot m^{-2}$.

The erythemally effective dose (EED) can be obtained from the following formula:

$$EED = 2000 \times (A + A^2 + 9A^3)\ J \cdot m^{-2}$$

where A = change in optical density at 330 nm. The average MED with this formula is 200 to 300 $J \cdot m^{-2}$. Compare the result obtained from the table with that from the formula. Then calculate the number of MEDs available during the day, and the number of MEDs received personally. Relate your personal experience to the results of the film badge test.

This method is very simple and, although the spectral characteristics of the film absorption change at 330 nm are not identical to the erythema, carcinogenesis or vitamin D

Table 4

Conversion of Δ OD_{330} of Polysulphone film to equivalent dose of 305 nm radiation in $W{\cdot}h{\cdot}m^{-2}$.

	0.000	0.002	0.004	0.006	0.008
0.01	0.04	0.05	0.06	0.07	0.08
0.02	0.09	0.10	0.11	0.12	0.13
0.03	0.14	0.15	0.16	0.17	0.18
0.04	0.19	0.20	0.22	0.23	0.24
0.05	0.25	0.26	0.28	0.29	0.31
0.06	0.32	0.33	0.34	0.35	0.37
0.07	0.38	0.39	0.41	0.42	0.43
0.08	0.45	0.46	0.48	0.50	0.51
0.09	0.53	0.54	0.55	0.57	0.59
0.10	0.60	0.62	0.63	0.65	0.67
0.11	0.68	0.70	0.71	0.73	0.75
0.12	0.76	0.78	0.79	0.81	0.83
0.13	0.84	0.86	0.87	0.89	0.91
0.14	0.92	0.94	0.96	0.97	0.98
0.15	1.00	1.02	1.04	1.05	1.07
0.16	1.09	1.10	1.12	1.14	1.16
0.17	1.18	1.19	1.21	1.23	1.25
0.18	1.27	1.29	1.30	1.33	1.35
0.19	1.38	1.40	1.42	1.44	1.46
0.20	1.49	1.51	1.53	1.56	1.58
0.21	1.62	1.65	1.67	1.70	1.72
0.22	1.75	1.77	1.80	1.83	1.86
0.23	1.88	1.90	1.93	1.96	1.99
0.24	2.03	2.06	2.09	2.12	2.14
0.25	2.17	2.20	2.22	2.25	2.28
0.26	2.30	2.34	2.37	2.40	2.44
0.27	2.46	2.49	2.52	2.56	2.59
0.28	2.63	2.65	2.69	2.72	2.75
0.29	2.78	2.82	2.86	2.89	2.94
0.30	2.97	3.02	3.07	3.11	3.15
0.31	3.20	3.25	3.30	3.35	3.41
0.32	3.47	3.53	3.60	3.67	3.74

To use the table, read down the first column to locate the value closest to, but not greater than, the OD difference that you have measured. Then move across the row to the column whose heading completes the OD difference determined. For example, if the the OD difference is 0.248, choose the row starting with 0.24, then scan across to the column headed 0.008, since 0.24 + 0.008 = 0.248.

photosynthesis action spectra, it provides a very good guide to the environmental levels of, and personal exposure to, the UVB radiation responsible for these cutaneous reactions.

There are some difficulties with the method. The film becomes saturated beyond about 0.3 absorption units of change. There is sufficient UVB in midsummer, at midday in Kingston, Ontario, Canada (latitude 44N; longitude 76W) to produce this change in about one hour. Badges recording available UVB must therefore be changed at hourly intervals in the midday period. If personal badges are worn during routine daily activity, the vertical disposition on the lapel region of clothing should make it possible for one badge per day to be used. However, if sitting or lying in the sun is a part of the activity, the personal badge may become saturated and will need to be changed.

The results obtained by reference to the table give an exposure which appears to be greater by an order of magnitude than that from the formula or, indeed, from considerations of the MED at 305 nm determined by phototesting. This difference is not easily explained and is a matter for discussion. It should be considered in any reference to publications which have used the film badge, and the method used to calculate the UVB fluence obtained should be clearly identified.

GENERAL DISCUSSION

The simple procedures set out in these photodermatology laboratories have been designed to allow the determination of normal skin responses to the sunburning UVB radiation and to two major forms of phototoxicity. In the process, it can be seen that relatively high doses of UVA and visible radiation alone have no effect on normal skin, and the tests demonstrated, therefore, allow some differentiation of abnormal photosensitivity in human skin. The mouse experiments should provide a basis for using mouse skin as a model for human skin reactions, both normal and photosensitized, and for the design of photocarcinogenesis experiments based on a knowledge of the severity of damage induced by a single exposure.

The use of the *Candida* test is perhaps the simplest of models for phototoxicity. It was developed specifically for photodermatology investigations and has proven value in various aspects of psoralen-induced photosensitization including the relation of cutaneous phototoxicity tests to the amount of psoralen in plants and perfumed materials. The *Candida* test is described in Experiment 19 of Chapter 12 and in Experiment 20 of Chapter 13.

The film badge method for monitoring environmental UVB is included since it has been adapted specifically for photodermatological studies for which it has provided much valuable data with a minimum of effort. If it is possible to phototest human skin as done with the UVB fluorescent tubes but with relatively constant sunlight around midday, the badge may well be used to monitor the exposure dose given in each test site.

SUPPLEMENTARY READING

Epstein, J.H. (1989) Photomedicine. In *The Science of Photobiology*, 2nd edition (Edited by K.C. Smith), pp. 155-192. Plenum Press, New York.

Fitzpatrick, T.B. (1974) Editor *Sunlight and Man*. University of Tokyo Press.

Frain-Bell, W. (1985) *Cutaneous Photobiology*. Oxford University Press, Oxford.

Harber, L.C. and D.R. Bickers (1989) *Photosensitivity Diseases: Principles of Diagnosis and Treatment.*, 2nd Edition. B.C. Decker Inc., Toronto, Philadelphia.

Magnus, I.A. (1976) *Dermatological Photobiology*. Blackwell, Oxford.

Regan, J.D. and J.A. Parrish (1982) Eds. *The Science of Photomedicine*. Plenum Press, New York.

Thomas, P. and P. Amblard (1988) *Photodermatologie et Phototherapie*. Masson, Paris.

REFERENCES

Baydoun, S.A. and A.R. Young (1987) An action spectrum for lethal photosensitization of *Candida albicans* by 8-MOP after low dose broadband UVA irradiation; an action spectrum for 8-MOP 4'5'-monoadducts. *Photochem. Photobiol.* **46**, 311-314.

Baydoun, S.A., N.K. Gibbs and A.R. Young (1989) Action spectra and chromophores for lethal photosensitization of *Candida albicans* by DNA monoadducts formed by 8-methoxypsoralen and monofunctional furocoumarins. *Photochem. Photobiol.* **50**, 753-761.

Black, A.K., M.W. Greaves, C.N. Hensby and N.A. Plummer (1978) Increased prostaglandins in E_2 and $F_{2\alpha}$ in human skin at 6 and 24h after ultraviolet B irradiation (290-320nm). *Br. J. Clin. Pharmac.* **5**, 431-436.

Daniels, F., Jr. (1965) A simple microbiological method for demonstrating phototoxic compounds. *J. Invest. Dermatol.* **44**, 259-263.

Davis, A., B.L. Diffey and T.K. Tate (1981) A personal dosimeter for biologically effective solar UV-B radiation. *Photochem. Photobiol.* **34**, 283-286.

Davis, A. and D. Gardiner (1982) An ultraviolet radiation monitor for artificial weathering devices. *Pol. Degrad. Stab.* **4**, 145-157.

Diffey, B.L. (1982) Ultraviolet radiation in medicine. *Medical Physics Handbooks* 11. Adam Hilger Ltd. Bristol.

Diffey, B.L. (1989) Ultraviolet radiation dosimetry with polysulphone film. In *Radiation Measurement in Photobiology* (Edited by B.L. Diffey), pp. 135-159. Academic Press, London, San Diego.

Diffey, B.L., R.J. Oliver and P.M. Farr (1984) A portable instrument for quantifying erythema induced by ultraviolet radiation. *Br. J. Dermatol.* **111**, 663-672.

Epstein, J.H. (1989) Photomedicine. In *The Science of Photobiology* 2nd Edition (Edited by K. Smith), 155-192. Plenum Press, New York.

Frain-Bell, W. (1985) *Cutaneous Photobiology*. Oxford University Press, Oxford.

Gibbs, N.K. (1987) An adaptation of the *Candida albicans* phototoxicity test to demonstrate photosensitizer action spectra. *Photodermatology* **4**, 312-316.

Holzle, E., G. Plewig and P. Lehmann (1987) Photodermatoses - diagnostic procedures and their interpretation. *Photodermatology* **4**, 109-114. See also P. Lehmann et al (1986) Lichtdiagnostiche Verfahren bei Patienten mit verdacht auf Photodermatosen. Zbl Haut-Geschl-Kr. 152, 667-682.

Johnson, B.E. and L.A. Mackenzie (1982) Techniques used in the study of the photodermatoses. *Seminars in Dermatology* **1**, 217. See also Magnus, I.A.(1976). Dermatological Photobiology. Blackwell, Oxford.

Johnson, B.E. (1984) Light sensitivity associated with drugs and chemicals. In *The Physiology and Pathophysiology of the Skin* Vol. 8, (Edited by A. Jarrett), pp. 2541-2608. Academic Press, New York.

Johnson, B.E., E.M. Walker and A.M. Hetherington (1987) *In vitro* models for cutaneous phototoxicity. In *Skin Models. Models to Study Function and Disease of Skin.* (Edited by R. Marks and G. Plewig), pp. 264-281. Springer-Verlag, Berlin.

Kavli, G. and G. Volden (1984) The *Candida* test for phototoxicity. *Photodermatology* **1**, 204-207.

Kligman, A.M. and K. H. Kaidbey (1982) Human models for identification of photosensitizing chemicals. *J. Natl.Cancer Inst.* **69**, 269-272.

Kripke, M.L. (1980) Immunology of UV-induced skin cancer: Yearly review. *Photochem. Photobiol.* **32**, 837-839.

Ljunggren, B. (1984) The mouse tail phototoxicity test. *Photodermatol.* **1**, 96-100.

Maguire, H.C., Jr. and K. Kaidbey (1983) Studies in experimental photoallergy. In *The Effect of Ultraviolet Radiation on the Immune System* (Edited by J.A. Parrish), pp. 81. Johnson and Johnson.

Morikawa, F., Y. Nakayama, M. Fukuda, M. Hamano, Y. Yokoyama, T. Nagura, M. Ishihara and K. Toda (1974) Techniques for evaluation of phototoxicity and photoallergy in laboratory animals and man. In *Sunlight and Man* (Edited by T.B. Fitzpatrick), pp. 529-557. University of Tokyo Press, Tokyo.

Norris, P.G., J. Morris, N.P. Smith, A.C. Chu and J.L.M. Hawk (1990) Chronic actinic dermatitis: an immunohistologic and photobiologic study. *J. Am. Acad. Dermatol.* **21**, 966-971.

Roberts, L.K., W.E. Samlowski and R.A. Daynes (1986) Immunological consequences of ultraviolet radiation exposure. *Photodermatol.* **3**, 284-297.

Seitz, J.C. and C.G. Whitmore (1988) Measurement of erythema and tanning responses in human skin using a tri-stimulus colorimeter. *Dermatologica* **177**, 70-75.
Thomas, P. and P. Amblard (1988) *Photodermatologie et Phototherapie.* Masson, Paris.
Wolff, K., F. Gschnait, H. Honigsmann, K. Konrad, J.A. Parrish and T.B. Fitzpatrick (1977) Phototesting and dosimetry for photochemotherapy. *Brit. J. Dermatol.* **96**, 1-10.

APPENDIX 1

The following is a sample Advised Consent Form. It is the responsibility of the experimentor to ensure that the consent form actually used, when performing the experiments given in this chapter, complies with all applicable national, state, local and institutional requirements. These vary from country to country and from region to region. Be sure to follow the appropriate regulations.

ADVISED CONSENT

Phototesting and photopatch testing.

The techniques of phototesting and photopatch testing have been explained to me.

I understand that small areas of skin will be exposed to ultraviolet or visible radiation, with or without previous application of a photosensitizing chemical.

The maximum area of skin involved at each exposure site will be 2×2 cm.

Skin damage liable to occur will vary from a mild reddening to blistering as in sunburn, confined to the small area irradiated, and this may be followed by some suntanning hyperpigmentation.

I understand that in some instances some prickling and burning sensation may be experienced.

I consent to involvement in the procedures so described.

Experiments Reprinted from *The Science of Photobiology*, First Edition

Kendric C. Smith, Editor

The following pages are a reprinting of about a dozen pages from the first edition of *The Science of Photobiology*. As noted in the preface, this lab manual has been designed to be a companion to the second edition of *The Science of Photobiology*. This second edition does not contain laboratory experiments. The first edition, on the other hand, contained several relatively brief sketches of experiments at the end of each chapter. Many of these are quite simple, straightforward, and instructive exercises that are still very useable. Therefore, with Dr. Smith's and Plenum's generous approval, we have reprinted those pages here.

The reprinted experiments cover many, but not all, of the sub-disciplines of photobiology presented in this book. The following list indicates the sub-disciplines covered and the author of the chapter in which the experiments appeared.

Section	Sub-Discipline	Author
4.6	Photosensitization	John D. Spikes
5.6	UV Effects	Kendric C. Smith
6.4	Environmental Photobiology	Howard H. Seliger
8.11	Chronobiology	Beatrice M. Sweeney
11.7	Photomorphogenesis	Walter Shropshire, Jr.
12.8	Photomovement	William G. Hand
13.12	Photosynthesis	David C. Fork
14.12	Bioluminescence	John Lee

4.6. EXPERIMENTS

One of the simplest demonstrations of photosensitization in biology is the photodynamic killing of paramecium (the same organism used by Raab in his discovery of photodynamic action). This experiment can be done in many ways; the following directions should be regarded only as a general outline. Prepare a 1×10^{-4} M stock solution of the dye rose bengal (MW = 974; ~10 mg/100 ml distilled water). Add 10 drops of a culture of paramecium (can be obtained at little cost from any biological supply house) to each of 3 small dishes or shallow vials numbered 1–3. Then, in dim light, add 10 drops of distilled water to dish No. 1 and 10 drops of rose bengal solution to dishes No. 2 and No. 3, and mix well. Immediately cover dish No. 3 and place in the dark to serve as the dark control. Place dishes No. 1 and No. 2 under a light source, and examine at intervals with a low-power microscope. If the dishes are placed ~5 cm from a 15-W cool white fluorescent bulb, a swelling of the cells, and a decrease in the swimming rate will be observed in ~1 min with *Paramecium caudatum;* by 2 min most of the organisms will be immobile or swimming erratically, and by 5 min the majority will be immobile, and some will have broken open. The reciprocal of the time of illumination (in minutes or seconds) necessary for the immobilization of 90 or 100% of the paramecia can be used as the rate of photodynamic immobilization. You should observe not only motility, but the morphology of the cell, the behavior of the contractile vacuole, etc. Dish No. 1 serves as a light control (without dye); when most of the animals are dead in dish No. 2, examine those in the dark control (dish No. 3). The dye solution may be too concentrated for some cultures of paramecia. If the reaction goes too fast, or if appreciable effects are observed in the dark control, try 3×10^{-5} M or 1×10^{-5} M dye.

This experiment is quite open-ended in that the rate of photodynamic immobilization, and the development of morphological changes can be measured as a function of dye concentration, dye type (methylene blue, neutral red, eosin Y, etc., can be used), light intensity, light color, temperature, etc. Also, the sensitivity of other organisms can be compared (other protozoans, *Euglena, Daphnia,* brine shrimp, nematodes, small tadpoles, small fish, etc.); bacteria could also be used, but samples would have to be plated out after successively increasing periods of illumination to determine inactivation. A very sensitive assay for carcinogenic hydrocarbon pollutants in air and water has been developed, based on the sensitization of the photodynamic immobilization of paramecia by such compounds.[75]

Another very simple, but spectacular experiment is the photodynamic hemolysis of red blood cells; after an appropriate period of illumination in the presence of a sensitizing dye the cells lyse, and the somewhat opaque tube of highly scattering erythrocytes becomes quite transparent. The experiment differs from that described above in that all solutions must be made up in isotonic saline to prevent osmotic effects on the cells. A detailed protocol for a student type of experiment on photodynamic hemolysis has been published.[76]

5.6. EXPERIMENTS

5.6.1. Photochemistry of Nucleic Acids

Hydrate Formation. Determine the absorption spectrum (200–300 nm) of uridylic acid (30 μg/ml). Irradiate at 254 nm (15-W GE germicidal lamp) in a quartz cuvet and follow the disappearance of absorption with time at 260 nm. When the absorbance at 260 nm is less than about 0.2, determine the full absorption spectrum. Adjust the irradiated solution to ~pH 1 with concentrated HCl (~0.1 ml), cover, and let stand at room temperature for 24 h, mix, and determine the absorption spectrum [R. L. Sinsheimer, *Radiat. Res.* **1,** 505–513 (1954)].

Thymine Dimer Formation. Determine the absorption spectrum (230–300 nm) of thymine (25 μg/ml). Place 250 μl of a solution of thymine (500 μg/ml) into each of several 10-ml beakers, freeze and irradiate (254 nm) in the freezer (you must start lamp while at room temperature and then put in freezer). Take samples at various times, thaw, and quantitatively transfer to a 5-ml volumetric flask. Determine the absorbance at 260 nm; when it has dropped to about 0.3, determine the absorption spectrum of the sample to confirm the loss of the characteristic absorption spectrum of thymine. Then irradiate this solution at room temperature and follow the return of absorbance at 260 nm with time. When the maximum increase in absorption has been achieved, determine a complete absorption spectrum to confirm the formation of thymine. (While frozen, the thymines are juxtaposed and the photochemical equilibrium between the formation and splitting of dimers favors dimer formation. In solution, however, the dimers are split by the reirradiation and then the monomers diffuse away and cannot be redimerized.)

The thymine dimers (after irradiation while frozen) can be chromatographed and separated from residual thymine [K. C. Smith, *Photochem. Photobiol.* **2,** 503–517 (1963)]. While the use of radioactive thymine would greatly facilitate these studies, the chromatogram can be cut into 1-cm strips (crosswise), and eluted in water, and the absorbance checked at 260 nm. The solutions suspected of containing the thymine dimer can be reirradiated as above to split the dimer, and yield thymine that will then absorb at 260 nm.

5.6.2. UV Survival Curves and Photoreactivation

Survival. Strains AB1157 (wild-type), AB1886 *(uvrA6),* AB2463 *(recA13),* and AB2480 *(uvrA6 recA13)* may be obtained from the Coli Genetic Stock Center, Department of Human Genetics, Yale University School of Medicine, New Haven, CT 06510. This experiment requires some expertise in microbiological techniques and a calibrated UV lamp at 254 nm. Approximate calibration can be achieved using 10^{-4} *M* uridylic acid (UMP) as a chemical actinometer (reference 8, p. 193). The UV fluence rate ($J\ m^{-2}\ s^{-1}$) is equal to

$$\frac{\left[\frac{A_I - A_F}{A_I}\right]\left[\frac{\text{moles (UMP)}}{\text{ml}}\right]\left[\text{ml (UMP)}\right]\left[\frac{11.9 \times 10^8\ \text{JE}^{-1}}{2537\ (\text{Å})}\right]}{[1.9 \times 10^{-2}\ \text{mol E}^{-1}\ (\Phi)]\ [(\text{sample surface area})\ \text{m}^2]\ [(\text{irradiation time})\ \text{s}]}$$

where A_I and A_F are the initial and final absorbances of UMP at 260 nm.

The survival curves in Fig. 5-12 can be approximated as follows: The cells were harvested after overnight growth in yeast extract nutrient broth and diluted 1:10 in buffer to give ~1×10^8 cells/ml. About 10 ml of cells in a 10-cm Petri dish on a shaker platform were UV-irradiated (8-W GE germicidal lamp at ~47 cm above the cells, giving a fluence rate of ~1 J m^{-2} s^{-1}) with various fluences, and appropriately diluted (estimated from Fig. 5-12) to yield about 200 survivors per plate when 0.1 ml is spread on a 10-cm Petri plate. Colonies may be counted after 24–48 h of growth.

Photoreactivation. Spread about 2×10^6 cells of strain AB2480 *(uvrA6 recA13)* per plate. Prepare 6 plates. Immediately UV irradiate 4 of the plates with 0.4 J m^2 (survival is approximately 7×10^{-6}). Place a Pyrex Petri dish lid full of water, as a UV and infrared filter, on top of 3 of the UV-irradiated plates and 1 non-UV-irradiated plate (without their plastic lids). Place two daylight fluorescent bulbs 5 cm above the agar surface of the plates and irradiate the UV-irradiated plates for 1, 5, and 10 min, and the non-UV-irradiated plate for 10 min. Incubate all plates at 37°C for 24–48 h and determine the effect of photoreactivation on survival.

6.4. EXPERIMENTS

It would be instructive to be able to measure one aspect of the growth rate of algae, the gross rate of uptake of inorganic carbon, by a radioactive tracer technique. The elements of the technique are described in pages 267–274 of reference 32. However, the following suggestions and simplifications should provide the student with a satisfactory grasp of the measurement.

Ten microcuries (μCi) of sodium bicarbonate, ^{14}C-labeled, is available from commercial suppliers under general license and can therefore be ordered for laboratory use without having to set up a specialized radioactivity facility. Most universities already have licenses for radioactive materials for research and the radioactive bicarbonate solution can be ordered through the Radiation Safety Officer. The student should have access to a liquid scintillation counter. However, if this is not available the samples that are collected on the Millipore filters can be placed in liquid scintillation vials containing 3 ml of dioxane and kept indefinitely. The vials can then be taken at any time to a laboratory where liquid scintillation counting facilities are available; any dioxane-based liquid scintillation solution can be added to the vials, and the samples counted. Since the half-life of ^{14}C is 5780 years, the decay of activity before analysis will be negiligible.

Any marine or estuarine natural water sample can be used as a source of phytoplankton. A laboratory culture of phytoplankton, if available, will also be perfectly satisfactory. As a rough rule of thumb, take a sample of the phytoplankton solution in a glass test tube into a dark room, wait 15 min to become visually dark-adapted, and irradiate the test tube with a "black light" (UV: 365 nm + 405 nm) lamp. If the red fluorescence of chlorophyll is visible, the sample will contain sufficient algae to give good results in the uptake experiments. Do not use this sample for the experiments as excessive UV radiation can kill some of the algae.

The radioactive bicarbonate should be dissolved in 0.22 or 0.45 μm (pore size) Millipore-filtered natural water or culture medium to a volume such that the activity is approximately 0.5 μCi ml^{-1}. We will be adding exactly 1 ml of this tracer bicarbonate solution to each glass bottle containing 100 ml of the algal sample. In order to keep the irradiation conditions constant, cool white fluorescent lamps should be used for the incubation. All sample bottles including the dark bottles should be filled with 100 ml of the phytoplankton solution, and kept at a fixed distance (~15 cm) from a bank of two fluorescent lamps for at least 1 h prior to addition of the radioactive tracer bicarbonate solution, in order to acclimate the phytoplankton to the maximum light intensity to which they will be exposed during the experiment. Different light intensities are achieved by wrapping one, two, and three layers of ordinary plastic fly-screening material (as a neutral density filter; Section 1.3.2.) around the "light" bottles. Samples irradiated at 0, 1, 2, and 3 layers of screening and the "dark" bottles should be run in triplicate. The normalization is obtained as follows: 0.1 ml of the tracer bicarbonate solution is added directly to a liquid scintillation vial to which is also added a 0.22 μm (pore size) Millipore filter through which 100 ml of the cold phytoplankton solution is filtered. This is done so that any "color" added to the liquid scintillation solution by algal pigments that may "quench" the scintillations produced by the ^{14}C beta rays can be corrected for. The incubation time should be between 1 and 2 h (the time of incubation is measured from the time of addition of the radioactive bicarbonate solution and should be the same for all samples). The samples should then be placed in dim light and filtered on 0.22-μm Millipore filters as soon as possible. Those filters containing the phytoplankton must be washed twice with 50 ml each of 0.22 or 0.45 μm-filtered sample water to insure that the only activity remaining on the filter is that contained *within* the algae. When the filters in the scintillation vials are counted in the liquid scintillation counter, the fractional uptake of radioactivity due to photosynthesis in the light bottle will be

$$f_L = \frac{\text{Net counts/min of incubated sample}}{\text{Net counts/min of normalized sample}} \times 10 \tag{6-17}$$

The factor of 10 is due to the addition of 0.1 ml to the normalization scintillation vial. This value must be corrected by the small amount of "dark" uptake so that the true fractional uptake is $f_L - f_D$. The true fractional uptake should then be plotted as a function of the number of absorbing screens, in order to verify that there is an approach to saturation of photosynthesis with light intensity.

For the more enterprising student the techniques described in reference 32 can be used to measure chlorophyll concentrations and carbonate alkalinity of the phytoplankton samples. If these are obtained it is possible to specify two additional parameters. The assimilation number is defined as

$$Z = \frac{\mu\text{g carbon taken up/h}^{-1}\text{ per liter}}{\mu\text{g chlorophyll } a \text{ per liter}} \tag{6-18}$$

Using a ratio of 50 (g carbon) /(g chlorophyll a) it is possible to estimate the gross growth rate constant of the phytoplankton

$$K = \frac{Z}{50}\,\text{h}^{-1} \tag{6-19}$$

8.11. EXPERIMENTS

Leaf "Sleep Movements" in the String Bean *Phaseolus vulgaris*.

Part I. Use a young (2–3 weeks old) bean plant with the first two leaves well expanded. Since the measurements of a circadian rhythm require more than a few hours' time, take the bean plant home with you, and put it on the window sill or in some propitious place with as much light as possible. Water this plant whenever the pot is dry. Turn it around every day or so in order to prevent too much phototropic bending if the light is unidirectional.

Measure the position of the two primary leaves as often as convenient for 2 days, but at least when you get up, around lunch time, and just before going to bed. Make the measurements as shown below, using a protractor or other device for measuring angles:

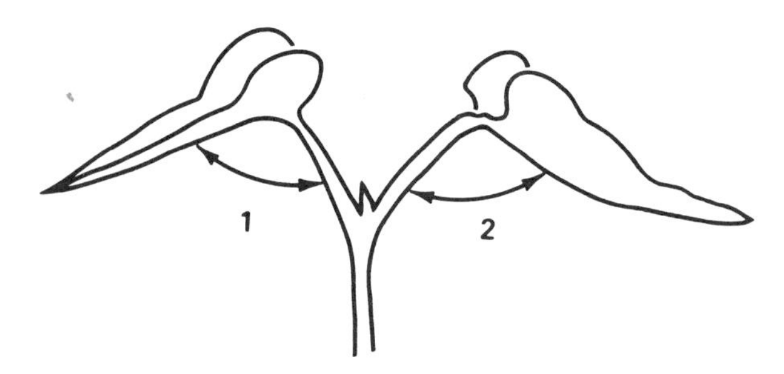

Part II. In the evening, at the end of the 2 days indicated in Part I, place the bean plant in constant darkness or in continuous artificial light for 2 days and continue measurements.

Part III. Try exposing your plant to sound or other environmental changes while in continuous light. Record and plot your measurements as a function of time in hours. Write a short report of your findings (see reference 8, pp. 1–12, 19–21 and reference 9, pp. 507–530).

11.7. EXPERIMENTS

11.7.1. Experimental Procedures

Experiment 1. The simplest demonstration of photomorphogenic responses is to compare young seedlings of beans or peas grown in complete darkness with those grown in the light. A number of differences, such as pigmentation, stem length, and leaf area, can be observed after several-days growth.

Fill small flower pots three-quarters full with water-soaked vermiculite, and allow to drain. On the surface, sow 6 to 12 seeds of bean or pea, cover with 1 cm of moist vermiculite, and maintain at ~25°C. Maintain a relative humidity greater than 80% by covering the pots with large inverted beakers. Place one set covered with black cloth in a darkroom to ensure no light exposure, and the other set in the laboratory under white fluorescent lamps (intensity > 100 footcandles) for 6 to 7 days.

Most of the differences observed are photomorphogenic responses. If dry weights of seedlings grown under the two conditions are compared, the differences are slight because seedlings have relatively large food reserves in the seed, and thus are not dependent on photosynthesis during this time period.

Experiment 2. The photomorphogenic responses may be measured quantitatively by using physiological responses such as hook opening or leaf disk expansion in bean seedlings.

If bean seedlings are grown in the dark, as outlined above, after 6 days a hypocotyl hook may be excised by cutting the stem portion just below the cotyledons into 2.5-cm-long sections under a dim green safelight. These sections are placed in Petri dishes on moist filter paper where they may be exposed to low irradiance red light (0.1 W m^{-2}) and the angle of hook opening determined by measuring with a protractor. Dark grown hooks at the beginning of the experiment have the short arm and long arm of the hook nearly parallel. After red light exposure, the cells on the inner region of the hook expand and the angle between the short arm and long arm is proportional to the fluence. For exposure to 1.0 mW m^{-2} for 20 h an opening of about 65° is observed.[55] If brief red exposures (1 kJ m^{-2} up to 30 min) are followed by exposure (0.3 kJ m^{-2}) to far-red energy, the hook opening is reduced.

If 5-mm disks of leaf tissue are cut with a cork borer from dark-grown leaves and placed on filter paper wetted with 5.8×10^{-2} *M* sucrose and 1×10^{-2} *M* KNO_3, they will expand. The diameters of the disks may be measured with a millimeter rule, or with a calibrated eyepiece in a dissecting microscope. The disks should be cut from both sides of the midrib of leaves, and after placing them in Petri dishes, exposed to red or far-red radiant energy for 10 min. They are then placed in metal cans (coffee cans work well), and kept in the dark at ~25°C for 3 days. Typical values obtained by students[56] are

Treatment	Diameter of disks (mm)
Far-red	6.47 ± 0.31
Far-red, red	8.01 ± 0.37
Far-red, red, far-red	6.77 ± 0.45

Both of these experimental materials demonstrate the red, far-red reversibility of the phytochrome system in a quantitative manner. Similarly, both are amenable to adding exogenous materials, such as auxins, nucleotides, or light-mimicking chemicals (e.g., cobalt ions), to study the interaction with light exposures.

Experiment 3. Seed germination also affords a striking system for demonstrating the control of the phytochrome pigment. Lettuce *(Lactuca)* and mustard *(Arabidopsis)* seed have most often been used, but other seed may be tested in the same manner, and many demonstrate similar responses. (*Arabidopsis* seed may be obtained from Prof. A. R. Kranz, Fachbereich Biologie (Botanik) der Universität, Siesmeyerstr. 70, D-6000 Frankfort/Main, West Germany.)

Seed are imbibed by placing them on filter paper moistened with 10^{-3} *M* KNO_3. (Typically three sheets of Whatman #1 filter paper, 10 ml 10^{-3} *M* KNO_3, and 50 *Arabidopsis* seed are distributed over the surface of 10-cm diameter Petri dishes.)[8] Wrap the dishes in black cloth and place them in a refrigerator (2–4°C) for 48 h. The dishes should be handled in complete darkness and exposed to red or far-red radiant energy either alone or in various sequences. Select your own fluence values or use comparable fluences to those used in Experiment 2. Seed are particularly good for measuring fluence-response curves by exposing for times ranging from 1 to 10,000 s. After exposure, keep the dishes in the dark for 4 days, in the case of *Arabidopsis*, and 2 days for lettuce at 25°C, and count the number of germinated seeds.

Experiment 4. Pigment production under the control of the phytochrome system may be demonstrated by measuring the amount of the red pigment anthocyanin produced in young mustard seedlings.[19] Mustard seed (Sinapis) are placed on water-moistened filter paper in glass storage jars (5.0 cm deep) or in small transparent plastic refrigerator storage boxes. Thirty-six hours after sowing and growing in the dark, the seedlings are exposed to radiant energy. An effective red radiance is 0.7 W m^{-2} and far-red is 3.5 W m^{-2}. Five-minute exposures give good responses when assayed 3 to 24 h after exposure. Select 25 seedlings and extract their hypocotyl and cotyledons in 30 ml of propanol–HCl–H_2O (18:1:81 v/v/v) by immersing extraction vials (50-ml glass scintillation vials with plastic snap caps work well) with seedlings in boiling water for 1.5 min. The capped vials are clamped in a metal rack to prevent the tops from blowing off and allowing excessive loss of solvent during boiling. The vials are allowed to cool at room temperature for 24 h to allow the extractions to proceed to completion, are centrifuged for 40 min at about 5000*g*, and the absorbance measured at 535 and 650 nm. A value for corrected absorbance is calculated as

$$A_{535} - 2.2A_{650} = \text{corrected } A_{535} \quad (11\text{-}13)$$

This system is very good for demonstrating the high-irradiance response. The time course of anthocyanin produced under continuous irradiation for both red and far-red exposures measured over a 12-h period is clearly a function of the irradiance levels used.[19]

This experiment may be modified for rapid qualitative results both by shortening the extraction times and filtering the extracts before measuring absorbances. Similarly, this system is a good one to demonstrate the validity of the reciprocity law (Bunsen–Roscoe) by determining the amount of anthocyanin produced by varying both time and irradiation fluence [(intensity) × (time) = k (a constant)].

Experiment 5. Photomorphogenesis under blue radiant energy may be demonstrated by measuring the photoinduction of the yellow carotenoid pigments in the fungus *Neurospora*. (Strains of *Neurospora crassa* can be obtained from the Fungal Genetics Stock Center, Humboldt State College, Arcata, California.) Wild-type cultures are maintained on 2% agar slants on Vogel's minimal medium.[57] A few drops of an aqueous suspension of conidia of the wild-type strain Em 5297a are added to each of several Erlenmeyer flasks (125 ml) containing 20 ml of Vogel's minimal medium supplemented with 0.8% Tween-80. The flasks are placed in the dark for 4 days at 25°C, or 6 days at 18°C. Under a red safelight (GE-BCJ 60-W incandescent lamp) the mycelial pads, which have grown from conidia, are poured out of the flasks, spread out on filter paper, filtered with mild suction on a Büchner funnel, and placed in 15-cm diameter Petri dishes with three pads per dish. The pads are floated on fresh medium (2–8 ml per pad), and equilibrated for 2 h in the dark at the temperature that is to be used during the subsequent light treatment. The pads are exposed to blue light (0.1 W m^{-2}), and allowed to grow at 25° or 6°C in the dark for 24 h. They are then extracted twice in 4 vol methanol per unit of fresh weight for 15 to 20 min. The methanol extracts are pooled. Then extract several times with four vol acetone until all pigment is removed from the mycelium. Combine all extracts, and add an equal volume of aqueous 5% NaCl. In a separatory funnel, extract the pigment by adding hexane (0.1 of the total volume). Pool the hexane extracts and dry over sodium sulfate. Decant off the hexane and measure the absorbance at 475 nm. The appearance of pigment occurs more rapidly at 25°C than at 6°C, but more total pigment is produced at the lower temperature during the 24-h development period.[32]

11.7.2. Source Materials and General Comments

For any photomorphogenic experiments, the sources of radiant energy are very important. For precise quantitative experiments, instrumentation for measuring the irradiance precisely is needed (radiometers, calibrated against standard lamps), as well as careful control of the spectral bandwidth employed (interference filters or optical dispersion systems, such as grating or prism monochromators). However, much information may be obtained by using broadband filters that are inexpensive and easily constructed. For example:

Red source: cool-white fluorescent lamps wrapped with several thicknesses of commercial red cellophane.[56]

Far-red source: incandescent reflector flood lamp mounted 15 cm away from several thicknesses of commercial red and blue cellophane. Heat can be removed from the system by passing the radiant energy through 10 cm of water in a large beaker before it passes through the cellophane filters. A simple wooden or cardboard box can be constructed to enclose the light source.

If a spectrophotometer is available, transmission of the filters should be measured. Good-quality red, far-red plastic filters may be obtained from Rohm and Haas, Chemische Fabrik, 61 Darmstadt, West Germany (U.S. Agent, B. E. Franklin, 421 Pershing Drive, Silver Spring, MD 20910) or if precise experiments are planned, interference filters should be purchased (see Section 1.3.2.).

12.8. EXPERIMENTS

12.8.1. Phototropism

Phototropism can be easily demonstrated using coleoptiles of *Avena*, the oat plant. Seeds may be procured from a biological supply house.

Germinate the seedlings in the dark for 24 h at 22–24°C prior to use. During the 12th hour of germination, subject the seedlings to a 5-min period of red illumination. After the 24-h period, transfer the seedlings to a beaker or tray lined with wet filter paper. Insert the seedlings through the filter paper, so that the paper holds the coleoptile erect, with the root down in the water. Allow the coleoptiles to continue to grow until they are approximately 30 mm high (3 to 5 days). It is important that this entire procedure be conducted in darkness or in a green safelight.

Subject the coleoptiles to differing periods (30 s, 1 min, 5 min, 15 min, 1h) of directional incandescent illumination (60-W lamp), and note the result. Devise a method of quantifying your observation, and plot the effect of illumination.

Further experimentation involving the causal agent (auxin) of phototropism may be conducted using the method devised by Went. For a detailed description of this experiment see A. Dunn, and J. Arditti, *Experimental Physiology,* Holt, Rinehart, Winston, New York (1968).

12.8.2. Phototaxis

A simple demonstration of photoaccumulation may be achieved using a young culture of the flagellate *Euglena gracilis*.

A unialgal suspension of *Euglena* may be prepared by obtaining a starter culture from a biological supply house and adding to 2–3 liters of Difco *Euglena* medium (broth). This suspension should be placed in a well-lit area for approximately 1 week until it turns a bright green. Pour some of the cell suspension into a large cuvet (20 × 20 cm) made from two glass plates cemented together with silicone cement. Illuminate the cuvet from one side with an image projected from a slide projector. Allow 5 min for the accumulation to take place, then turn off the slide projector. The image projected should be "outlined" in *Euglena.* The image will then fade as the cells randomize.

In order to determine the spectral sensitivity of the response, color slides may be introduced, these being made by photographing a spectral chart from a physics or chemistry textbook. Note where the greatest accumulation occurs, correlating it directly with the position of the spectral display on the cuvet.

In order to determine whether the accumulation is due to phototaxis or photokinesis, a direct examination of the cells with a microscope is advised. The basic procedures are discussed in Sections 12.3. and 12.4.

13.12. EXPERIMENTS

13.12.1. The Separation and Identification of the Major Leaf Pigments

In this experiment, the major photosynthetic pigments are extracted from leaves, separated from each other, and identified by their absorption and fluorescence characteristics as well as color and position on the chromatogram.

Homogenize about 2 g of leaves and a "pinch" of calcium carbonate in about 10 ml of chilled (refrigerator) 100% acetone in a Waring blender or in a mortar and pestle. Filter the homogenate on a Buchner funnel.

The chromatographic separation of the pigments can be conveniently done on chromatography paper (about 4 cm wide and 20 cm long) or on a cellulose thin layer plate such as Eastman Kodak Chromatogram Sheet for thin layer chromatography (13255 Cellulose without fluorescent indicator, No. 6064, 20 × 20 cm).

Prepare the chromatography plate or paper by allowing 8% vegetable oil in petroleum ether (50–105°C) to run until the front is about 1 cm from the edge. Dry the plate or paper thoroughly.

Streak the acetone extract with a small pipet or medicine dropper repeatedly along the edge of the chromatography plate that is free of the vegetable oil. Let the acetone evaporate between each application. Perform this operation in dim light to avoid light degradation of the pigments, and repeat this procedure until a dark band is obtained.

Develop the chromatogram in a methanol, acetone, water solvent (15:5:1) in darkness. For this purpose the solvent can be put in the bottom of a 25-cm high thin-layer chromatography tank or other suitably sized beaker or glass container. The end of the chromatography paper containing the green band of pigments can be dipped into the solvent (but not so deep as to cover the green band) and held upright by hanging the paper or leaning the chromatography plate against a wire taped in place across the top of the tank or beaker. Cover the tank with a glass plate or the beaker with a Petri plate, and develop the chromatogram in darkness until good separation is obtained (about 45–75 min).

The chromatogram may be examined for fluorescence of the pigments under long-wavelength UV radiation (360 nm). What pigments are fluorescent? Explain why the fluorescence is red.

Note the colors and relative amounts of the green and yellow bands. The pigment bands can be identified by their position on the chromatogram,[83] and can be isolated by scraping off or cutting out the band and eluting the pigment in about 10 ml of acetone.

If a suitable spectrophotometer is available the absorption spectra of the isolated pigments can be measured. If desired, the pigments can be transferred from acetone to other solvents by means of a separatory funnel, and spectra compared to published curves.[6]

(The details of this experiment were provided courtesy of Dr. Joseph Berry and Ms. Susan Reed.)

13.12.2. Hill Reaction With DCIP

Wash and blot dry about 10 g of spinach leaves. Tear off and discard the large central midribs, and cut the remaining parts of the leaves into several small pieces so that they can be ground up in a chilled mortar and pestle with about 60 ml of chilled 0.4*M* sucrose, 0.05*M* KH_2PO_4-Na_2HPO_4 buffer pH 7.2[84] and 0.01*M* KCl.

Filter the green juice through a layer of "Miracloth" (Cal Biochem) or through 8 layers of cheesecloth, and centrifuge at about 200*g* for 2 min. Centrifuge the supernatant at about 1000*g* for 10 min (a clinical centrifuge can be used at full speed for this spin). Discard the supernatant and resuspend the pellet (choroplasts) in about 10 ml of the buffer solution. Choroplast preparation should be performed at ~ 0°C.

The reduction of DCIP can be measured by adding about 0.5 ml of the chloroplast suspension and 0.5 ml of a $5 \times 10^{-4} M$ stock solution of DCIP (0.145 mg DCIP/ml water) and enough of the buffer solution used to prepare the chloroplasts to bring the volume to 10 ml. These solutions can be added to colorimeter tubes or to test tubes and exposed to a 100-W bulb at a distance of about 15 cm. They may also be conveniently illuminated by placing the samples in the beam from a slide projector. To serve as the dark control, one tube can be covered with aluminum foil to exclude light. All the tubes can be put in a large beaker of water to maintain the desired temperature (usually about 20°C).

The disappearance of the blue color results as a consequence of the reduction of DCIP by the electrons that are obtained from the oxidation of water by the operation of photosystem 2 of photosynthesis (Section 13.6.2.2.). The Hill reaction can be measured quantitatively by determining the change in optical density at 600 nm in a spectrophotometer. The extinction coefficient of DCIP at this wavelength is 21.8 mM^{-1} cm^{-1}[85] (i.e., at 600 nm a 1-m*M* solution of DCIP in a 1-cm pathlength cuvet will produce an optical density of 21.8). If the concentration of chlorophyll is determined,[86] then the activity of chloroplasts can be expressed as micromoles of DCIP reduced per milligram chl per hour.

The effect of an "uncoupler" of phosphorylation on electron transport (Section 13.7.2.) can be demonstrated if a substance such as methylamine (neutralized with HCl) is added to the reaction mixture to a concentration of about 20 mM.

The effect of the herbicide DCMU (Fig. 13-5) on the Hill reaction can be investigated by adding this compound to the reaction mixture before illumination. For this purpose a 10^{-3} M stock solution of DCMU can be made by dissolving 0.23 mg/ml of DCMU in ethyl or methyl alcohol. Adding 0.1 ml of this solution to the 10 ml of the reaction mixture should give strong inhibition of the Hill reaction.

The effect of temperature on the Hill reaction can be investigated by running the reaction in water baths held at various temperatures, and, if a means of measuring the relative or absolute light intensity is available (photocell, light meter or thermopile), the light curve (at one or different temperatures) showing the rate of DCIP reduction as a function of light intensities can be investigated. It is also possible to determine the effect of different wavelengths of light on the Hill reaction. For this purpose, a slide projector can be adapted to accept square colored glass or interference filters. If heat production is not too great, it is possible to use Kodak Wratten gelatin filters mounted in glass to provide actinic light of appropriate wavelengths.

14.12. EXPERIMENTS

A number of simple demonstrations can be carried out without access to elaborate laboratory facilities. The firefly reaction can be used to demonstrate the ATP assay, the oxygen requirement, and the pH effect on the spectra.[14,19] With the luminous bacteria, an oxygen effect and the NADH reaction can be demonstrated. The light emission can be seen by the dark-adapted (10 min) eye, but for more quantitative measurements some sort of photometer should be used, such as a fluorometer with the excitation lamp turned off, or a scintillation counter operated in the out-of-coincidence mode. Quantitative experiments can be developed if the student has access to a refrigerated centrifuge and a pH meter.

Fireflies may be easily captured, or dried whole specimens may be obtained from several chemical supply companies (Worthington Chemical Corp., Freehold, New Jersey; Sigma Chemical Co., St. Louis, Missouri). Take about 10 dried fireflies, cut off the tails, and grind them with about 1 g of clean sand and 3 ml of buffer solution. This solution is made by dissolving 0.4 g glycine and 0.1 g ammonium bicarbonate in 100 ml of distilled water. It should have a final pH of about 7.6. Grinding should be done for about 10 min, and then the solution should be filtered (or better if available, centrifuged). The filtrate is the crude extract. All these procedures should be carried out under ice-cold conditions. Divide the extract into about three portions in test tubes. Make up a solution of ATP by dissolving 0.1 g in 100 ml of buffer, and add to this 0.01 g $MgSO_4$ (or other soluble Mg salt). All the light reactions must be carried out at room temperature. To a portion of extract, rapidly add 1 ml of the ATP solution. A flash of yellow light will be observed.

Bubble nitrogen gas for about 5 min into a 1-ml sample of ATP solution and also a second portion of the firefly extract, and then pour the solutions together rapidly while still bubbling with nitrogen. The nitrogen removes oxygen from the solutions and only after agitation for a few seconds in air will there be enough oxygen returned to induce the light emission. Add one drop of 10% concentrated acetic acid to 1 ml of the ATP solution, and then add this to the last portion of extract. A weak red glow should be seen (see Fig. 14-2).[14]

A culture of luminous bacteria can be purchased from the American Type Culture Collection, 12301 Parklawn Drive, Rockville, Md. 20852; ATCC 11040, *Photobacterium phosphoreum*, grows well. Alternatively a wild strain can be isolated from the surface of a decaying fish or squid. Standard microbiological techniques can be used to transfer and grow the bacteria on a Petri plate containing sterile media of the following composition (100 ml): sodium chloride 2.5 g, sodium phosphate tribasic 0.5 g, potassium phosphate monobasic 0.2 g, ammonium phosphate dibasic 0.05 g, magnesium sulfate 0.01 g, Bacto-peptone ("Difco") 1 g, glycerol 0.3 ml, Bacto-Agar 1 g. The final pH of this mixture should be around 7.3. About 3 days after the inoculation, the surface of the plate should be covered with the growth of the luminous bacteria. If a stream of nitrogen is passed into the Petri plate, keeping the lid as tightly closed as possible, the light will dim perceptibly, and then return when reexposed to air.[17]

Bacterial luciferase may be purchased from the above-mentioned chemical companies. Make a solution of reduced nicotine adenine dinucleotide (NADH) by dissolving 0.1 g in 10 ml of buffer, another solution of flavin mononucleotide (FMN) by dissolving 0.05 g in 1000 ml of water, and then a suspension of dodecanal in water by taking 1 drop of the pure dodecanal and shaking it vigorously in 10 ml of buffer for 1–2 min to obtain a cloudy suspension. To 2 mg of bacterial luciferase dissolved in 1 ml of buffer in a test tube, add one drop of the dodecanal suspension, then 1 drop of the FMN. In a darkened room, add to this 1 drop of the NADH solution. A blue flash will be seen, not as bright as the firefly light, but it then settles down to a steady glow that continues for many minutes.

REFERENCES

Section 4.6

75. Epstein, S.S., M. Small, J. Koplan, N. Mantel, and S.H. Hutner (1963) Photodynamic bioassay of benzo[a]pyrene with Paramecium caudatum. *J. Natl. Cancer Inst.* **31**, 163-168.

76. Hoar, W.S. (1968) Some effects of radiation on cells. I. Photodynamic action. In *Experiments in Physiology and Biochemistry*, Vol. 1. (Edited by G.A. Kerkut), pp. 132-135. Academic Press, London and New York.

Section 6.4

32. Strickland, J.D.H. and T.R. Parsons (1968) *A Practical Handbook of Seawater Analysis*, Fish Res. Bd. Canada Bull. No. 167.

Section 8.11

8. Sweeney, B.M. (1969) *Rhythmic Phenomena in Plants*, pp. 1-12, 19-21. Academic Press, London.
9. Withrow, R.B. (Editor) (1959) *Photoperiodism and Related Phenomena in Plants and Animals*, Publication No. 55, AAAS, Washington, D.C.

Section 11.7

8. Shropshire, W. Jr., W.H. Klein and V.B. Elstad (1961) Action spectra of photomorphogenic induction and photoinactivation of germination in *Arabidopsis thaliana. Plant Cell Physiol.* **2**, 63-69.
19. Lange, H., Shropshire, W. Jr. and H. Mohr (1971) An analysis of phytochrome-mediated anthocyanin synthesis. *Plant Physiol.* **47**, 649-655.
32. DeFabo, E.C., R.W. Harding and W. Shropshire, Jr. (1976) Action spectrum between 260 and 800 nanometers for the photoinduction of carotenoid synthesis in *Neurospora crassa. Plant Physiol.* **57**, 440-445.
55. Klein, W.H. (1963) Some responses of the bean hypocotyl. *Am. Biol. Teach.* **25**, 104-106.
56. Powell, R.D. (1963) A simple experiment for studying the effect of red and far red light on growth of leaf disks. *Am. Biol. Teach.* **25**, 107-109.
57. Vogel, H.J. (1956) A convenient growth medium for *Neurospora* (medium N). *Microb. Genet. Bull.* **13**, 42.

Section 13.12

6. Vernon, L.P. and G.R. Seely (Editors) (1966) *The Chlorophylls.* Academic Press, New York and London.
83. Goodwin, T.W. (Editor) (1965) *Chemistry and Biochemistry of Plant Pigments.* Academic Press, New York and London.
84. Dunn, A. and J. Arditti (1968) *Experimental Physiology.* Holt, Rinehart, and Winston, New York.
85. Armstrong, J. McD. (1964) The molar extinction of 2,6-dichlorophenol indophenol. *Biochim. Biophys. Acta* **86**, 194-197.
86. Arnon, D.I. (1949) Copper enzymes in isolated chloroplasts. Polyphenoloxidase in *Beta vulgaris. Plant Physiol.* **24**, 1-15.

Section 14.12

14. Seliger, H.H. and W.D. McElroy (1965) *Light: Physical and Biological Action.* Academic Press, New York.
17. Harvey, E.N. (1952) *Bioluminescence.* Academic Press, New York.
19. Strehler, B.L. (1968) Bioluminescence assay: Principles and practice. *Methods Biochem. Anal.* **16**, 99-181.

APPENDIX A

The following tables of symbols, units and conversions have been derived from those published in *Photochemistry and Photobiology*. They were compiled by the current editor, Pill-Soon Song, and are printed here with his generous permission. SI units are considered to be the standard, and conversion factors are provided for common units not part of the SI system.

NAMES AND SYMBOLS OF SI UNITS

Physical quantity	Name of unit	Unit symbol
Photometric units		
luminous intensity	candela	cd
power, radiant flux	watt	W (J/s)
fluence	kilojoule per sq meter	kJ/m^2
fluence rate	watt per sq meter	W/m^2
luminous flux	lumen	lm (cd·sr)
luminance	candela per sq meter	cd/m^2
illuminance	lux	lx (lm/m^2)
wave number	reciprocal meter	m^{-1}
radiant intensity	watt per steradian	W/sr
Basic units		
length	meter	m
mass	kilogram	kg
time	second	s
electric current	ampere	A
temperature	kelvin	K
amount of substance	mole	mol

Derived units

area	square meter	m^2
volume	cubic meter	m^3
frequency	hertz	Hz (s^{-1})
density	kilogram per cubic meter	kg/m^3
speed	meter per second	m/s
angular speed	radian per second	rad/s
acceleration	meter per second squared	m/s^2
angular acceleration	radian per second squared	rad/s^2
force	newton	N ($kg \cdot m/s^2$)
pressure	pascal	Ps (N/m^2)
kinematic viscosity	sq meter per second	m^2/s
dynamic viscosity	newton-second per sq meter	$N \cdot s/m^2$
work energy, quantity of heat	joule	J (N·m)
electric charge	coulomb	C (A·s)
voltage, potential difference, electromotive force	volt	V (W/A)
electric field strength	volt per meter	V/m
electric resistance	ohm	Ω (V/A)
electric capacitance	farad	F (A·s/V)
magnetic flux	weber	Wb (V·s)
inductance	henry	H (Wb/A)
magnetic flux density	tesla	T (Wb/m^2)
magnetic field strength	ampere per meter	A/m
magnetomotive force	ampere	A
entropy	joule per kelvin	J/K
specific heat	joule per kilogram kelvin	J/(kg·K)
thermal conductivity	watt per m kelvin	W/(m·K)
activity (radioactive source)	becquerel	Bq (s^{-1})
absorbed dose of ionizing radiation	gray	Gy (J/kg)

Supplementary units

plane angle*	radian	rad
solid angle	steradian	sr

CONVERSION FACTORS

To convert from	To	Multiply by
Photometric units		
foot·candle	lumen/meter	1.076×10^1
lux	lumen/meter^2	1.00

Other units

acre	meter^2	4.046×10^3
angstrom	meter	1×10^{-10}
atmosphere	Pa (newton/m^2)	1.013×10^4
calorie (thermochemical)	joule	4.184
centipoise	newton·second/m^2	1.0×10^{-3}
curie	disintegration/second	3.7×10^{10}
day (mean solar)	second (mean solar)	8.64×10^4
degree (plane angle)	radian	1.745×10^{-2}
dync	ncwton	1×10^{-5}
electron volt	joule	1.602×10^{-19}
erg	joule	1.0×10^{-7}
erg/(cm^2·s)	watt/m^2	1.0×10^{-3}
farenheit (temperature)	kelvin	$T_K=(5/9)(T_F+459.67)$
faraday	coulomb	9.649×10^4
foot	meter	3.048×10^{-1}
gallon (UK liquid)	meter^3	4.546×10^{-3}
gallon (US liquid)	meter^3	3.785×10^{-3}
gauss	tesla	1.0×10^{-4}
inch	meter	2.54×10^{-2}
kayser	reciprocal meter	1.0×10^2
kilocalorie (thermochemical)	joule	4.184×10^3
millibar	Pa (newton/meter^2)	1.0×10^2
millimeter of mercury (0°C)	Pa (newton/meter^2)	1.333×10^2
minute (plane angle)	radian	2.909×10^{-4}
pint (US liquid)	meter^3	4.732×10^{-4}
poise	newton·second/m^2	0.10
psi	Pa (newton/m^2)	6.894×10^3
rad (radiation dose absorbed)	Gy (joule/kg)	1.0×10^{-2}
roentgen	coulomb/kilogram	2.5798×10^{-4}
second (plane angle)	radian	4.848×10^{-6}
torr (0°C)	Pa (newton/m^2)	1.333×10^2
watt/cm^2	watt/m^2	1.0×10^4
yard	meter	9.144×10^{-1}

* Use of degrees, minutes, seconds of arc, or revolutions (r) is permitted with SI units.

Participants

AGATI, Dr. Giovanni
Istituto di Elettronica Quantistic
CNR, Via Panciatichi 56/30
50127 Firenze
ITALY

ALARIE, Mr. Romain
MRC Group in the Radiation
Sciences, Faculty of Medicine
Université de Sherbrooke
Sherbrooke, Québec
CANADA J1H 5N4

ANDREONI, Dr. Alessandra
Dept. of Biology, Cellular
and Molecular Pathology
2nd Faculty of Medicine
Via Sergio Pansini, 5
80131 Napoli
ITALY

ARNASON, Dr. Thor
Dept. of Biology
University of Ottawa
Ottawa, Ontario
CANADA K1N 6N5

ARRIAGA, Dr. Edgar A.
Dept. of Physiology
University of Kansas Medical Center
39th and Rainbow Boulevard
Kansas City, KS 66103
U.S.A.

AUCOIN, Mr. Richard R.
Department of Biology
University of Ottawa
Ottawa, Ontario
CANADA K1N 6N5

AVERBECK, Dr. Dietrich
Institute Curie
Section de Biologie
26 Rue d'Ulm F-75231
Paris Cedex 05
FRANCE

BAYRAKCEKEN, Dr. Fuat
Ankara University
Dept. Eng. Phys.
06100 Ankara
TURKEY

BEAUREGARD, Dr. Marc
Max-Planck Institut
Strahlenchemie
Mülheim/Ruhr
D-4330 FRG

BEDWELL, Ms. Joanne
National Medical Laser Centre
The Rayne Institute
5 University Street
London WC1E 6JJ
UNITED KINGDOM

BERTHOMIEU, Mlle Catherine
Service de Biophysique
Dépt. de Biologie
Centre d'Etudes Nucléaires de Saclay
91191 Gif-sur-Yvette
CEDEX, FRANCE

BLAKEFIELD, Ms. Mary K.
Dept. of Biology
University of Kentucky
800 Rose St.
Lexington, KY 40536
U.S.A.

BOWMAKER, Dr. Jim K.
Dept. of Biological Sciences
Queen Mary College
University of London
Mile End Road
London E1 4NS
UNITED KINGDOM

BOYLE, Dr. Ross W.
MRC Group in the Radiation
Sciences, Faculty of Medicine
Université de Sherbrooke
Sherbrooke, Québec
CANADA J1H 5N4

BRENNAN, Dr. Thomas
Associate Professor of Biology
Dickinson College
Carlislc, PA 17013
U.S.A.

CONN, Ms. Pauline Frances
Chemistry Department
Paisley College
High St.
Paisley, Renfrewshire
U.K. PA1 2BE

CROKE, Dr. David T.
Department of Biochemistry
Royal College of Surgeons in Ireland
123, St. Stephen's Green
Dublin 2, IRELAND

CUNNINGHAM. Ms. Sharon
CADNA Medical Diagnostic Inc.
138 Sunrise Avenue
Toronto, Ontario
CANADA M4A 1B3

DELLINGER, Dr. Marc
Muséum National d'Histoire Naturelle
Laboratoire de Biophysique
INSERM U-201
43, rue Cuvier
75231 Paris CEDEX 05
FRANCE

DOUGLAS, Dr. Ron
Dept. of Optometry and Visual Science
The City University
311-321 Goswell Road
London EC1V 7DD
UNITED KINGDOM

FAHMY, Dr. Karim
Institut für Biophysik und
Strahlenbiologie der
Universität Freiburg
Albertstrasse 23
D-7800 Freiburg, FRG

FOULTIER, Dr. Marie-Thérèse
Nantes Hopital La Vaennec BP1005
44035 Nantes, CEDEX 01
FRANCE

FUSI, Dr. Franco
Istituto di Elettronica Quantistic
CNR, Via Panciatichi 56/30
50127 Firenze
ITALY

GARBISU CRESPO, Mr. Carlos
Dpto. Bioquimica y Biologia
Molecular
Facultad de Ciencias
Univ. Pais Vasco
APDO 644J, 48080 - BILBAO
SPAIN

GATTO, Dr. Barbara
Dept. Organic Chemistry
University of Padova
Via Marzolo, 1
I-35131 Padova
ITALY

GRAGG, Mr. Richard D.
College of Pharmacy and
Pharmaceutical Sciences
Florida A&M Univ.
Tallahassee, FL 32304
U.S.A.

GROSS, Mr. Eitan
Dept. of Physics
Bar-Ilan University
P.O.B. 90000
Ramat-Gan
52900 ISRAEL

HENRIKSEN, Dr. Dennis Bang
University of Copenhagen
H.C. Orsted Institute
Chem. Lab. II
DK-2100, Copenhagen
DENMARK

HIENERWADEL, Dr. Rainer
Institute für Biophysik und
Strahlenbiologie
Albertstr. 23
D-7800 Freiburg
FED. REP. GER.

HIROTA, Mr. Jungi
Dept. of Bioengineering
Tokyo Institute of Technology
Meguro-ku, Tokyo 152
JAPAN

HOTT, Mr. John L.
GA Tech School of Chemistry
Box 51
Atlanta, GA 30332-0400
U.S.A.

JOHNSON, Dr. Brian E.
Dept. of Dermatology
Level 8, Polyclinic Area
Ninewells Hospital
Dundee DD1 9SY
UNITED KINGDOM

KENNEDY, Dr. James C.
Department of Radiation Oncology
Queen's University
Kingston, Ontario
CANADA K7L 2V7

KOZOREZ, Dr. Alexei
Biophysics Lab Kurchatov
Institute of Atomic Energy
142092, Troitsk
Moscow Region
U.S.S.R.

LEARN, Dr. Doug
Schering-Plough
3030 Jackson Avenue
Memphis, Tenn. 38151
U.S.A.

LEE, Dr. Ki-Hnan
Centre de Recherche on Photophysique
Univ. Québec, Trois Rivières
C.P. 500
Trois Rivières, Québec
CANADA G9A 5H7

LEE, Dr. John
Department of Biochemistry
The University of Georgia
Boyd Graduate Studies Research Center
Athens, Georgia 30602
U.S.A.

LERCARI, Dr. Bartolomeo
Dipartimento di Biologia della Piante Agrarie
Sezione di Orticoltura e Floricoltura
Universita Degli Studi di PISA
Viale Delle Piagge, 23
PISA 56124
ITALY

LILGE, Dr. Lothar
Wellman 2
Massachusetts General Hospital
32 Fruit Street
Boston, MA 02114
U.S.A.

MARCOTTE, Mr. Louis
Dept. of Chem. and Chem. Eng.
The Royal Military College of Canada
Kingston, Ontario
CANADA K7K 5L0

MARROT, Dr. Laurent
Laboratoires de Rechercche Fondamentale
L'OREAL
1, Avenue Eugene Schueller
Boite Postale No. 22
93601 Aulnay-Sous-Bois
CEDEX, FRANCE

MATHESON, Ms. Melissa Jean
Faculty of Veterinary Science
University of Sydney
N.S.W. 2006
AUSTRALIA

MATHIS, Dr. Paul
Service de Biophysique
Départment de Biologie
CEN Saclay
91191, Gif-sur-Yvette
CEDEX, FRANCE

McCULLOUGH, Dr. John J.
Organic Chemistry Dept.
Queen's University
Belfast BT95AG
NORTHERN IRELAND

McDONAGH, Dr. Antony F.
Department of Medicine
University of California in San Francisco
San Francisco, CA 94143
U.S.A.

MONTEITH, Ms. Cherilyn A.
School of Optometry
University of Waterloo
Waterloo, Ontario
CANADA N2L 3G1

MUA, Mr. John
Chemistry Dept.
University of Papua New Guinea
P.O. Box 320, University, N.C.D.
PAPUA NEW GUINEA

NADEAU, Dr. Pierre
Dept. of Chem. and Chem. Eng.
The Royal Military College of Canada
Kingston, Ontario
CANADA K7K 5L0

NAVARATNAM, Dr. S.
The North East Wales Institute
Deeside, Clwyd
UNITED KINGDOM

NULTSCH, Dr. Wilhelm
Lehrstuhl für Botanik
Fachbereich Biologie
Der Philipps-Universität Marburg
D-3550 Marburg/Lahn, den Lahnberge
Deuchland (GERMANY)

O'HARA, Dr. Julia A.
Norris Cotton Cancer Center
Hanover, NH 03756
U.S.A.

PEDERSEN, Ms. Marit
Dept. of Physics
AVH, University of Trondheim
College of Arts and Science
N-7055 Dragvoll
NORWAY

PENNING, Dr. Louis C.
Dept. Med. Biochem., Sylvius Laboratory
Faculty of Medicine, University of Leiden
Wassenaarseweg 72
2333 AL Leiden
THE NETHERLANDS

PFAU, Dr. Jürgen
Fachbereich Biologie
der Philipps-Universität
Marburg, Dept. Botanik
Lahnberge, 3550 Marburg
FED. REP. GER.

PLATOU, Ms. Trine
Institutt for Kreftforskning
Avdeling for biofysikk
Postadresse: Montebello 0310
Oslo 3
NORWAY

POLJAK-BLAZI, Dr. Marija
Rudjer Boskovic Institute
BiJenicka 54
Zagreb
YUGOSLAVIA

POTTIER, Dr. Roy
Department of Chem. and Chem. Eng.
The Royal Military College of Canada
Kingston, Ontario
CANADA K7K 5L0

RAVANAT, Dr. Jean-Luc
Centre d'Etude Nucléaire Grenoble
DRF-Laboratoires de Chimie
85X, 38041 Grenoble
CEDEX, FRANCE

ROBERTI, Dr. Giuseppe
University of Naples
Biology Dept.
II Fac Med.
Via Sergio Pansini 5
I-80131 Napoli
ITALY

ROTOMSKIS, Dr. Ricardas
Vilnius University
Laser Research Center
Sauletekio ave 9, corp. 3
Vilnius 232054
Lithuania
U.S.S.R.

RUSSELL, Dr. David A.
Dept. of Chem. and Chem. Eng.
The Royal Military College of Canada
Kingston, Ontario
CANADA K7K 5L0

SENGE, Dr. Mathias
Department of Chemistry
University of California
Davis, CA 95616
U.S.A.

SERRANO, Dr. Aurelio
Instituto de Bioquimica Vegetal
y Fotosintesis
CSIC/Univ. de Sevilla, Facultad Biologia
Apartado 1113
41080 Sevilla
SPAIN

STUMPE, Dr. Hugo
Deptartment of Organic Chemistry
University of Linz
A-4040 LINZ-AUHOF
AUSTRIA

SULLIVAN, Mr. Eric
Department of Chemistry
Queens University
Kingston, Ontario
CANADA K7L 3N6

SWAIRJO, Ms. Manal
Physics Department
Boston University
590 Commonwealth Avenue
Boston, MA 02215
U.S.A.

THIBODEAU, Dr. Danièle
Service de Biophysique
Dépt. de Biologie
Centre d'Etudes Nucléaires de Saclay
91191 Gif-sur-Yvette
CEDEX, FRANCE

TOON, Mr. Stephen P.
Research Tech., Solar Energy Reserch Inst.
University of Denver
1617 Cole Blvd.
Golden, CO 80401
U.S.A.

VALDUGA, Dr. Giuliana
Department of Biology
University of Padova
Via Trieste 75
35121 Padova
ITALY

VALENZENO, Dr. Dennis Paul
Department of Physiology
University of Kansas Medical Center
39th and Rainbow Boulevard
Kansas City, KS 66103
U.S.A.

VAN DER WAL, Dr. J.
Department of Medical Biochemistry
Faculty of Medicine, University of Leiden
Wassenaarseweg 72
2333 AL Leiden
THE NETHERLANDS

VAN LIER, Dr. Johan E.
MRC Group in the Radiation Sciences
Faculté de Médecine
Université de Sherbrooke
CHUS
Sherbrooke, Québec
CANADA J1H 5N4

WANG, Ms. Tian P.
The University of Illinois at Chicago
Department of Chemistry (M/C 111)
4500 Science and Engineering South
Box 4348
Chicago, Illinois 60680
U.S.A.

WEAGLE, Mr. Glenn
Dept. of Chem. and Chem. Eng.
The Royal Military College of Canada
Kingston, Ontario
CANADA K7K 5L0

WILSON, Dr. Brian C.
The Ontario Cancer Treatment
and Research Foundation
Hamilton Regional Cancer Centre
Henderson Clinic
711 Concession Street
Hamilton, Ontario
CANADA L8V 1C3

WOOD, Mr. Simon
Department of Biochemistry
The University of Leeds
Leeds LS2 9JT
UNITED KINGDOM

Index